WILLIAM ROXBURGH

The Founding Father of Indian Botany

William Roxburgh. Engraving by Charles Turner Warren, first published with an obituary in the Transactions of the Society for the Encouragement of Arts, Manufactures, and Commerce *in 1815. This, the source for all later portraits, was based on a miniature then owned by his widow, Mary Roxburgh, the present whereabouts of which is unknown.*

WILLIAM ROXBURGH

The Founding Father of Indian Botany

Tim Robinson

PHILLIMORE
IN ASSOCIATION WITH ROYAL BOTANIC GARDEN EDINBURGH

2008

Published by
PHILLIMORE & CO. LTD
Chichester, West Sussex, England
www.phillimore.co.uk

ISBN 978-1-86077-434-2

The author and publisher would like to thank the trustees of
the Orcome Trust and the Sibbald Trust for their significant
contribution towards the cost of producing the colour plates
that appear on these pages.

The Sibbald Trust is an independent charitable trust which
gives financial support to many different facets of the
botanical and horticultural interests of the Royal Botanic
Garden Edinburgh.

Printed and bound in Great Britain at
CAMBRIDGE UNIVERSITY PRESS

CONTENTS

Monogram used by William Roxburgh as a
seal on his letters, taken from a letter from
Roxburgh to Dr James Edward Smith,
dated 20 August 1794. Linnean Society,
Smith Letters, vol. 8 f188.

LIST OF FIGURES

LIST OF TABLES

ACKNOWLEDGEMENTS

For a work of this size, there are a number of people and organisations which have made it possible, by answering queries and providing me with information as well as allowing me access to a wealth of primary sources. I would like to thank: Henry Noltie, of the Royal Botanic Garden Edinburgh, who first suggested that I should work on the Roxburgh Collection of drawings held by the Garden and has been an invaluable support and mentor during so much of the research, with his wealth of botanical and Indian knowledge; Nick Phillipson, of the Edinburgh University History Department, who guided me through the Enlightenment and much historical background; Mrs Jane Hutcheon, Librarian at the Royal Botanic Garden Edinburgh, who introduced me to the Roxburgh Collection of drawings held by them and showed me some of their outstanding books and periodicals; Debbie White, who photographed some of the material held in the library of the Royal Botanic Garden Edinburgh; Neil Chambers, of the Banks Archive Project, at The Natural History Museum, London, who first introduced me to the wealth of correspondence that has led me all over the world; Malcolm Beasley, the Librarian of the Botany Library at the Natural History Museum, London, for allowing me to use so much of their archival material which was so relevant to sorting out aspects of William Roxburgh's life, and the members of his staff for all their assistance; Iain Milne, Librarian of the Royal College of Physicians of Edinburgh, for allowing me access to their important collections which were so crucial in unravelling the medical qualifications of the late 18th century as well as showing me the archives relating to John Boswell; Martha Whittaker, at the Sutro Library, California State Library, for the large collection of letters involving William Roxburgh which they hold and organising copies of them; Louise Anematt, at the Mitchell Library, State Library of New South Wales, for solving problems in accessing the material on their website; Alison Lewis, of the American Philosophical Society Library, Philadelphia, for clearing up some of the problems of the correspondence with America, and introducing me to the elephant's skeleton; Andrew Wilson, at the Special Libraries and Archives at Aberdeen University, for confirming the date of Roxburgh's graduation; Gina Douglas, Archivist and Librarian at the Linnean Society of London, for allowing me access to the archives which sorted out Roxburgh's relations with a number of contemporary scientists; the staff at the National Library of Scotland, who have helped me find so many of Roxburgh's publications as well as the numerous books and periodicals which I kept requesting; Arnott Wilson, Archivist, Edinburgh University Library, for helping to elucidate William Roxburgh's education; the staff of the Special Collections of the Edinburgh University Library, who have been so helpful and supportive in finding so much contemporary material; the staff at Edinburgh Public Library, George IV Bridge, Edinburgh, who have been willing to find a wealth of background reading; Susan Bennett, Curator for the Royal Society for the Encouragement of Arts, Manufactures and Commerce, who showed me the manuscript sources held by them; David Blake and the staff of the British Library who found and carried vast tomes from their enormously extensive archives; Kate Pickard, Archivist at the Royal Botanic Gardens, Kew, who has been willing to search for relevant material held by them, and Marilyn Ward, who showed me the Kew Roxburgh drawings collection; Elizabeth Mackay, Librarian at the Edinburgh Academy, who cleared up problems over William Roxburgh's youngest son's education; David Brown of the National Archives of Scotland, who found some interesting relevant manuscript material; Derek Alexander, Archaeologist for the West of Scotland with the National

Trust for Scotland, who answered queries about the pavilion at Culzean; Leander Wolstenholme, of the Botany Department of the Liverpool Museum, who helped sort out problems from their herbarium material; John Rourke and Ted Oliver, of the National Botanic Institute, Kirstenbosch, who helped solve some of the problems relating to Roxburgh's work in South Africa; David White, at the College of Arms in London, who gave me details of the Roxburgh grant of arms; Drs H. J. Chowdhery, Prasanna, P. Laksminarasimhan, Ansari, S. S. Hammed and Sanjappa of the Botanical Survey of India and its Indian Botanic Garden at Howrah for all their time and effort in showing me so much of the wealth of material at the Calcutta Botanic Garden; and various descendants of Roxburgh's who are working on their families' genealogies, who have always been most interested and supportive of any problems that I have discussed with them. Finally, due acknowledgement must be given to my wife, Anne, for all her help and support to see this work through research to publication.

ILLUSTRATION CREDITS

The author and publisher would like to thank the following individuals and organisations for permission to reproduce material. Every effort has been made to contact copyright holders, but if any errors or omissions have been made, we would be happy to correct them at a later printing.

British Library, By permission of the British Library, 28, 29, 42, 70, 93, 94, 95

Brock, K. J., 53, 72, 73, 121, 123

Central Edinburgh Public Library, By courtesy of Edinburgh City Libraries, 20, 34, 48, 49

College of Arms, Reproduced with the kind permission of the College of Arms, 129 (MS Grants 50, p. 403)

Edinburgh University Library, Special Collections, By kind permission of Edinburgh University Library, 12, 14, 17, 18, 19, 21, 26, 27, 52, 65, 66, 67, 69, 89, 109, 111, 112, 113, 118, 119, 120

Indian Botanic Garden, Howrah. Reproduced by the kind permission of the Indian Botanic Garden, 51, 76

Linnean Society of London, Reproduced with the kind permission of the Linnean Society of London, 81

National Library of Scotland, Reproduced with the kind permission of the Trustees of the National Library of Scotland, 55, 58, 59, 63, 75, 117

National Trust of England, Reproduced with the kind permission of the National Trust, 38

Natural History Museum, Botany Library, © The Natural History Museum, London, 22, 60, 41, 62

Royal Botanic Garden Edinburgh, Reproduced with the kind permission of the Royal Botanic Garden Edinburgh, frontispiece, 5, 6, 7, 8, 9, 16, 23, 24, 33, 35, 57, 64, 71, 77, 78, 79, 80, 83, 84, 86, 87, 88, 91, 92, 96, 97, 98, 104, 105, 106, 107, 108, 110, 114, 125, 126, 128

Royal Botanic Garden Kew, Reproduced with the kind permission of the Royal Botanic Gardens Kew, 47, 99

Royal College of Physicians of Edinburgh, Reproduced with the kind permission of the Royal College of Physicians of Edinburgh, 2, 45, 46

Royal College of Surgeons of Edinburgh, Reproduced with the kind permission of the Royal College of Surgeons of Edinburgh, 3

Victoria and Albert Museum, © V&A Images/Victoria and Albert Museum, 25, 74, 82

PREFACE

Interest in the pioneers of botanical exploration has never been greater, with recent years seeing a steady stream of books detailing their lives and their legacies to science and gardening. This book on the life of William Roxburgh fills an important gap. Aptly subtitled, the Founding Father of Indian Botany, it provides a rich source of insight and information on one of the foremost figures of 18th-century botany and the first to attempt the systematic classification of the plants of India.

Dr Tim Robinson has pursued his research into Roxburgh diligently and whereas many publications on the botanical pioneers present an overly romanticised account of their adventures, this is a scholarly account based on careful research and analysis. That research formed the basis of a thesis on Roxburgh for which the University of Edinburgh awarded the author the degree of Doctor of Philosophy.

Roxburgh's legacy extends far beyond the 2,000 species and the 50 genera of plants that he described as new to science or the thousands of drawings and herbarium specimens preserved today in collections across Europe. In addition to his remarkable contribution to Indian botany, Roxburgh made substantial contributions to knowledge of the Cape Flora and the plants of St Helena. He also ranged far beyond botany into meteorology, mineralogy and zoology, looking at the world with a practical eye and a keen interest in the well-being of its inhabitants. The author attributes Roxburgh's humanitarian behaviour to influences that shaped his character during his early years in Scotland. He lived with Dr John Boswell, the President of the Royal College of Physicians, studied under Dr John Hope and came into contact with other leading members of the Edinburgh Enlightenment. From them, and perhaps particularly from Hope, he developed an experimental approach to his investigations of nature that came to the fore during his time in India. There he investigated the properties and cultivation of many economically important plants that were in use as crops, fibres, spices or sources of dyes. But he also sought new plant resources by investigating related species and others in which he saw potential. His concern for fellow humans led him to consider several local palm species as sources of food in times of famine and to devise schemes for drainage and irrigation to maximise the production of crops during times of drought or flooding. His work was, therefore, immensely practical as well as academic and should serve as a reminder to us today that human well-being and economic prosperity ultimately depend upon the wise use of the natural resources of our planet. Congratulations to the author on an excellent and much-needed publication.

STEPHEN BLACKMORE
Regius Keeper
Royal Botanic Garden Edinburgh

INTRODUCTION

Arriving at Madras in May 1776, William Roxburgh was immediately at the centre of so many strands that affected Indian history in the nigh on 40 years before he left in 1813. There was the local politics of Fort St George, where the struggle for control of the Council led to the indictment of the Governor Pigot. When Roxburgh moved to Nagore two years later, he with all the East India Company employees was chased out at very short notice by Haider Ali's army and leaving all his things in the escape. Once at Samalcottah, he once more lost everything that he had collected when a cyclone washed all his belongings including his herbarium, botanical drawings and his library into the sea. At this point, it was still not only acceptable but encouraged by the likes of Warren Hastings and Sir William Jones to discover as much as possible of Indian culture and civilisation, which were both held in high esteem. How different was the atmosphere by the time he had been in Calcutta for a few years, under the influence of the Governor Wellesley and the support of the Evangelicals. These 37 years, from 1776 to 1813, covered so many crucial elements that were to mould Indian history over the next two centuries – and William Roxburgh was at the heart of this change. His life, therefore, is a mirror to the changing *mores*, from the rationalism of the Enlightenment to the approaching jingoism of high Victorian colonialism. A study of his life offers a spotlight onto this important period. As the first person systematically to describe the botany of India, he is pivotal in understanding the development of the plants of the sub-continent. Both as a field botanist and as the Superintendent of the Calcutta Botanic Garden, he was at the centre of the growing knowledge, and used his positions to spread this knowledge across the world.

A number of conundrums remain about his life, of which the greatest must be regarding his origins, for which there appears no trail to follow. Once in India, and he had had his first son almost certainly by a native woman, should one assume that this led to his having an understanding of Telegu and/or Tamil? Certainly it looks as though he was loyal to his Telegu artists and one should expect that he had to learn their language to be able to tell them the style required of his drawings (even if Sir Joseph Banks was not good at offering specific help on this) and the need to be able to communicate with the local people when he went on his plant-hunting expeditions.

The greatest sadness must be the loss of his notebooks which contained, one imagines, following Dr John Hope's guidance, very detailed records of where, when and the conditions of his plant collecting, as well as his full meteorological records. Equally, the letters that do exist give few insights into the man himself: there is plenty about his botany and the wider scientific interests, but as a character he remains elusive with few examples of his own feelings about events, people and conditions.

However, the remaining archives show a man of enormously wide interests and contacts, a man brought up in the seat of the Scottish Enlightenment of Edinburgh, a capital city that was still looking for a new role after the union of the Parliaments in 1707. From Edinburgh, Roxburgh took with him a sound scientific knowledge and interest in experimental botany; he also took a network of contacts on which he built to develop a world-wide circle of correspondents. This produced a legacy of enormous value in both his taxonomic work on the flora of India, South Africa and St Helena. Allied to this was the stunning collection of botanical drawings and there still remain some traces of his herbarium material. Of his more practical, economic experiments, some were well ahead of his time, others reflected the passing of a bygone age: he pointed out where tea should be found and grown, and the usefulness of jute as a fibre; but his work on dyes was superseded by the coal-based aniline colours within a few decades of his death.

* * *

For consistency of style, a number of conventions have been used. The main problem area has been deciding what to do about spellings, and the practice followed has been that the original spellings, capitalisation, abbreviations and punctuation have been used in quotations. Deciphering some of the manuscript handwriting has not always been easy, and there will, inevitably, be some problematic transcriptions that have not been solved.

In the case of place-names, those that were common parlance at the time have been used; thus Madras rather than Chennai, Calcutta rather than Kolkata, and Ceylon rather than Sri Lanka, for instance, except where there might be confusion when the modern name has been given as well.

The third area which has caused considerable problems has been the names of plant species. The practice in this case has been to use the name that Roxburgh gave, or that was used at the time, provided it was published in some periodical, such as his *Flora Indica* or the *Asiatick Researches*, in which case it can be traced to the modern name; where the modern name is well established, this has also been given. However, there have been some occasions, particularly in Appendix 4, when it has not been possible to find any known species that fits the manuscript, in which case the species name has been preceded by a question mark.

PART ONE

THE LIFE OF
WILLIAM
ROXBURGH

CHAPTER 1

Origins and Education 1751-1776

The terrace of early Georgian houses, now nearly a hundred years old, gave no indication of the lives of its inhabitants on this late November afternoon, already dark in the northern capital of Edinburgh. To keep out the cold, the shutters were closed and barred, and smoke rose from the chimneys: justly had Edinburgh earned the nickname of Auld Reekie. One of these houses contained a large family, the elderly husband was now very sickly, his wife some 23 years younger was looking after their children and those of his from his second marriage, helped by a maid while the cook busied herself preparing their evening meal. The old man, Dr William Roxburgh, was in his study, trying to put pen to paper, writing one of his many letters to Sir Joseph Banks, the President of the Royal Society in London.

> When I left London, it was my intention to return in October, which I was prevented from doing by the severity of my complaints. At this moment I am rather better, tho satisfactory signs of my recovery are still remote. Some time ago I wrote to Brown respecting our East India Botany, & wish much to be favored with a letter from him in reply. I trust you have received the rest of my drawings & descriptions from the India House, agreeable to the request I made to the Directors through their Secretary before I left London. You might then be able to make the last two numbers of the 3rd vol. of Indian plants richer .... Some time past, but since my arrival in Edinburgh (where the Thermometer has been already as low as 20°) I learned that you had sent one or two men to the east coast of South America, &c, to collect for Kew. Permit me, therefore to beg a small portion of their labour may be bestowed on such subjects as are still wanted for our Asiatick empire ... Whether I am ever able to return to India, or not, my anxious wish, to the last hour of my existence, will be to add all in [my] power to the vegetable riches of our possessions there.[1]

The old man looked up at the whickering candle as he put his quill pen on its stand, staring with screwed-up eyes behind his small round framed glasses. He rubbed his shivering hands, covered in thick woollen gloves that were almost hidden by his jacket and heavy overcoat. Behind him, the coal fire glowed red in the grate, radiating out its heat that could not warm him enough. To keep out the blistering cold outside, the windows were shut, the shutters closed and the heavy curtains drawn. One wall was covered with shelves of leather-bound volumes, many showing the stains of dampness acquired during numerous Calcutta monsoons. There were his books on travel, including those of Captain Cook and John Forster which both covered their circumnavigations of the globe 40 or so years earlier; a number of tomes covering different aspects of science, from the manufacture of salt and a kaleidoscope of dyes, through the growth and care of important economic crops like tea and sugar, to various floras, including John Lunan's recently published *Hortus Jamaicensis* that lay on his desk and that had set him off on this letter. Volumes such as this, on plants, occupied the pride of place, especially the first two volumes of his own handsomely produced *Plants of the Coast of Coromandel* that his recent employer, the East India Company, was taking such an age to complete under the guidance of Sir Joseph Banks. How William Roxburgh longed to get his own flora of India published, if only he had the time and the health.

As he sat in his study, he wondered why he had striven for nearly forty years to make his fortune only to return to Edinburgh to die of cold with so much still to do. He was stronger now than three months ago when he was so ill that he could hardly write, and then only with a trembling hand, still evident to this day in the copy of the letter that he wrote to Dr J. E. Smith, President of the Linnean Society in the middle of August, and held in their archives now. Before picking up his pen, he put his hands to his ears: if only he could be rid of the continual noise in his head that had plagued him for years, how much clearer his thoughts would be. Only with concentration and busy-ness could he lessen this distraction. He dipped his pen into the inkwell and finished off the letter with his easily recognisable signature.

> I trust you & the Ladies continue to enjoy good health, & that the weather may be better than it is at Edinburgh, where every variety is to be experienced in the course of 24 hours ... we have fortunately got an excellent well built house, in a warm situation on the south side of Edinburgh, & rather without than within it.
> I remain, with much regard
> My Dear Sir
> Yours most faithfully
> W. Roxburgh

William Roxburgh had a very wide circle of correspondents and among them were a number of real friends who cared for this eccentric botanist who had spent so long in India. They were solicitous of his health and when he visited them often gave him produce from their estates: on his previous visit to Britain in 1807[2] when he had visited Smith in Norfolk, he wrote that 'Such a Turkey as yours was never I saw. Many thanks from Mrs Roxburgh & myself for it.' But the letter of November 1814 was not the last, as two more still exist from him, the first to Robert Brown who worked closely with Banks, and the other to another of his great friends and mentors, Dr James Edward Smith. Shortly after he received this letter, written in the middle of January, Smith would have discovered that its author had died and was buried next to his daughter and in his wife's family grave in Greyfriars Kirkyard in Edinburgh, hardly 100 yards from the house in which he had died.

The obituaries now started pouring in, some brief as in the papers in Edinburgh, to which he had returned only about eight months previously. Others were much longer, reflecting his importance in the expanding world of global botany as well as the wider circles of the scientific community. He was heralded as a man known on all five continents, with a long list of publications and who had made a collection of Indian botanical drawings that are still of renown for their enormous artistic merit and their botanical importance. How different was this ending to a life, with letters flying round the world bemoaning the passing of what one writer referred to as the greatest botanist since Linnaeus, to his origins. Of these, virtually nothing is known except what can be gleaned from these obituaries and his will.

EARLY YEARS

For a man who in his own life corresponded with many of the great of his day,[3] the facts about William Roxburgh's early life are very scant. The detective work to find even his date of birth hit a number of dead end trails: there is no mention in any of the parish registers of any William Roxburgh being born around 1751 which is the year that emerges from his various obituaries. It is indeed from these obituaries that most information can be traced of his early years. Disentangling the fact from myth leaves out some gloriously purple passages, likening him to Robert Burns who was born eight years later in nearby Alloway:

> His early years passed away rapidly, amidst the romantic scenery that seems to have inspired the Muse of his countryman Robert Burns, and conferred both grace and energy on the poetical labours of an humble ploughman.[4]

What does seem certain from these sources is that he was born in June 1751, but there is doubt as to whether the actual date was the 3rd or the 29th, as both appear in his obituaries. The date of his death can be cleared up easily, for the *Edinburgh Advertiser* for Tuesday, 28 February 1815 states in the list of deaths: 'At Edinburgh, on the 18th instant, WILLIAM ROXBURGH, M.D., F.L.S., Chief Botanist to the Hon. East India Company ...'

The corollary of this is that there is no firm evidence for where he was born. However, the various obituaries all give the county as Ayrshire, and most give the parish as either Craigie or Symington (which appeared as Lymington in some, due almost certainly to a misreading of the script) at Underwood. Both of these are possible (see Fig. 1), if one takes the fact that the Boswell home at Auchinleck was nearby, and this proximity explains the reason for Roxburgh having stayed with Dr John Boswell when he studied at Edinburgh University and retained a strong connection with the family all his life. There is also a description of the house of Underwood in the *Kilmarnock Standard* of 4 July 1931, where it is described as being very derelict:

> In 1785 this estate was purchased from the creditors of Messrs Alexander, merchants, Edinburgh, by John Kennedy, the only surviving son of Robert Kennedy, of Green, near Ayr, who was a descendant of the Cassillis family .... When the estate came into their possession, the remains of an old baronial castle, with a moat, stood upon it, but in such a state of decay as to be irreparable. This was taken down and the present commodious and comfortable residence was erected on the same site, ... Close to the farmhouse of Underwood are the remains of a small clachan, on the joiner's shop in which may still be seen a lintel bearing the date 1742.[5]

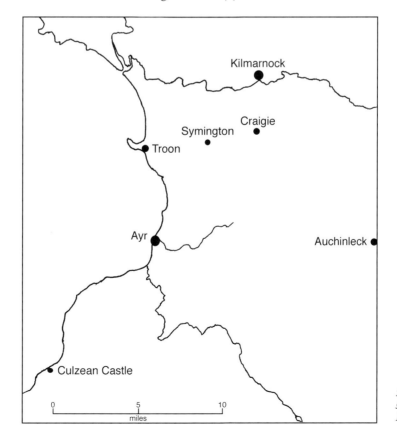

1 Map of part of Ayrshire, showing the area in which Roxburgh was brought up.

There are references to Roxburgh being brought up in the country, so it is possible that his father worked in some capacity on the Underwood estate, corroborated by the statement that 'his family was not in affluent circumstance'.[6]

More information, though, can be gleaned from Roxburgh's will, which was recorded on 9 July 1815, and is held in the Scottish Record Office.[7] This states that at the time of writing the will, in July 1814, Roxburgh left bequests to a half-brother, Thomas Parkhill, of Ayr, a half-sister, Isabella Humphrys of Symington, and William Parkhill, of Glasgow, son of Roxburgh's late half-brother, John Parkhill. There was also a John Roxburgh who matriculated at Edinburgh University for the three academic years 1775/76 to 1777/78 and who could have been a younger brother.[8] Added to this is the fact that Sir Joseph Banks, in a letter to Roxburgh, stated that Roxburgh's son had an uncle, Mr Orme(s).[9] All this leads to the conclusion that William Roxburgh had at least one sister who married a Mr Orme(s), possibly one brother as well as another set of siblings after his father died and his mother remarried a Mr Parkhill. What is strange is that no records for any of these events appear to exist in any of the Ayrshire parish records, although there are a number of Roxburgh families in the Kilmarnock area, and there is one William Roxburgh born in Kilmarnock on the 5th and baptised on 10 November, 1751, first child of Adam Roxburgh, a 'Baxter' (a baker) but no wife is mentioned.[10] It is unlikely that this is the subject of this study as there is no record of Adam Roxburgh's death during the next ten years nor marriage of a Parkhill between 1750 and 1780.[11] The other possibility is that Roxburgh was the illegitimate son of a well-connected family, which could have been the Boswells or that family was asked to look after him. Although this theory fits the Boswell connection, it does not agree with Roxburgh having a full brother unless the John Roxburgh is a coincidence which is, of course, perfectly possible. His half-brothers would then have been the children of his mother after she married a Mr Parkhill elsewhere and came to settle in the Ayrshire region, to be near Auchinleck, the Boswell family home. This theory also ties in with the nepotism common in Edinburgh and the care that the Boswell family took of Roxburgh for the rest of his life.

Another idea must be that, with the proximity of the Kennedy family home and the connections between them and the Boswells, that Roxburgh was the son of a Kennedy and Roxburgh's own mother. There are so many alternatives that it is impossible to be certain, but I do like the idea of William Roxburgh being the illegitimate son of one of the larger landed gentry from Ayrshire: it fits so many of the imponderables. The concept of landed gentry having illegitimate children was not unusual, nor was the fact that they were then cared for and their education and careers established by their father. This was, indeed, the subject of a number of contemporary novels and plays. This is further highlighted by the fact that Roxburgh himself did just this for his own natural son, John. Thus the possibility of William Roxburgh being illegitimate is strong, reinforced by a variety of factors, ranging from the lack of evidence of his origins to the close relationship he had with the Boswell family to the extent of marrying as his third wife Mary Boswell, who could have been a cousin of some sort: a look at any number of families at this time shows the frequency with which even first cousins married. This is further suggested by the fact that Edinburgh men of science were generally better born, with fathers who were better connected with the brokers of patronage: the ease with which Roxburgh fitted into the Edinburgh scientific and medical society certainly suggests that he had some family connection with it, as was his future in India eased by powerful letters of introduction.

2 *Dr John Boswell (1707–80), President of the Royal College of Physicians of Edinburgh. Portrait attributed to William Millar, who painted his wife.*

3 *Professor Alexander Monro, secundus (1733–1817). Portrait by John T. Seton.*

EDUCATION AND EDINBURGH

Roxburgh must have attended at least the local village school, unless he had the use of a private tutor, benefiting in the former case from the fact that Ayrshire had always been an educationally well-provisioned county: over half the parishes had appointed a schoolmaster by the 1660 Restoration.[12] There is a delightful piece in his obituary which gives an insight into this education: 'The happy facility, and comparative ease, with which knowledge is obtained in Scotland, soon pointed out a learned profession as an object of laudable ambition to his parents.'[13]

His education almost certainly followed the norm, with a strong emphasis on the classics, for his knowledge of Latin does seem to have been at least adequate: he was later able to carry on a correspondence with the Rev. John in Tranquebar in Latin, as well as writing some of his descriptions in Latin.[14] Another of the pillars of this early schooling was the relationship between the good and the useful, a fact that bore deeply on the thinking of Roxburgh, as shown by so much of his scientific work. Similarly, by looking forward to his later achievements, further light can be thrown on this stage of his education: from the work that he did in mapping the Circars in the early 1790s, it would seem that he benefited from the teaching of geography and mathematics, which were commonplace elements of the Enlightenment education. The strength of the mathematics was also reflected in his interest in standardising the volume of bales for indigo.

Otherwise, we know nothing of this stage of his education or any other details of his life until he matriculated at Edinburgh University for the session of 1771/72, to study anatomy and surgery under Dr Alexander Monro.[15] Here Roxburgh would have experienced the emphasis given to anatomical dissection and post mortem examination, that made the Edinburgh medical school one of the most progressive. One of the effects of this teaching was the stress given to empirical observation, which can be seen with Roxburgh in the care he devoted in later years in his taxonomic descriptions and his work on alternatives to hemp ropes.

Unfortunately the class lists for Dr John Hope are missing for 1772 to 1775, but all the later evidence points to the fact that Roxburgh also studied under him, although it is not clear exactly when he did so.[16] As mentioned above, Roxburgh lived with Dr John Boswell who was one of the pillars of Edinburgh Enlightenment society, being President of the Royal College of Physicians as well as James Boswell's uncle.[17] With these contacts, and meeting and living with so many people who were at the centre of Edinburgh Enlightenment society, it is not surprising that he appears with so many of

the attributes of the educated Enlightenment man of science. One interesting example of the closeness of this social group was the painting, made in 1786, which showed the meeting between Walter Scott and Robert Burns in the house of Adam Ferguson and in the presence of James Hutton, Dugald Stewart, John Home, Adam Smith and Joseph Black.

It is worth at this point giving some background to the intellectual life in Edinburgh during the third quarter of the 18th century, for it helps explain Roxburgh's approach to science and particularly botany. The combination of the European Enlightenment, which signalled the ability of man both to understand and control his social and natural environment,[18] and the determination of Edinburgh Town Council to re-establish the city as a major player in at least Scottish life following the Union of the Parliaments in 1707, produced a flowering of intellectual life. For example, by 1767 plans had been adopted for the building of the New Town which had reached completion as far west as Hanover Street by 1785.[19]

4 Professor John Hope (1725–1786) talking to his gardener John Williamson in the Royal Botanic Garden, Leith Walk, Edinburgh. Drawn and etched by John Kay, 1786, reproduced in A Series of Original Portraits and Caricature Etchings *(Edinburgh, 1792).*

The intellectual centre of the city was the University which was predominantly clerical and professorial,[20] and its success can be gleaned from the fact that by the turn of the 18th century, textbooks, treatises and histories by its professors were being used in an astonishing number of universities at home and abroad.[21] But of crucial importance was what Phillipson referred to as 'the alternative academic culture', centred round such societies as the Philosophical Society, in which the professors of divinity, philosophy and medicine met lawyers, ministers, men of letters, and local noblemen and gentry to discuss moral, natural and medical philosophy: it was stated, indeed, in the rules of the Philosophical Society that a third of its members should be 'gentlemen who do not make Philosophy or Physick their particular Profession.'[22] The basis of much of these discussions was 'rational enquiry',[23] based on the writings of such men as David Hume, Adam Smith and Lord Kames. On these two pillars of the Edinburgh Enlightenment were built both the rationalism of the empirical approach to science which followed the teachings of such people as Boerhaave[24] but also questioned these teachings as well; and developed a very close-knit society which was very prepared to help each other and their students. This experimental method was also based on the approach favoured by such established English scientists as Sir Isaac Newton and Francis Bacon, which had abandoned the *a priori* arguments of the Middle Ages, in which a given cause produced certain effects, in favour of the most scrupulous observation of phenomena. This was in contrast to such earlier writers as Montesquieu, who derived from the ancient writers Tacitus and Aristotle, an approach that applied logic in the absence of fact, to a chain of cause and reasonable effect. As a result, practical advances in both medicine and agriculture stimulated theoretical science in turn. Against this must be seen the belief in a wise and good Nature propounded by James Hutton in his *The Theory of the Earth* (though not published until 1785), supported by the stoicism of Smith's ideas, against the earlier catastrophic view of nature that had plunged Dr Johnson into gloom on his tour of the Highlands – it is possible that Roxburgh may have met Hutton, for from 1767 to 1774 Hutton served on the management committee of a scheme to link the Forth and Clyde rivers by a canal. This scepticism of previously accepted thinking meant that education was based on a thorough understanding of what had been written plus a reliance on observation and where there was doubt as to which should take precedence, this was given to the results of clearly defined observations and experiment.

An example of this small circle is gleaned from the fact that there were only 163 members of the Edinburgh Philosophical Society between 1737 and 1783. Among those who were also Fellows of the Royal College of Physicians of Edinburgh in 1739 included Charles Alston, John Boswell, John Clerk, Andrew Plummer, William Porterfield, Sir John Pringle, Andrew St Clair and John Stevenson; and other names which appear on the admission role for 1782 include Joseph Black, Henry Cullen, William Cullen, Andrew Duncan, James Gregory, James Hamilton, Francis Home, John Hope, James Lind, Alexander Monro *secundus*, Donald Monro, Daniel Rutherford (Hope's successor as Professor of Botany) and Charles Clark.[25] As will be shown in the case of William Roxburgh, this network of eminent Scottish academics was crucial to his career.

Furthermore, Roxburgh will almost certainly have come across the work of men such as Maclaurin, the Professor of Mathematics, Joseph Black, the Professor of Chemistry, as well as men such as Matthew Stewart. These men, both directly and indirectly, would have emphasised a number of skills that Roxburgh took with him into later life. One of these was the quest for precise quantification, shown particularly well in his experiments on hemp and on black-body radiation. There was another side to this coin, which was that specialist terms of description could not be applied to these men who were at home in a broad range of the sciences and intellectual pursuits: an educated man was expected to be able to contribute to a number of fields.

It must be remembered, however, that for all the 'politeness' of society, the conditions under which even these pillars of society lived, was not as polite as may be imagined. Sir Walter Scott, writing *Guy Mannering* in 1815, gave a graphic account of conditions for those living in Edinburgh in the 1770s:

> The period was near the end of the American war. The desire for room, of air, and of decent accommodation, had not as yet made very much progress in the capital of Scotland. Some efforts had been made on the south side of the town towards building houses *within themselves* as they are emphatically termed; and the New Town on the north, since so much extended, was then just commenced. But the great bulk of the better classes, and particularly those connected with the law, still lived in flats or dungeons of the Old Town .... The extraordinary height of the houses was marked by light, which, glimmering irregularly along their front, ascended so high among the attics, that they seemed at length to twinkle in the middle sky. This *coup d'œil*, which still subsists in a certain degree, was then more imposing, owing to the uninterrupted range of buildings on each side, which, broken only at the space where the North Bridge joins the main street, formed a superb and uniform Place, extending from the front of the Luckenbooths to the head of the Canongate, and corresponding in breadth and length to the uncommon height of the buildings on either side.[26]

This environment occupied by Roxburgh was especially important as power in Edinburgh at this time was very much in the hands of these same academics and the scientific audience which listened to and followed them. Added to this, since its re-establishment in 1727, a significant part of Edinburgh science took its origins from and was centred on the activities of the university Medical School, and it was through the work of the medical professors and other physicians that many of Edinburgh sciences, such as chemistry, physiology and botany rose to eminence.[27] This is further reinforced by the fact that when Charles Elliot set up his bookshop in Edinburgh's Parliament Square in 1771, he specialised in medical books, importing them from Europe and London as well as printing translations of foreign editions and publishing Edinburgh authors.[28] This intertwining of Edinburgh society is further reflected in the membership of the Aesculapian Club, founded in 1773 as a supper club, which appeared to have as one of its aims the promotion of medical research, and Alexander Monro *secundus*, Andrew Duncan, John Hope and Daniel Rutherford were all members.[29] Living with John Boswell, attending the course of Alexander Monro, and studying under John Hope, Roxburgh was at the very centre of this intellectual ferment and it had a long-lasting effect on his attitudes towards botany and wider science, as will be shown further in Chapter 7.

Not only were these people involved with the discovery of new areas of science, but they were also very eager to promote the application of science in particular practices. This was shown in the slightly later case of Sir George Mackenzie of Coul (1780-1848).[30]

DR JOHN HOPE'S INFLUENCE

Dr John Hope was to be a seminal influence on Roxburgh. Like the Boswells, Hope was connected to the Scottish nobility, being the grandson of Lord Rankeillor, Judge of Session under William and Mary, and his father was a well-known Edinburgh surgeon. At Edinburgh University during the 1740s, he studied anatomy under Alexander Monro *primus*, botany and materia medica with Charles Alston and probably natural philosophy (physics) under Colin Maclaurin. His botanical interest would have been fired by attending the lectures of Charles Alston, Professor of Botany and Materia Medica, the joint Chairs that Hope succeeded on Alston's death in 1761, when he was appointed by the Crown as King's Botanist for Scotland and Superintendent of the Royal Garden in Edinburgh, at the same time the Town Council appointed him to the University Chairs. By this time he had already established a very successful medical practice, which he retained until his death, treating many of the Enlightenment characters in Edinburgh.[31] These two topics of botany and materia medica were of essential importance to a physician as most remedies used at this time were plant based, and indeed, botany was part of the medical course at Edinburgh University until the second half of the twentieth century. Many doctors had their own recipes developed from plants which many grew themselves, so a knowledge of botany was crucial for the simples that were given to patients. This led on to a tradition by which many doctors became botanists.

Hope's lectures were based on three important concepts: critical reading of the established texts; fieldwork which was studied both from the angle of the plants and the conditions in which they grew, as well as the inculcation of keeping full and careful notes; and thirdly, that experiments should be rigorously executed, with proper scientific controls and the ability to repeat the experiments and obtain the same results. This clarity of methodology was well shown in a series of experiments on the relative strengths of the effects of light (phototropism) and gravity (geotropism) showing the dominance of light over gravity on the direction of plant growth.[32] With the aid of diagrams such as these, he devoted an entire lecture (Number 28) to the conflicting forces of these two factors on plant growth.[33]

He emphasised the importance of practical fieldwork and varied his lectures accordingly:

> As the season is uncommonly late, I find if I proceed immediately to Classification I shall run short of Flowers for your dissection, and therefore for some days I shall be employed in considering whether or not there is such a thing as the distinction of sexes in Plants, on the supposition of which the celebrated Linnaeus founded all his order of arrangement.[34]

This is further emphasised in a series of plant lists which Hope collected in 1765, which 'show us also that at the period referred to considerable attention was given to the flora of Scotland, and that field Botany was a definite part of the teaching of Botany by John Hope.'[35]

He performed a set of experiments in which he demonstrated, by the use of growing plants in such a way as to show that they would grow towards the light, even if this meant growing downwards; and when he grew plants upside down they

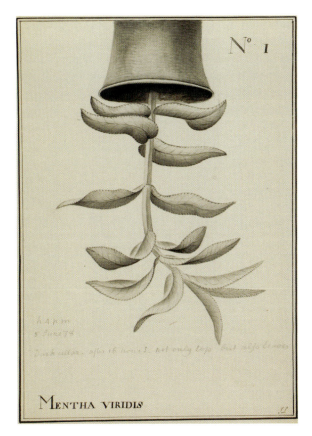

5 *Hope's experiment on the effect of gravity on the growth of spearmint (now* Mentha spicata*). Ink drawings by John Lindsay demonstrating negative geotropism under both light and dark growing conditions. The four drawings are captioned: I 'h 4 pm 5 June 78. dark cellar, after 16 hours, not only top but also leaves'. II 'h 4 pm 5 June 1778. light room 16 hours'. III 'Dark cellar 2 days 15 hours observe top leaves'. IV 'Light room 2 days 15 hours'.*

curved upwards providing that there was no directional light to interfere with the effect of gravity. Although his students would have been aware of this work, Hope never published it, but it pre-dates 'by nearly a century any similar experiments on the interaction of plant tropisms.'[36] Additional support for this emphasis on clarity of exposition and scientific methodology is shown by the importance that Hope gave to plant anatomy, devoting in the 1777/78 session five lectures to the structure of flowers and sex in plants, emphasising in lectures 30 to 32 the importance of really careful experiments and showing the weakness of those of Alston. After one lecture on hybridisation and an introduction to classification, lectures 34 to 40 and 57 to 64 were devoted to describing the various classes and orders, according to the Linnean sexual system, relying on the use of microscopes to give basically mechanistic explanations: when he was in doubt, Hope gave explanations of the varying theories and left it up to his students to decide, such as the function of the anther in lecture 56.[37]

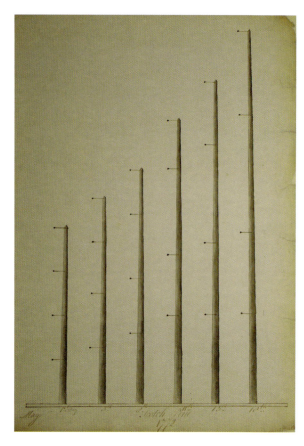

6 *Hope's experiment on the rate and position of growth of a root tip of Scots pine* (Pinus sylvestris)*. Ink drawing by one of Hope's assistants. These experiments were performed in May 1772, with recordings taken on days 1, 5, 8, 11, 13 and 18, and the results appear to have been included in about his 8th lecture each year during at least the later 1770s.*

Hope had been trained by Alston in the Boerhaavian methods of reading yet always questioning the theories of the past against his own observations. Thus, he started his series of lectures with the importance of relying on one's own senses: 'I believe Gentlemen that we derive more knowledge from the senses, viz. the taste & smell, than from all books together. I can say thus far for myself, that I got more knowledge from these in the Materia Medica, than in books.'[38] He then continued his series with four lectures on taste. The importance of observation and dissection is gleaned from the number of lectures which he devoted to this latter subject: during the session 1777/78, he devoted Lectures 34 to 40 and 46 to 50 on dissecting flowers to show how they could be separated into their various classes, having spent Lectures 24 to 28 dissecting flowers to show sex in plants, starting with the simple tulip before moving on to a further five species.[39] This Boerhaavian attitude was reflected well in his own teachings: for instance, in his twelfth lecture, he referred to the work of Sir John Hill, who had described the cuticle as consisting 'of Longitudinal vessels'.[40] This emphasis on continual questioning as well as relying on first principles is further exemplified by his comments on the sexual system of classification set out by Linnaeus.

Hope also appears to have been well aware of the importance of the role of doctors in the developing second British empire following the independence of America, both as physicians but also using their knowledge of materia medica as a starting point to advance knowledge of the natural history of those areas which were beginning to be studied and visited.[41] Nearly five per cent of the students on his class lists were to be doctors in India, but as a number of his students did not graduate from Edinburgh, of whom Roxburgh himself was a good example, this figure must be seen as a minimum.[42] This international aspect of his teaching is also reflected in the number of foreign students who attended his lectures: for the ten years in which the country of origin is given in his Class Lists, 55 of the 813 students did not come from Scotland (nearly seven per cent), the majority of these 55 came from Ireland (33), but others came from as far afield as America (six), the West Indies (three), Sweden (three) as well as Russia, Spain and Italy who all provided one. This attitude of Hope's is confirmed by his comment in one of the existing sets of his lecture notes, although sadly, and inevitably as is the case with these things, the undated set:

7 Plan of the Edinburgh Botanic Garden, Leith Walk, as developed by Hope, drawn in 1777 by William Crawford. It shows very clearly the intended circuitous paths that were to prove so popular to visitors, as well as the rhubarb field, off the top left corner.

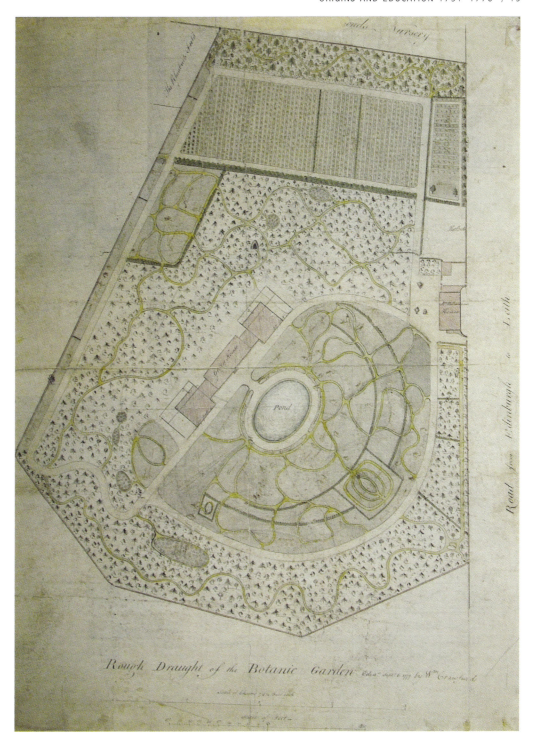

Rough Draught of the Botanic Garden

I mention this [the fact that roots can be propagated when seeds are difficult or impossible to ripen, and he gave as an example the case of the potato] to you gentlemen (as you will probably be in different parts of the globe) to encourage you to try to introduce into colder countries useful plants of warmer especially if the virtue is in the root which will ripen tho the plant will not bear proper seed.[43]

One of the main characteristics that emerges from a study of the various surviving versions taken from Hope's lectures is the value he put on showing examples of as many plants as possible.[44] Where possible, he showed living plants but was aware of the problems of fieldwork during bad weather: 'the 3 & last part of the course consists in demonstration, in this part I have made some improvement & in others I have given it up entirely, for a N° walking thro' the Garden especially in bad weather I found to be a great injury to it.'[45] What gave this aspect of his teaching great value was that Hope had a very wide circle of correspondents as was common amongst the scientific community of the time, and thus received plant material from the Americas, West Indies and elsewhere. Thus in the same second lecture he referred to plants coming from 'Mr Masson, who went out with Captain Cook', from Jamaica and from Spain, so that 'the number of Interesting plants in this Garden is daily increasing.' This meant that his students became aware of the discoveries of new plants as they were first sent to Britain. It also meant that on leaving Edinburgh, these embryonic botanists had a wealth of knowledge of all parts of the world so that they could develop further botanical discoveries.[46]

The importance of fieldwork was also stressed by Hope, who gave prizes for the best *hortus siccus* produced by his students at the end of their course. This stress on work in the field was reflected in the work that Hope did on the Scottish flora, and his influence on Roxburgh in this area is shown in his *Flora Indica*, which 'may be considered a specialised [*sic*] sort of "survey", no doubt partly inspired by Hope's pioneering (though largely unpublished) work on the Scottish flora'.[47]

Hope's own original work was also shown by his lectures on classification. As a starting point, he based his system on 'a continual gradation in point of sense &c. System and Classification are used by us in order to our understanding this gradation, for it holds good with regard to the vegetable & Fossil Kingdoms also but in nature I believe there is no such thing as Classification, we use it only as an assistant to our senses.'[48] This was a major advance from earlier systems, which viewed plant classification as ordered by God into three groups: trees, shrubs and herbs, of Caesalpino and Ray. Although he used some sexual characteristics in his system, as did Linnaeus, Hope's lectures stressed the natural system: 'There followed many lectures of dissections, &c. Such methods are natural as make many circumstances the foundation of the Classes …. The first objection to a sexual system is that he [Linnaeus] rests the foundation of the first eleven classes solely upon number, than which nothing is more variable as you have seen in numerous examples.'[49]

Although Roxburgh retained the sexual system for his own classification, as appears, for instance, in his *Hortus Bengalensis*,[50] he was probably aware of Hope's advances in thinking. This is because he came into close contact with Francis Buchanan, whose lecture notes they are (of 1780), and also Roxburgh was in contact with Hope up to the latter's death in 1786. The reasons for not using Hope's system were probably twofold. First, the Linnean system was easy to use and had become widely accepted, and so would be understood by most people reading the descriptions. Secondly, Hope never published his system, therefore to use such a system would not have been generally understood. In his lectures of 1780, he said, 'My design was to define the different classes And this I have at last accomplished and have reduced them into a sort of system which I intend soon to publish. This system I shall first give you and then present you with the plants belonging to each.' There follows a series of tables giving his system; however, Hope's system still needs to be fully analysed.

One of the important parts of Hope's course was the dissection of plants, so that students could actually see, handle and get to know the enormous variety of plants. To this end, he was always adding to the range, from all over the world:

The number of plants in this Garden is daily increasing, Coffee, Scammony, Camphor, the Indian pink & the most wonderful of all, the moving plant [*Mimosa pudica*] are in

8 Codariocalyx motorius, the telegraph plant (Bengali: burrum chundalli). *Drawing by an unknown Bengali artist. This drawing was sent to Hope by Dr James Kerr from Bengal around 1774, so could not have been seen by Roxburgh; but Hope must have shown his students similar botanical illustrations (prints and originals) by Western artists. The care with the botanical details is shown by the enlarged drawings at the bottom and was a practice followed by Roxburgh.*

great perfection …. Since last year I have received a considerable addition to my Stock, all the plants that Mr Anderson Collected on his voyage he sent me; I have also got a large collection from Mr Boswell who is lately arrived from India. This is the largest I have ever got.[52]

This leads to an interesting side of Hope's botanical teaching: how he included material from overseas, thus stressing the importance of exploration and plant collecting, a number of his former students going overseas as doctors and scientists. Over and over again, he mentioned the plants he acquired from overseas, as stated above. However, he also expanded on the gardens and collectors with whom he was in contact:

> The attention of his Majesty is not alone in contributing largely for the Botanic Gardens, but also in sending Missionaries in distant parts of the world for procuring herbs and seeds. His Majesty has also Missionaries to gather all plants of a rare kind. Mr Masson, who went with Captain Cook &c. Came home loaded with plants, & is now out on the same business.[52]

> The Buddlya globosa was brought to me by Mr Anderson, a surgeon to Captain Cook's ship who promised to be an eminent naturalist but suffered with his captain [on the fateful 1776-79 expedition].[53]

More examples were given in his 27th lecture in 1780. There are also two specific mentions in Roxburgh's letters of plants to be forwarded to Hope in Edinburgh: 'There are also two parcels of most of them [a list of 17 plants], one is directed for Dr. Hope,'[54] and 'In the chest N° 3 there is a small parcel of seeds for Dr. Hope, which I request of you to forward to him.'[55]

A study of the Indian Medical Service in the late 18th century with Hope's student lists gives an interesting result (Table 1). Bearing in mind that the numbers of doctors

Year	Number of students	Doctors in India
1761	55	2
1762	61	2
1763	72	6
1764	51	3
1765	27	2
1766	37	1
1767	73	2
1768	41	1
1769	55	3
1770	66	1
1771	84	0
1776	97	6
1777	97	6
1778	80	5
1779	61	9
1780	67	5
1781	77	5
1782	76	5
1783	96	0
1784	98	3
1785	90	2
1786	76	1
Total	1537	70
Percent		4.6

Table 1 Numbers of students attending John Hope's annual botany courses, and the number of students who can be identified for each year who joined the Indian Medical Service after graduating from Edinburgh. The class lists for 1772 to 1775 are missing.[56]

recorded in this Table are only those found practising in India, and that a number also went to the West Indies and probably others to America, the fact that as many as 4.6 per cent of Hope's students went to India is a significant proportion. This is particularly true as it does not include those who were ships' doctors nor does it include men like Roxburgh himself who was a botanist rather than a doctor after 1785. There is also the fact that a number of Hope's students never graduated from Edinburgh (again Roxburgh is an example). All of these factors will increase the percentage of doctors from Edinburgh in the Indian Medical Service but not appearing as Edinburgh graduates in the right-hand column.

A similar picture emerges when the class lists are examined with reference to the number of students who attended Hope's classes from overseas (see Table 2). In this case, there are a number of years in which no origin for the students was given, hence they have been omitted. Although the majority of what have been referred to as overseas students came from Ireland (33), other countries represented include America (six), West Indies (three), Sweden (three), and Russia, Spain and Italy each providing one, and there are a number whose names suggest that they were foreign, such as de Chair and Engelhart. There was also one whose origin was given as 'French prisoner'. One person who was probably a contemporary under Hope was Archibald Menzies who studied at Edinburgh University for the sessions from 1771/72 to 1774/75 and had been one of Hope's gardeners. Menzies went on to become the Government Naturalist in the *Discovery* under Captain Vancouver's voyages on the Pacific coast of America from 1790 to 1795, when he returned to high acclaim and when he introduced the Monkey Puzzle tree (*Araucaria araucana*).

Year	Number of students	Overseas students
1763	72	8
1777	97	8
1778	80	4
1779	61	5
1780	67	5
1782	76	4
1783	96	8
1784	98	9
1785	90	2
1786	76	2
Total	813	55
		6.8%

Table 2 The number of students attending John Hope's botany courses in years in which the country of origin has been given for those not from Scotland. Years in which this data has not been given have been omitted.[58]

It is interesting to note in this connection that Hope made specific mention in his lectures to the botanic gardens in Jamaica and Spain. In both cases, he gives a glimpse of the growing importance of botanic gardens and the collections which they were able to grow:

In Jamaica there are 2 Botanic Gardens which are very large, tho' in General when we speak of a Botanic Garden, we mean only a few acres of ground in which as many plants as can be procured are placed, yet these two I was speaking of are very large, one of 50 acres, the other 70, one is placed in the warmest part of the Island; the other in the coldest. Thro' the warm one runs 2 small rivulets, which supply it with water, & the Cold one is placed 3,600 feet above the level of the sea; this is higher, I believe than any ground in Scotland. There is at Madrid a very considerable Botanic Garden. There is no Kingdom in the world so fit for Botanic studies as Spain, its Kingdom is so extensive.[57]

Although not mentioned in Morton's short biography of Hope[59] as one of his correspondents, Hope did give the name of the botanist at Jamaica as Dr James Clerk.[60] It is thus possible that they were in touch, which would explain why students should come from the West Indies to Edinburgh. In Lecture 27, he also mentions plants from Pennsylvania (*Dionea muscipula*) and from Mr James Cherr (*Hydisarum movens* and *Crotalaria* sp.).[61]

Richard Grove stated that 'the process of botanical garden making was highly imitative even in the colonial period. The pattern and influence of Leiden and Amsterdam, for example, exercised an extraordinary organising power, at Paris, at the Cape, at St Vincent, at Calcutta.'[62] The analysis of Hope's lecture notes does seem to suggest that people in such influential positions as Hope may well have exercised more

9 *Carpogon capitatum by one of Roxburgh's Indian artists.*

direction in the development of these gardens than was previously thought. Certainly in the case of Roxburgh at Calcutta, converting it into a fully fledged botanic garden with research facilities with over 3500 species rather than the small collection of some 350 species that he inherited, may well have been instigated more by what he learnt at Edinburgh than from the patronage and global network of Sir Joseph Banks. Although Banks may have engineered a number of placements for people such as Roxburgh at Calcutta and Alexander Anderson in the West Indies,[63] the training of these men was a necessary precursor on which Banks could build. Although many of the plants which were coming into the country were beginning to find their way into the collections of private individuals, the main driving force for their collection was still scientific though collections such as Ashton Lever's museum 'which from 1774, when he established his museum in London, until its dispersal in 1806 served the dual function of popularizing the unusual and providing a resource for such well-known naturalists as Thomas Pennant (1726-98), John Latham (1740-1837), and George Shaw (1751-1813).'[64]

It is also worth stressing at this point that Hope had found the original garden too small and had persuaded the Town Council to allow him to move it in 1765. The new site of five acres, in what is now at the top of Leith Walk, was considerably larger and gave Hope the opportunity of laying the plants out in the Linnean system and other areas where they were planted according to their medicinal use but also making the design such that there were new vistas as one walked round the paths. Thus, writing in 1779, Hugo Arnot said that 'the botanical garden is generally visited by strangers, and considered one of the ornaments of the city of Edinburgh ... and so to place those [plants], that at least one of every species may be seen, in the course of the serpentine walk around this division.'[65] Thus, Roxburgh was introduced to the idea of a botanical garden having the dual role of containing examples of as many species as possible, and yet being the centre of botanical research.

One further point that is worth raising here is some advice given by Hope to James Edward Smith when he was studying in Edinburgh under Hope ten years after Roxburgh: Smith attended the university for the sessions 1781/82 and 1782/83. In a letter to his father, Smith stated that Hope 'recommends me, above all things, first to make myself master of Latin.'[66] This very practical guidance was so important in the medical and botanical world of the Enlightenment, when so much of the international scientific writing was still written in Latin. It would seem reasonable to assume that a similar attitude of Hope's would have applied to Roxburgh, the fruits of which were evident in his correspondence with the Rev. John at Tranquebar.

It does appear from this initial study of Hope that he did indeed adapt his teaching to prepare students for colonial science and especially botany. By extrapolating from a study of the Scottish flora and adding the examples he acquired from collectors, he showed the way in using the Linnean system in describing the new species that were being discovered at such a fast rate. The influence on Roxburgh can be seen from the number of genera and species that Roxburgh described.

A final comment about the medical rather than botanical training which students received prior to arriving in India is given by William Huggins, writing in 1824:

> Medical men form a branch of the Company's establishment, receive salaries, and are distributed among the different districts like civil servants. These gentlemen, after having taken their diplomas in England, and with, perhaps, a very moderate degree of experience, go out to India to practise their profession. Every country has its peculiar diseases, and a man must be practically acquainted with them, in order to treat them properly; between those of England and India the greatest difference exists, and a different treatment is, of course, necessary, the knowledge of which can only be acquired by experience and a residence in the country, so that, generally speaking, medical men on their first arrival are very unfit for the duties they undertake.[67]

This would certainly have been true of Roxburgh. He left Edinburgh after two years, spent the next thirty months as Surgeon's Mate on the India run before his appointment at Madras in 1776. It was not until 1790 that he was awarded his M.D., from Aberdeen, *in absentia*, as he was still at Samulcottah at the time, by which time he was virtually a full-time botanist.

Before leaving Roxburgh's time at Edinburgh, it is worth saying something about his time with Alexander Monro, studying anatomy and surgery. Like Hope, Monro stressed the importance of first hand experience, dissecting, observing and drawing in the advancement of medicine in Edinburgh. The quality of this work is beautifully shown in his *The Structure and Physiology of Fishes Explained and Compared with those of Man and Other Animals* that was published in Edinburgh in 1785. Once more, the emphasis on detailed observation and hands-on experience that Monro reinforced was to prove so important to Roxburgh.

The contacts Roxburgh made at Edinburgh were to be a gateway through which he gained access to a wider scientific sphere. The active role taken by his patrons is shown in a letter from Hope shortly before he died to Banks, in which he recommended one of his ex-pupils, Archibald Menzies, who 'was employed in the Edinburgh Botanical Garden while he studied for the Medical Profession', a position that Roxburgh may also have occupied.[68] Building on his friendship with Hope and Boswell, he met Sir John Pringle who was then President of the Royal Society, probably in 1772.[69] It was Pringle who read Roxburgh's first published article, to the Royal Society, and Roxburgh wrote in a letter: 'my most valuable friend the late Sir John Pringle' and in a later letter 'As far back as twenty nine or thirty years ago, I sent to Sir John Pringle, the seeds of various plants preserved in the hardened mucilage of gum arabic, which he wrote me reached him in very excellent conditions.'[70] The importance of Pringle lay in the fact that he was, amongst other things, President of the Royal Society, and in a London that viewed Scots with distrust, he consciously advanced his compatriots in the metropolis. The importance of the scientific contacts that Roxburgh made in these early years will be further developed in Chapter 7.

ARRIVAL IN INDIA

William Roxburgh must have left Edinburgh in the late summer or early autumn of 1772, for his next appearance is as Surgeon's Mate of the East India Company's ship *Houghton*, whose log for this journey starts at Deptford on 9 November 1772.[71] The *Houghton* had been launched in 1766 and this was its third voyage; at just over 700 tons, it ranked at this period towards one of the larger ships of the fleet; it was nearly 140 feet long with a beam of 34 feet 4 inches and a hold of slightly over 14 feet; but the height between decks was only 5 feet 10 inches. The total crew for this journey consisted of a Commander, five Mates, Purser, Surgeon, Surgeon's Mate, and 112 others, plus 73 soldiers and seven passengers of whom four were 'waiters' and one a 'Black Servant'. These servants would probably have belonged to the Commander, Surgeon and First Mate, and, as there were two unaccounted passengers, one or two of the servants may have belonged to them. A number of ships' Surgeons used their spare time to study the natural history of the land surrounding the ports of call on their voyages, and many had draughtsmen as their servants.[72] Roxburgh joined as assistant (Surgeon's Mate) to Richard Ballantyne, ship's Surgeon. After sailing from Deptford on 27 November for Gravesend, which was reached on 1 December, they sailed on towards The Downs a fortnight later and remained there a month. Having tried to leave on 23 January 1773, in the company of the frigate *Southampton* and Indiaman *Northington* as well as several smaller vessels but had to return two days later, the small fleet finally left on 29 January for Madeira, where they stayed from 17 February to 2 March.

They then took the normal route, following the South East Trade Winds which, as one contemporary put it, 'oblige ships bound round the Cape of Good Hope into the

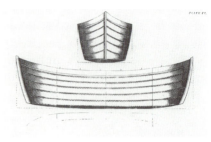

10 'Madras Embarking'. Hand-coloured aquatint by C. Hunt, May 1856, after a watercolour by J.B. East. Passengers are carried into a massulah boat before being ferried to a ship standing off in the Roads, and the dangers of a soaking for passengers and cargo going through the surf is evident.

11 Plan of a Massulah boat, as used chiefly at Madras for loading and unloading cargoes, and carrying passengers to and from the Roads. Engraving from John Eyre from 'Description of the various Classes of Vessels ....', Journal of the Royal Asiatic Society of Great Britain and Ireland (page 9, 1834).

Indian Ocean to stretch down to the coast of Brazil before they begin their easting, so Trinidad was an important landfall and departure.'[73] From Madeira, the *Houghton* sailed on to the Cape, arriving at False Bay on 4 May which they left on the 11th. The next call was Madras where they remained from 8 July until the 29th. This would have introduced Roxburgh to the delights of landing at Fort St George, through the surf in massulah boats, vividly described by William Hickey and referred to by almost all who arrived there.[74] Spavens's description of landing is equally graphic for someone of his poor education: 'There is scarce any bay, but it is almost like riding in the open sea; and at that fine time of the year, the surf runs so high and impetuous, that it is not practicable for a boat to land with any degree of safety, except the masula-boats which are built on a very peculiar construction, and of such materials as render them flexible, and so very pliant as to yield to the beach when they strike on it, so that they are able to bear those shocks that would dash any other boat to pieces.'[75]

After the short journey to Bengal, they arrived at Culpee on 8 August and remained there until 7 January 1773. It was here that Roxburgh was introduced to what appears a fairly frequent practice and one that he appears to have followed two years later, of members of the ship's company 'running' (i.e., jumping ship), in that the entry for 30 December, states against Richard Ballantyne 'Run December 30th 1773 at Bengall', three other crew members also 'ran' at Bengal. As there is no mention of anyone replacing Ballantyne as Ship's Surgeon, it must be presumed that Roxburgh took on this job but did not receive any increase in pay for it. Although the voyage was comparatively uneventful, an incident at Calcutta, on 12 December 1773, does highlight one of the fears of the crew:

> The Northington hailed us & told us that a ship which we saw coming was the Dolphin Man of War, upon which our people applied for leave to go on shore as they were afraid of being pressed; on my [William Smith, the Commander] refusing them, they said they were sorry to do any thing that was disagreeable to their commander & Officers, but at the same time they must Consider their own Liberty, & that they would take a boat & go on shore. Consequently they hauled up the Cutter, & went forcibly into her. Soon after the Gunner came and told me, that many of them had gone into the Gunroom & took upwards of 20 Cutlasses from him. Swearing they would defend themselves from being pressed. They said at going off, if the ship was in any danger and a signal was made, they would come off immediately to her assistance. The Number gone is 33 petty officers included.

This does suggest that Smith was deemed a good Commander but their fears were far from groundless, for some months later seven men were transferred to His Majesty's ship *Buckingham*, on 15 July 1773, while at sea. The fear of the press gang hardly needs expansion, for the conditions in the navy at the time were enough to generate a mutiny at Spithead in 1792. With eight of the crew dying at sea, one of whom fell overboard, there was a total loss of 19 out of the 121 who started, and of the 73 soldiers, ten died at sea and one went missing at the Cape.

The return journey was slightly shorter, lasting from 8 January 1774 until 4 June, covering a total of 13,410 miles in 134 sailing days, compared with 15,789 miles in 152 sailing days on the outward journey (averages of 100 miles per day on the homeward voyage against 104 miles per day on the outward). On this journey, the *Houghton* stopped at the Cape, for ten days from 6 March, took 13 days to reach St Helena where they remained for a week (leaving on 4 April) and the only incident of interest on the journey home appears in the log for 19 April:

> At about ½ past 2 pm or midnight was Alarmed with a sudden Appearance of Shoal Water close under our lee Bow. The Helm was put instantly a lee but the Mizon & Mizon staysail Clewlines being both up on account of a squall coming on, the ship did not come too very fast, so that we were in the white water before we lost our way & perceived it was only arising from a quantity of Fish collected together. I was the more Alarmed as there is a Look Out inserted in the Charts near our then Latitude for which I has ordered a good look out to be kept but it rained so hard we could see very little way. It appeared in patches after we had passed through it & the water was full of shining particles astern more than usual, which confirms me in my Opinion for we had not time to sound.
>
> Had we passed a little to windward or to leeward of this we might very possibly have determined it to be a shoal (unless I had sent a Boat to it) & such things have doubtless given rise to some which are laid down in common tracks, which if they existed surely must be seen by some as there are now usually so many more ships passing than formerly and this too convinces me how little a dependance there is to be layed on a look out when any real danger is in the way confirmed by the narrow escapes of the King William in 1738 & Waber in 1748 among the shoals off the South West Part of Madagascar & others I have seen in Journals.

There is now an unaccounted gap in Roxburgh's movements, but unlike his early years this time for only about four months, as he next appears Surgeon's Mate on the *Queen*, whose log for the relevant voyages opens at Deptford on 19 October 1774.[76] The log gives more insights into life on board, and the problems with the weather before leaving, as there were gales and snow storms and frosts, all of which must have made life below decks very uncomfortable. There is also information about private trade being taken on board, for the Captain and officers who had strict quotas laid down as to how much they could each have, giving one possibility on how these men could make some money from each trip over and above their pay: Roxburgh, as Surgeon's Mate, was due £1 10s. per month, payable at the end of the complete voyage (one way the East India Company tried to stop crew from jumping ship during the voyage). The Captain, George Stainforth, seems to have had more of a problem with discipline than William Smith on the *Houghton*, for there are a number of occasions when crew and soldiers were 'punished on the gangway' or lashed for drunkenness and/or 'sticking' one of the midshipmen.

After leaving the Downs, off the Kent coast, on 26 December, the journey took them first to Madeira, where they stayed from 12 to 15 January, but because of strong gales had to return on 24 until 31 January. By the middle of June they had rounded Ceylon and arrived at the Madras Roads on 23 June, after 144 days at sea: after so long and such heat, the state of the water and food does not bear contemplating. Here they stayed until 4 August, during which time at least one crew member died, six were 'transferred to the King's Service' and one to the Indiaman *Salisbury*, meanwhile they took on board three seamen. A number were also sufficiently ill for the fact to be entered that on 8 July 'we have now 19 sick'.

From Madras, they sailed across the Bay of Bengal towards the Malay peninsula and then on up to Macau which they reached on 10 October, reaching Wampoa (the port for Canton) on 21 October. Here the ship remained until 2 January 1776, but without Roxburgh, for he 'ran at Canton 1 January 1776'; probably he was one of the 26 who went to Canton in the liberty boat on 26 December and failed to return, forfeiting his pay for the whole voyage in the process, for the East India Company paid off the crews of the ships on their arrival back in England. This action by Roxburgh, of foregoing a year's pay, must be seen as an extraordinary step to take unless he had some alternative security such as an open letter of introduction or source of income.

12 A View from the King's Barracks, Fort St George. Hand-coloured aquatint from Francis Blagdon's A Brief History of Ancient and Modern India *(plate II, London, 1805).*

These three-plus years and two voyages confirm Roxburgh's statement that 'he had made two voyages to the East'.[77] They introduced Roxburgh to many of the places that would become important to him in later years: St Helena, the Cape, Madras, Bengal and the East Indies. With Hope's teaching and the influence of Pringle behind him, he would have travelled with his eyes open to the botanical opportunities that these places gave and also met some of the important scientists within the network that these men would have developed. The range of climates and the differences of the seasons would also have been highlighted by these voyages, and with the interest of the Royal Society in recording meteorological data, it may have concentrated Roxburgh's mind on doing some of this himself, leading to his first two published papers, particularly as the first was read by Pringle although by the time the second was read, Pringle had retired and Sir Joseph Banks had succeeded as President of the Royal Society.[78] It must also be remembered that at this time it was usual for the ship's doctor to be a naturalist, who would use the voyages to further his own scientific knowledge and make collections of natural history, often supported by detailed diaries and drawings.

REVIEW OF ROXBURGH'S FIRST TWENTY-FIVE YEARS

By the time Roxburgh reached Madras in 1776, he had gained a sound knowledge in botany and an interest in wider sciences, some of which will be revisited in Chapter 7, and an awareness of the importance of performing and describing reproducible experiments, examples of which are described in Chapters 7, 9 and 10. This botanical knowledge was based on the Linnaean sexual system of classification and nomenclature, a system that he used for the rest of his life although aware by the end of its weaknesses. It was supported by the importance that Hope had given to accurate field work as well as the value of descriptions already published, although Hope had emphasised the need to treat these critically. Hope had also introduced Roxburgh to the wider range of plants which grew in the emerging world, as more and more parts were opened up and the discoveries reported back to Britain.

Already he had gained the friendship of Sir John Pringle, President of the Royal Society, as well as that of John Hope and the family of John Boswell. During his years travelling between Britain and the East, he had visited St Helena and the Cape of Good Hope as well as spending time in the islands of the East Indies and visiting China. Not only had this introduced him to the natural history of these regions, and an indication of the range that was still to be discovered and described, but also to the problems which plant collectors could face, and which are further discussed in Chapter 6.

CHAPTER 2

Coromandel Coast Years 1776-1793

MADRAS AND NAGORE

After five months' travel, Roxburgh arrived at Madras and was appointed as Assistant Surgeon at the General Hospital at Fort St George on 28 May 1776, interestingly without anyone acting as sponsor.[1] It was an appointment that suggests that he must have carried with him some letter of introduction from one of his patrons (Hope, Boswell or Pringle), because it was necessary for anyone joining the East India Company to have two sponsors and there is no record of this in the archives: this was one way the Company had of restricting who entered India, as we shall see later with Roxburgh's friend William Carey. It acted as a form of censorship as well, as the Company could rescind the employment of anyone if they broke the patterns of business or behaviour that was acceptable. Having said that, once established in India, the limits of behaviour were very broad! This need will be highlighted below in the case of Roxburgh's eldest son, William junior, for whom Sir Joseph Banks was requested to be the nominee.

The Madras that Roxburgh entered was in a state of political turmoil, for the Governor (President of the Council), Lord Pigot, was removed from his post in August 1776.[2] This was

> because never in the whole history of the Company had there been, or was there to be, a worse scene of general fatuity than had been displayed by the Company's agents at Madras at the time that Hastings joined them. In every possible way affairs had been grossly mismanaged. Not content with mere negligence in providing the Company's investment, the Madras service had become thoroughly corrupt, and not content with being merely corrupt, it had involved its employers in a needless and ruinous war which was the principle cause of the virtual bankruptcy of the Company .... Peace [with Haidar Ali] had at length been dictated to the thoroughly cowed and terror-stricken councillors at the very gates of their city only a few weeks before Hasting's arrival.[3]

Warren Hastings had been appointed as Governor in 1771 but by 1774 was operating as a minority faction within the Council, and tried to set about improving the situation, partially by a radical change of attitude, because he understood that India 'was another world with a set of values, ideas and institutions entirely different from those of his own country.'[4] The wars against Haider Ali and Tipu Sultan with their effects on British rule in India did affect Roxburgh, but the results were not of major consequence except in so far as they determined his postings, as will be shown below and also probably the ease with which he was granted an appointment at the General Hospital in Madras on his arrival. The important point is that from the 1750s the French had been assisting the Indian princes with both training and armaments while the British had hung back even though they had signed a treaty with Haidar Ali in 1769. As a result, Haidar Ali and then Tippu Sultan, his son, joined forces with those of Mysore, which he was controlling, against the Marattas, in a series of wars that lasted until Tippu's death at the fall of Seringapatam in 1799.

One of the great interests of the Royal Society at this time was collecting meteorological data, so that the climates of those parts of the world that were being explored and opened up to trade could be better understood. As President of the Royal Society, Sir John Pringle was very involved with this and thus Roxburgh would also have been made aware of its importance, both intrinsically and if he wanted to be accepted amongst the metropolitan scientific world. It is not surprising, therefore, to find that the first two of Roxburgh's published articles were on climate and were read by the Presidents of the Royal Society, first Pringle then Banks. Roxburgh's work on climate and meteorology is studied in more detail in Chapters 7 and 11.

Roxburgh's first article, published in the *Philosophical Transactions of the Royal Society* in 1778, also shows that he made excursions away from his house, for there are no readings for the period 28 January to 19 February 1777, where he stated that he was 'in the country about 6 miles W' and then 'in town again.' Later, some of these absences were on botanical trips, because by the following year at the latest, Roxburgh had met Johann Gerhard König, a Dane who had come to Tranquebar as surgeon and naturalist in 1768 then entered the service of the Nawab of Arcot in 1774 before his appointment by the

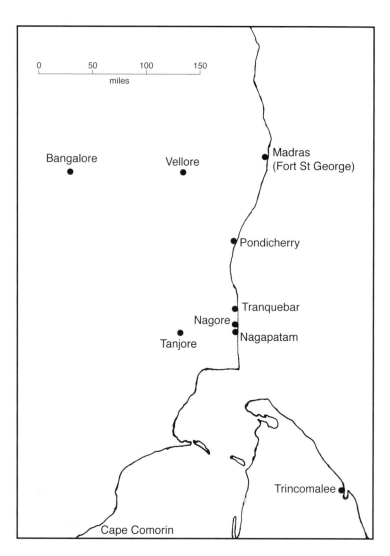

13 Map of Southern India.

East India Company as Naturalist on the Madras Establishment on 17 March 1778.[5] König stayed in Madras, certainly at first, with Andrew Ross who also became a great friend of Roxburgh's,[6] but of equal importance to Roxburgh, he had been a student under Linnaeus, thus reinforcing the botanical teaching of Hope in Edinburgh. The importance of the Danish Mission of the Society of United Brethren at Tranquebar emerges again and again over the decades following König's arrival there, through the work of Benjamin Heyne, Johann Gottfried Klein and John Peter Rottler, the first of whom took over from Roxburgh at Samulcottah in 1793.[7] The spread of the botanical network from this outpost in southern India reached Sir Joseph Banks, who as early as 1777 had been receiving plants from them, helped, no doubt, by the proximity of their headquarters in London, at Lindsey House, not far from the apothecaries' Physic Garden in Chelsea.[8] There is, therefore, a double link between Roxburgh and Banks: through Pringle as President of the Royal Society to his successor Banks, and via the botanist König at Tranquebar.

From his very early days in India, Roxburgh must have been sending plant material home, for in the first extant letter of his, to Sir Joseph Banks in March 1779, he stated 'I have looked over all the copys of the collections I have sent home.'[9] This letter suggests that he had recently come from Madras and that he did not expect to remain there long: 'Nagore the place of my abode at present, ... My removal from Madras has put me out of my usual way a little.' It also gives one of the few indications of the impact of the Mysore War, for he suggested to Banks that specimens of seeds and plants would be safer sent in Danish or Dutch ships. The letter is also the first

14 Tipu Sultan (1750–99). Hand-coloured aquatint from Francis William Blagdon's A Brief History of Ancient and Modern India *(frontispiece, London, 1805).*

indication of the friendship between Roxburgh and König, for he stated that 'Dr König is now on a voyage to the islands in the Straits of Mallacca, and Siam. I expect him daily at this place on his way to Madras.' One final point which emerges from this letter is that it is possible that it was Banks who initiated the correspondence, for Roxburgh started the letter by saying

> About a month ago I was honored with your very agreeable Letter dated March 25th 1778. Till then I durst flatter myself that any collection of seeds, or specimens of Plants I could make would be half so acceptable as you say, or I would not have waited for your orders, for I wish to send such things to every person that will only pay proper attention to them, and not let them be lost.

These last few lines give a deep insight into Roxburgh's views of plant collecting, a practice that was thoroughly grounded under Hope in Edinburgh. Roxburgh was obviously not prepared to waste his time sending plants to people who would not give them 'proper attention' and equally, he would not spend unnecessary time despatching plants if they were liable to get lost by their recipient: he was aware from a very early stage in India that care had to be taken over the way the plants were sent home and he devoted much care and energy to ensuring that they arrived in a viable state. Banks was not good at letting Roxburgh know the state of the plants on arrival, and there are a number of letters in which Roxburgh expressed dismay that no information was given about this: perhaps as well, as many arrived dead, as will be described later (see Chapter 6).

Another insight from this letter is Roxburgh's lack of confidence, a characteristic that reappears over the decades and in spite of the high regard in which people held him. The presence of Warren Hastings at Madras when Roxburgh arrived was also important, for Hastings was instrumental in setting up the Asiatic Society in Calcutta, with its interest in Indian culture, science and religion, and was crucial to the academic milieu of that city when Roxburgh arrived there in 1793. To have met Hastings would therefore have been a great help and would have been assisted by his own connections with the scientific hierarchy of London.

CORINGA

By his own reckoning, Roxburgh was not a good correspondent at this stage, for after he was driven out of Nagore by Haider Ali in about March 1781, it took him until November 1782 to write that he had 'been drove from Nagore by Haider Ally above 20 months ago'.[10] However, the letter also stated that 'About the end of last year, I got appointed Surgeon to this small out Garrison [Samulcottah],' which makes one wonder whether he spent the early part of 1781 on active service; it also confirms his promotion from Assistant Surgeon to Surgeon. These early letters also give some indication of the character of Roxburgh which is reinforced over the years in many of the letters that followed. The first point is that he was very much an experimental and field botanist and scientist who supported his work by reading the earlier publications and referring to them in his letters: thus he wrote about 'my own experiments' and 'I have made no further experiments on the Galls'; 'here I am generally at a loss for Paper to put my specimens in' and 'Since I am to the Northward, I have observed a Plant, which is common hereabouts, but never saw it to the Southward, which I take to be a new species of Convallaria'; he mentioned that Nagore was 'called Nasux in Jeffreys map' and 'In Burman Thesaurus Zeylandica (which is the only Book I am possessed of that mentions its purgative virtues).'[11]

As regards his character, he comes over as a person who was a stickler for detail but also lacking in self-confidence: the former is shown by his explanation and description of the medicinal uses to which Indian plants are put, the descriptions of his experiments with dyes and fibres, and his meteorological diaries (on the latter

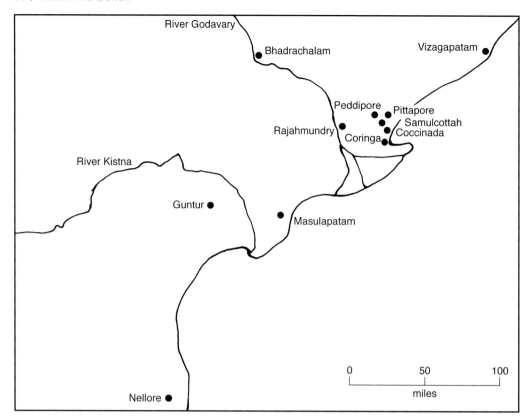

15 *Map of the Coringa area.*

it is interesting to note that although these were published for the years 1776-78, he never received copies of them and added the diaries for 1779-80; these aspects of his character are dealt with in more detail in Parts 2 and 3. The matter of confidence, already alluded to, is reflected in his comment, 'I shall now exert my frail abilities to render my letters more deserving of your attention. I am much pleased with the last part of your letter which in some degree removes a great deal of diffidence which I laboured under, regarding composition or language.'

The move from Nagore was the first disaster which Roxburgh suffered, for in December 1784, he wrote 'The beginning of 1781 we were obliged to evacuate Nagore in a great hurry, by which I lost my Thermometers & many things of value.'[12] Although he did not again start keeping his meteorological diary immediately, this had been kept for some time by January 1793, and he was referring to the rainfall measurements in 1791.[13] The second disaster was even more devastating for Roxburgh: a hurricane swept through the Guntur District in May 1787 with sufficient force and causing such damage that Patrick Russell wrote to Banks 'You will from public accounts hear of the dreadful hurrican[e and] inundation that happened last month at Coringa, Ingeram, &c ... Poor Roxburgh with his family made their escape with the utmost difficulty, his house at Cochinara was totally destroyed.'[14] Roxburgh's losses, however, were not all from the hurricane, for there is an interesting series of records in the archives regarding a theft of 12 candy loads of cotton cloth from Roxburgh after the storm.[15] The occurrence of cyclones was not uncommon, it being stated that 'the usual months for such storms are May and November.'[16] What was unusual in this case was the severity of the storm, highlighted in a letter to Andrew Ross after the event:

> From the 17th May, it blew hard from the North East; but as bad weather is unusual at such a Season, we did not apprehend that it would become more serious, but, on the 19th at night it encreased to a hard gale; and on the 20th in the morning it blew a perfect hurricane, insomuch that our Houses were presently untiled, our Doors and windows beat in, and the railing and part of the wall of our inclosures blown down. A little before eleven it came with violence from the Sea, and I presently perceived a multitude of the

Inhabitants, crouding towards my House, crying that the Sea was coming in upon us! I cast my eyes in that direction; and saw it approaching with great rapidity, bearing much the same appearance as the Bore in Bengal River. As my House was situated very low, I did not hesitate to abandon it, directing my steps towards the Old Factory, in order to avail myself of the Terraces; ... but before we attained the Place of our destination, we were nearly intercepted by the torrent of Water .... I think the Sea must have risen fifteen feet above its natural level.

William Parsons continued: 'The Fishermen, a most useful body of People, inhabiting chiefly by the Sea-side, have been almost totally extirpated': the size of the disaster can be imagined when one remembers the catastrophic effect of the tsunami that hit this same coast in 2004. In another letter, Roxburgh himself wrote that

On the 20th, about daylight it [the wind] veered to NNE and NE and began to blow and rain exceeding hard, the Wind encreasing to a degree of violence I cannot describe: I was with my Family in a House which I had, near the sea, at Cockanara: By 9 o'clock the Sea had risen 6 feet above its usual high-water level: I then with my family, with utmost difficulty, reached a large Gentoo House that stood high. By Noon the Sea was 12 or 13 feet above its high-water level, and had extended upwards of Eight miles inland ...[17]

Roxburgh continued:

The loss in lives is reckoned about 15,000 drowned; My House was entirely washed away, and everything I had in it to the value of about 10,000 Pagodas: all we saved was ourselves, and just what we had on our backs: I had not even time to carry off Mrs R's jewels; what I regret most, was a most valuable Botanical Library, all my Manuscripts, Drawings, preserved specimens of Plants, &c which I had been collecting ever since I came to India.

16 Xylia xylocarpa (Mimosa xylocarpa). Hand-coloured engraving from Roxburgh's Plants of the Coast of Coromandel (plate 100, London, 1798), based on a drawing by one of Roxburgh's Indian artists (Roxburgh drawing 482).

This extract raises some interesting questions, for a young Surgeon in the East India Company would not have been able to build up assets of 10,000 Pagodas (equivalent then to about £4,000) in a period of ten years from salary savings. The Medical Regulations of 1785 had fixed the pay for surgeons as the same as captains, which in 1796 was the equivalent of £27 13s. 6d. per month when in garrison or cantonment duty, rising to £37 10s. on field duty, but little of Roxburgh's time was at the latter level.[18] It does suggest that he may well have been a very successful trader during these early years, and to have 12 Candy loads of cloth stolen, over 2½ tons, suggests that this was more than he might need for private consumption! This may also explain why he employed Mr Amos as his agent at Madras, a fact again mentioned in many of his letters, and he also had the advice of Andrew Ross, one-time Mayor of Madras with whom he struck up a strong friendship. Later, in correspondence with the Rev. John at Tranquebar, there is frequent mention of Roxburgh supplying various goods to John and his colleagues.[19] To get an idea of the size of this trading, when Macartney led his embassy to China in 1793, he noted that the value of the

cotton imported into Canton in 1792 was worth the equivalent of over £400,000 and
that the profit margin was in the region of 100 per cent.[20] He also stated that 'Our
settlements in India would suffer most severely by any interruption of their China trade
which is infinitely valuable to them, whether considered singly as a market for cotton
and opium, or as connected with their adventures to the Philippines and Malaya.' It
is thus easy to understand how a well-connected member of the community could
soon earn a considerable fortune: we shall see later the level that Roxburgh reached
by his death.

As far as Roxburgh was concerned, the greatest calamity was the loss of his drawings,
his herbarium collection and his library. The former could be remedied, by setting his
artists to start again, but first he had to build up his plant collection. His losses were
such that 'he wrote for some necessaries' to a friend to supply.[21] Russell continued that
although the floods were not bad at Vizagapatam, 'our trees and hedges were utterly
stripped of their leaves, presenting as bleak an appearance as a Greenland winter.' As
far as his library was concerned, Ross managed to get Roxburgh some books 'at Mr
Davidson's Sale & otherways'.[22] However, Roxburgh continued to complain for some
time about this loss, writing, for example, to N. Kindersley, Dr J. E. Smith's nephew,
that he had 'but a very poor Botanical Library'.[23] By the time he left India, though,
he does appear to have built up a good collection, as shown by the List of Books in
Roxburgh's Library in Appendix 6.

Again the closeness of Roxburgh and König is shown by the fact that Roxburgh
was with his friend when he died and was, indeed, his executor. When Patrick Russell
wrote to Banks in July 1785, he stated that he had been ill since before returning from
Calcutta at the beginning of May, when 'Dr Roxburgh had interdicted an immediate
application to business'. By the end of that month, he was much weaker and both
König and Roxburgh were apprehensive of the consequences. By 6 June he had
made his will and when Roxburgh saw him two days before his death, he had made
final arrangements for all his papers. König had planned to set out for Madras to
prepare for a journey to Tibet or Bhutan,[24] but contracted dysentry so severely that
Russell 'was so much alarmed that he sent for Roxburgh the Surgeon General to
his assistance. Both of them loved König.'[25] Roxburgh probably took König's papers
down to Madras that September before forwarding them to Banks, for 'Dr Roxburgh
having acquainted me [Patrick Russell] with his intention of transmitting these papers
immediately to Madras, in order to be sent to you by the first conveyance, I took the
liberty of suggesting whether it might not be more advisable to wait a more favorable
season.'[26] Russell immediately followed König as Company Naturalist on the Carnatic
with an allowance of 60 Pagodas per month, confirmed in November 1785.[27] However,
by December 1788 Russell was packing up to come home and recommending that
Roxburgh should succeed him, 'he is the only person on this Coast in any degree
qualified, and I make no doubt of his being appointed'.[28] This was, indeed, confirmed
in April 1789.[29] However, an interesting letter was written by Banks in 1787 that
suggests that he was not the unbiased patron of Roxburgh that the latter presumed.
Banks wrote that 'Swartz [the Danish botanist] is the best botanist I have seen since
Solander. I have hopes of getting him to supply Koenig's place in India'.[30] I have
not been able to follow this up to see what steps Banks took to try to effect this
placement, but how different Indian botany would have been had he succeeded.

Samulcottah stands on the edge of a hilly region possessing a very interesting flora
that Roxburgh explored with great ardour,[31] and it does appear as though Roxburgh
enjoyed his years there. For example, there is a delightful letter from the Rev. John
at Tranquebar, where he refers to 'your description of your northern paradise is very
tempting.' The letter also has a lightness which is equally appealing, for John starts the
letter 'My Dear Adam' and ends with wishes for 'Eve' as he referred Roxburgh's wife,
Mary.[32] Equally, 'The soil of this country is in general fruitful, but better about Corcondah than to the southward, in the vicinity of the Godavary, where it is more sandy. It

is, however, by art and industry, brought almost to the state of an earthly paradise.'[33] Conversely, shortly after Roxburgh arrived at Calcutta, he was bemoaning the effect of the Bengal climate on his already poor health: 'I am now from home on account of a bad state of health which I have been more or less afflicted with ever since I left the Coast but more so for these last two months.'[34] On a recent visit back to this area, I noted that both cotton and tobacco are grown extensively: both requiring rich soil.

However, the fact that Roxburgh included the phrase 'should anything remove me from hence' in his letter of August 1790, does suggest that he was beginning to think of leaving. This is reinforced by a number of factors. He had arranged with Marischal College in Aberdeen to be awarded his MD, which was granted on 12 January 1790;[35] and at two meetings at the Royal College of Physicians of Edinburgh in February of the same year, Roxburgh's election as a Fellow was accelerated, to take place within a fortnight rather than the customary three months (it is interesting to note that he was seconded by Joseph Black, a friend of Hope's).[36] Added to these, he was elected a non-resident fellow of the Royal Society of Edinburgh a year later, on 4 January 1791, which may have spurred him to send them a collection of dried plants that is now in the herbarium of the Royal Botanic Garden Edinburgh.[37]

This preparation for leaving India and to gain the necessary qualifications to practice medicine in Britain must have been instigated in late 1788, for in his letter of 20 January 1790 to Patrick Russell, he started by saying, 'When I last wrote you by the Dublin, my mind was very unsettled, and otherwise unhappy; which thank god is not the case now; and I have made up my mind to a longer stay in India than I ever intended.' This change of heart was almost certainly due to the death of his first wife, referred to by both Andrew Ross ('the sincere & serious concern which I felt at the intimation I rec'd of the irrepairable loss which you met with – & I heartily condole with you on the Occasion') and Patrick Russell ('I shall resign my appointment, and recommend Roxburgh as successor. I imagine he had been about leaving the country, but having lately lossed his wife, he intends remaining some years longer'), and the latter's reason for suggesting Roxburgh as his successor as Company naturalist on the Carnatic.[38] It must also be remembered that life expectancy in India was short and Roxburgh had already survived there for over twelve years. There is, in addition, the fact that although he had lost one fortune a few years earlier, he may have accumulated sufficient to look after his family in Britain, particularly if he could develop a medical practice there.

Although the timing may appear odd, it must be remembered that letters could take anything from six to twelve months to arrive in England. Roxburgh would have approached both Marischal College, Aberdeen, and the Royal College of Physicians Edinburgh, in the autumn of 1788 with it arriving in autumn of 1789, which makes his elections for January/February 1790 reasonable. Thus, Roxburgh wrote in August 1790, 'I am afraid all the ships of the season are arrived & no account of the Books you mention, how cruel it is to tantalize me so. a chest of Books & Diplomas from Edinburgh is in the same predicament.'[39]

Also by this time, Roxburgh had married his second wife, Mary Huttemann, probably the sister of George Huttemann who ran the Free School in Calcutta. The connections with the Nonconformists is strengthened that the marriage was solemnised by the Rev. Gericke who was one of the missionaries from Tranquebar.

COMPANY NATURALIST

When Patrick Russell retired from India, Roxburgh was appointed his successor as Company Botanist on the Madras coast, in April 1789, when the position was described as 'making Researches in Natural History on the Coast.'[40] (The terms Company Naturalist, Company Natural Historian and Company Botanist all appear to have been used synonymously.) When Russell had been appointed, the job was described to render the Botanical improvments of the late Dr König [who held the post previously] more immediately subservient to the uses of Life, either with regard to Medicinal Applications or to Arts & Manufactures – And you are to require from Dr Russell, so long as he shall continue to receive the Company Pay, an Annual Communication of all his discoveries and Improvements which is to be regularly transmitted to Us that we may lay the same before the Royal Society.[41]

This has a strong indication that Banks, as President of the Royal Society, was behind this requirement of reporting back to London, a condition that does not seem to have applied to König who died only a short time after Banks became President.

This position of Company Naturalist on the Carnatic had been originated so that the Company could make use of the knowledge and skills of Gerhard König, who had been a student of Linnaeus in 1757 and arrived in India as a physician in 1768, determined to make his name as a naturalist. Not being able to finance botanical field trips on this salary, he took the post of Naturalist to the Nawab of Arcot in 1774, which enabled him to travel more extensively, both to the nearby hills as well as further afield, including to Ceylon. Passing through Madras, he would have met Dr James Anderson and after he arrived in 1776, William Roxburgh, confirmed in the latter's letter of 1779. König's aim was an appointment with the Company, which he achieved in 1778. An idea of the breadth of his brief is given when his successor was being looked for: 'not botany only but zoology, mineralogy and chemistry.'[42] Patrick Russell, who was his successor, had tried to persuade him to publish some of the results of this research, and through the good offices of Banks had won the agreement of the Company to help. This would end up as Roxburgh's *Plants of the Coast of Coromandel*, for which 'Curtis's Flora Londinensis be taken as the Model'.[43]

As the title suggests, the position of Company Naturalist in the Carnatic broadened into more than just botanical: Roxburgh's predecessor, Dr Patrick Russell, had a great interest in snakes and produced a pioneering work on Indian snakes,[44] and Roxburgh gave him some help on this, just as Russell helped Roxburgh over his plants, indeed, 'spent his remaining years checking Roxburgh's descriptions'.[45] Roxburgh himself took some interest in this subject, for he sent descriptions of snakes to the Rev. John at Tranquebar to identify, and to Russell on the Arnu Boa which had killed two sepoys at Rajahmundry.[46] In a further letter, Russell stated that

> I have only to regret that it has not been more in my power to execute all I intended, & ardently wished to perform. Leaving matters in this unfinished state, it is natural to wish that some researches may continue to be prosecuted with equal Zeal, but with superior abilities, & this I hope will be admitted as an apology for presuming to suggest a Surgeon. The Gentleman to whom I allude is Mr Roxburgh of Samulcotah, of whose skill and practice in Indian Botany, I have unequivocal proofs & this conjoined with his local knowledge, acquired in the course of a Long residence in the Country induces me to think him singularly qualified to be of essential service to the Company ... more with a view to immediate utility, than to the extension of Pure Botany.[47]

Russell, however, had been far more outspoken in Roxburgh's praise to Banks a couple of months earlier:

> Great industry and much practice in India Botany has enabled him to repair in a great measure the late loss of his large collection; and his local knowledge render him a very fit person to prosecute the views of the Directors. He has described, and got coloured drawings of a considerable number of plants, and has favoured me with some specimens for your inspection. He is the only person on this Coast in any degree qualified, and I make no doubt of his being appointed.[48]

As mentioned before, Roxburgh's first wife had died towards the end of 1788 prior to which Russell was under the impression that 'he had been about leaving the country, but, having lately lossed his wife, he intends remaining some years longer'.[49] Roxburgh remained unsettled for some time after his wife's death, and it appears that something

17 Coluba stolatus (*native name* Wanna Cogli), *a snake that was sent to Dr Patrick Russell by William Roxburgh in July 1788. One specimen was sent from Rajamundry and one from Samulcottah. Hand-coloured aquatint from Patrick Russell's* An Account of Indian Serpents (*plate XI, vol. 1, 1796*).

*18 Squirrel on a tamarind.
Hand-coloured engraving
by William Hooker after a
painting made by James Forbes
in 1769, from Forbes's* Oriental
Memoirs *(page 354, vol. 2,
1813).*

happened in late 1790 or early 1791 which nearly made him resign, for Ross referred to 'the troublesome business which is the subject of your last letter is determined, and respect to that I can only say that the good Man Dr Duffin gives me hope that the Business will be managed in such manner as to relieve you from the necessity of removing or of making any resignation, on the plea of your Complaints.'[50] This was followed by a letter two days later, when 'Mr Gerick called upon me soon after, & on canvassing the embarrassing circumstances of your present situation he seemed to be decidedly of opinion that the forfeiture of your rank was a more convenient alternative than an absolute resignation.'[51] However, a week later all appeared to be resolved, for having spoken to Sir Charles Oakeley, the Governor at Madras and well known to Roxburgh, Ross was able to write that 'there is no occasion for the removal of yourself, or of the other Gentleman in the Physical line to the Northward who were desired to repair to the Army, so that I may congratulate you now on the probability of your stay there you [*sic*] being permanent.'[52]

This problem could have been caused by a clash between his two roles, of Surgeon and Company Naturalist. These dual roles certainly did pose problems for Roxburgh, and by early 1792 he was looking into ceasing his medical practice, but his pay was related to his rank in the Medical Service. This was first raised in an undated letter but from its position in the archive would appear to be March 1792, when Andrew Ross gave him advice as to how to set about achieving his aim, and was restated in July, when Ross ended a letter with

> I am prompted by Dr Berry (from a motive of friendship) to suggest to you – that it is now proper that you should consider seriously what line it may be most eligible and convenient for you to pursue in respect to your future views & advantage in the Companies Service as there is reason to believe that those who are your Juniors in rank will now – (on the near departure of Dr Duffin) do all that they can to benefit themselves by the apparent diffidence which you lately expressed of taking an Active part in the Medical or Chirurgical line and the seeming desire which you expressed of being left to pursue your separate Study of Natural History.[53]

This debate about his future was also carried on with the Rev. John at Tranquebar, for there is one of the latter's long, flamboyant letters at the end of the year, persuading Roxburgh to remain in India:

> For Gods & mankinds sake don't leave before 10 or 20 Years your Place. Europe does not want you so much as India. Shall your Plantations all go to Ruin. Can you make so many Discoveries for the Benefit of the World in England which is now in the greatest Cumotions by Payn's Right of Man, than you have done & will do here. You will be torn in pieces with all your Plants, Manuscripts & Drawings & you shall never have my Blessing in such a Degree as you have & shall have as long as you remain in India.[54]

Further consideration of Roxburgh's botanical work while at Samulcottah is treated in Chapters 8 and 11.

19 *Spotted kingfisher on* Aponogeton natans. *Hand-coloured engraving by William Hooker after a painting made by James Forbes in 1784, from Forbes's* Oriental Memoirs *(page 270, vol. 2, 1813).*

THE CORCONDAH LEASE – BACKGROUND

From the mention of 'your plantations' in the letter from the Rev. John mentioned above, it does appear that Roxburgh had already taken out the lease of some land shown by the fact that he had felt able to consider retiring to Britain in 1788, even though his fortune had been lost in the hurricane of 1787. In the beginning of January 1793, with the help of Andrew Ross, he set about renting an estate, at Corcondah, and the study of the negotiations for it are a good example of the procedure that was followed for this.

The exact relationship between Roxburgh and Ross is never clearly explained in their correspondence. However, Ross, an employee on the Revenue Board and one-time Mayor of Fort St George, was a man of position, but nevertheless was prepared to go to enormous effort to help Roxburgh in achieving this lease: over a period of ten months, he wrote nearly seventy extant letters either to Roxburgh or about his business interests in connection with the Corcondah lease, and Roxburgh appears to have sent a similar number to him, but sadly these are now mostly missing. One insight is from a letter of June 1793, in which Ross wrote 'that the application for Corconda will succeed to our wish' which does suggest that there was some common interest in the matter.[55]

One insight into the relationship comes from Ross's awareness of Roxburgh's humanitarian views. This was shown by his worries over the deaths caused by the devastating famine that had lasted for three years, and his plans to introduce sago palms and other crops to obviate the future occurrence of such disasters:

> I now give you a communication of very important papers passed in the Revenue Board – relative to the Administration of the Northern Circars – which you will peruse with the best attention & make such remarks on every part of them without reserve as your knowledge of the several circumstances noticed, & your judgment may direct you & particularly with regard to the conduct of the different Landholders & others in power towards the inferior Classes of the people & to the poor at the commencement of & during the scarcity.[56]

After the Peace of Seringapatam after the Third Mysore War in March 1792, the Guntur Circar was formally occupied and Company Collectors were appointed to survey rental assessments for five-year periods. These were based on village records as settled by Tipu Sultan at the time his revenue was at its greatest. The problem here, as will be shown below, was that these did not take into account the terrible famines which lasted from 1790 for three years to 1792. Added to this, the judicial side of the rent-collecting was rudimentary, the Government believing that a permanent settlement of the revenue was necessary before a system of courts, judges and judicial regulations could be introduced.[57] This opening up of the Guntur Circar may have provided the opportunity for Ross and Roxburgh, and a number of others, to apply for newly available plantations.

This idea of renting land was still fairly novel, and the fact that other people were doing the same, Ross felt to be a help, and in his letter of 28 March 1793, quoted three people as precedents: Mr Cochrane who produced silk, sugar and indigo; Mr Hamilton; and Mr Popham who cultivated 1500 acres of cotton. The value of this tenanting had reached such a level that by April, the agents of Roebuck and Abbott (a trading partnership based at Madras) had been granted 'several Districts'.[58] Ever the man for ensuring that Roxburgh's interests were being looked after, Ross also suggested in the letter of 28 March that he should put in for 'an addition to your monthly as Botanist & what you Profit in the Natural History way & therein state the length of time the labour, the attention, the expence – risk to your health – & other inconveniences to which you have been exposed.' This avuncular care of Ross for Roxburgh is also seen in the mild admonishment given, 'When I said I wished you to make your Observations, what Grounds might be fit for the cultivation of Sugar, it was not in my mind, that you should Manufacture it; but only that there might be an apparent advantage in knowing it; to those who might hereafter be able to procure Long Leases or Purchases of such Grounds.'[59]

The letter which Roxburgh wrote setting out the reasons for renting land, particularly the purgunnah of Corcondah, is extensive and detailed running to several pages.[60] It reflects so many aspects of his life in Samulcottah: his work on experimenting with new and potentially economically useful crops, such as sugar and indigo; his knowledge of the countryside, from his trips collecting botanical discoveries; his humanitarian sympathies for the state of the natives and his attempts to improve their lot; his ability to get on well with the local chiefs, rajahs and zemindars; and his awareness of the political necessity of approaching people in the manner most likely to bring about a satisfactory conclusion.

The short introductory paragraph gave a broad economic reason for developing crops that would improve the trade between India and Great Britain, particularly sugar and indigo. He then continued, explaining that his knowledge of these two crops depended on his experiments over a number of years, by implication largely at his own expense: this is confirmed in his letters, for instance to Banks about sugar, and his early publication on indigo (and see Chapter 9 for more information on his work on indigo).[61]

Roxburgh continued with an explanation of why he had been unable to rent a property for this purpose before. Tenancies were traditionally inherited by the sons of the previous tenant on his death but cases did arise when there was no heir: these were described as havally lands. In those parts that were controlled by the East India Company, they took over the arrangement for the new tenant. Local zemindars, who dealt with this problem in native run territories, were not keen on letting to Europeans, who would not have accepted the same level of control, an attitude that Roxburgh confirmed.

The letter went on to show Roxburgh's familiarity with the area, from personal experience, as it was accompanied by a sketch map. By implication, Roxburgh must have been an intrepid traveller, for he refers at one point to 'the wild independent Purgunnah of Rampah' and then later that he had been in friendly correspondence with the Rajah of Rampah. The wildness of this country was emphasised by a description of four villages that he had visited, that had been 'reduced to ashes & the cattle carried off'. This visit to the area had also persuaded Roxburgh as to which areas could grow the various crops that he had in mind:

> culture of Sugar Cane will only be found here & there scattered throughout the whole, & for Indigo on the lighter soils along the skirts of the Mountains immediately on the banks of the Godavary, the latter situation will also prove favourable for the growth of the Mulberry, which I shall be glad to extend, if it meets with the wishes of Government.

The final part of the letter concentrated on the terms of the lease, which Roxburgh pleaded for ten years, to give time for him to make it worthwhile to invest sufficient to make 'such improvements as may promise a speedy change on the face of the Country for the better'. This was reinforced by the fact that the 'hostile neighbouring Polygar of Rampah' would render the area less profitable, and therefore should be let at a lower rental than normal. The work that Roxburgh did in preparing the financial security of the project is further highlighted in a letter to Andrew Ross dated 4 July 1793, in which he stated that he had 'made every possible enquiry about the collections [rents] of that country & find that during the best of times before the Famine when the country was full of Inhabitants, it would not do more than pay that revenue (20,000 pagodas) after deducting charges for collections'. After querying how anyone could offer to take the land for 24,000 pagodas ('they must take unwarrantable means in drawing anything like that sum.'), his keenness for the land was then stressed,

> nothing would be more agreeable to me than the possession of that country, but the terms are at least a third more than I will give, & even then I log my account with only just clearing myself during the first two years or till by encouragement I can bring in People to cultivate even the lands that were formerly in cultivation.

The care with which he approached the whole project is further developed towards the end of this letter, when he stated that with between half and a third of the population having died during the famine just over, and the tanks broken by the heavy rains that ended the drought, he had taken advise from 'Mr Evans who knows the state of the country well [and] thinks I might give 16,000 [pagodas]'. As a result, he felt that 'it is better to stop even now than to involve myself in a troublesome ruinous concern.' All this suggests that he had produced quite a detailed business plan.

Roxburgh, as he stated in a letter printed in *Oriental Repertory*, had travelled in the hills surrounding the Godavary estuary, while searching for *Nerium tinctorium* [a dye plant], and had obviously noted the lie of the country as well as remembering the comments given to him some years earlier about the richness of the soil.[62] As a result of this, he must have mooted to Ross the idea of renting 'a large spot of ground in the Corconda Zemindary',[63] which lies 'near the foot of the Badrachelam Hills. The soil is good, well cultivated, and affords ... a tolerable profit.'[64]

This was followed by a suggestion from Ross first, then supported by Major Beatson, that Roxburgh should help in surveying the Godavary with a view to canalising a part

20 *Peons of the sort that Roxburgh requested to guard his plantations against marauding hillmen. These ones were from Mysore, painted by H. Salt and engraved by Pollard for Lord Valentia's* Voyages and Travels to India *(London, 1809, vol. 1, facing page 407).*

of it. The reason behind this was to provide irrigation and reservoirs so that the famine caused by the drought of 1790-92 would not be repeated.[65] An exploratory survey was carried out with Mr Denton in late February and by the end of March Ross sent a theodolite 'which with its Frame or Stand I bought at Coll. Ross Sale for Pgs 7 after Major Beatson had valued it at from 15 to 20',[66] but the perambulator had to be specially made at a cost of just over Pgs 20 and was despatched to Roxburgh in early May.[67] As so often, meetings at one stage in his life produced benefits later, for when he visited St Helena on his return to England in 1813, Beatson was Governor of that island. Further discussion of Roxburgh's work on proposals for watering the Circars is given in Chapter 13, and his work on St Helena in Chapter 6.

The first indications of the process of gaining the lease came in April, with Ross calming Roxburgh's fears that the negotiations could take as long as a year, and also pointing out that the leases obtained by Roebuck and Abbott near the mouth of the Godavary were for '10 to 12,000 Ps Annualy, which is an encouragement to you.'[68] He therefore suggested that Roxburgh should apply for a lease immediately. The letter written the next day gave the exact figures, and described the fact that 'it is publickly understood that the whole Business is for the account of Messrs R & A & that the Native is introduced as a convenience.'[69]

Ross wrote that he had spoken to Sir Charles Oakeley about taking 'a long Lease of an extensive piece of Ground – for the cultivation of Sugar to which he did not by any means give a refusal.'[70] This is the start of a flurry of letters: among the extant archives of Roxburgh material, there are over the following eleven months 74 remaining letters and references to about a further 60 which are missing. One effect of this was that Ross urged Roxburgh to learn the native (Gentoo) language [presumably Telugu, but Roxburgh may have already have had some knowledge of either Telegu and/or Tamil, from the mother of his eldest son, John] and in the process gather 'any Mss or other writings of curiosity & value in that language …. This is an important & Curious Subject if you can pursue it to advantage, you would much gratify your learned Friend Sir Wm Jones & Coll: Kyd.'[71] (The relationship between Roxburgh, Jones and the Asiatic Society is considered in more detail in Chapter 7.) Interestingly, Ross felt that 'the French have been far superior to the English in this respect & that very many of their Comp[ys] Serv[ts] & others, Civil & Military had (& some still do) improved themselves in this way.' There were, however, a number of major problems which Roxburgh with Ross's assistance, had to overcome, from local Company policies which were against giving leases to Europeans, and politics and jealousies between the British officials, to those connected with the local Zemindars who had learnt to play the various factions against each other. Once these had been overcome, further detailed problems arose concerning the negotiations for the land, its protection and rent, and ending with the need for a manager.

On 9 May, Roxburgh wrote again to Haliburton, the Secretary to the Revenue Board at Fort St George, repeating much of the letter and adding that the recent famine and its concomitant depopulation of the area had reduced the value of the Zemindary.[72] World politics took a part at this point: the arrival of news via the 'fast route' via Cairo rather than round the Cape, of the declaration of war by France against England, meant that Ross was delayed in bringing the matter to the attention of Sir Charles Oakeley. However, Ross felt sufficiently confident that he suggested 'that you may make every preparation, but without making communications to others.'[73]

THE LEASE – NEGOTIATIONS

Ross worked indefatigably behind the scenes, talking to Mr Cockburn, the new member of the Madras Revenue Board, followed by a long letter to Haliburton in which he asked what more needed to be done to get a more acceptable basis for reckoning the level of rental, as that proposed had not been deemed satisfactory

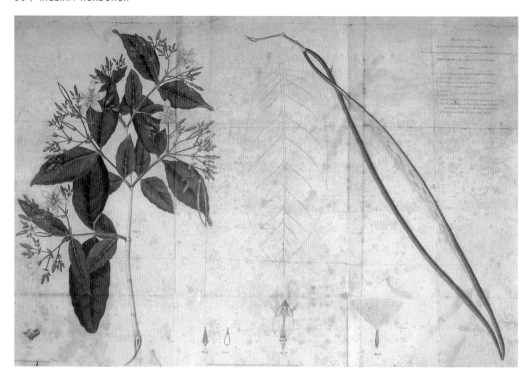

21 Wrightia tinctoria (Nerium tinctorium). *Engraving published by Roxburgh with a description of indigo in Dalrymple's* Oriental Repertory (*vol. 1, p 39, 1793*), *from a drawing by one of Roxburgh's Indian artists (Roxburgh drawing 18).*

and emphasising again the calamitous effects of the famine: 'In many places where populous villages formerly stood, there is at present neither vestige of Man or Beast, & not a village in the country, which does not exhibit the most melancholy marks of depopulation and decay.'[74] This very detailed letter explaining the advantages of the proposal was followed the next day with a further long letter, with a description of all that Roxburgh had done to enhance 'the prosperity of the Comp[ys] Possessions on this Coast' and thus was deserving of the lease, the rent of which should also be reduced because of the lawlessness of the neighbouring district.[75] This letter then had to be seen by three members of the Revenue Board 'separately at home' which had been done by 24 June when 'the Revenue Board, at this days sitting, have Agreed to recommend your proposals to Government, without reference to the Chief & Council of Masulipatam.'[76] The reason for the latter comment was because of the 'shameless practices of the C[hief] of C[ounci]l with several persons to whom they wished the Corconda District to be let.'

Four days later, sadly in a letter that is badly mutilated and therefore large parts are not decipherable, Ross appeared to set out some of the politics and the fact that the Revenue Board had put a value of Pgs 20,000 as the rental, being Pgs 4,000 less than the offers made through the Collector.[77] Next day the news was even better: 'The time is 5 years certain, at Ps 18,000 per an. Gross Revenue renewable for 5 years more at Ps 20,000 which [will] no doubt give you satisfaction;' this was confirmation of the Minutes of the Revenue Board meeting a week earlier which Ross sent to Roxburgh early in July.[78] However, Roxburgh felt that, after the deprivations caused by the famine, which killed between a half and two-thirds of the population, he could not pay more than Ps 15,000 but was advised by Mr Evans 'who knows the state of the country well thinks I might give 16,000.'[79]

The cowle, or lease, under the terms given by Ross on 28 June, was agreed by the President and Council, and two days later Ross wrote recounting some of the shenanigans that had been going on between the Collector and the Chief and Council, to try to ensure that the application by Mr Roebuck would go through: as Ross put it 'that the C[ollecto]r has more than a voluntary kindness' for those who hoped to rent the land! The later letter, of 10 July, seems to suggest that the 'voluntary kindness' may have been in the region of a full year's rental.[80] An indication of the problems of communication on such a fast moving set of negotiations can be gleaned from the fact that Roxburgh's letter of 4 July took over a week to reach Ross, who then had the progress of the cowle stopped. By the time Ross had had a chance to discuss this with his contacts on the Revenue Board, the issue had been clarified to the extent that these friends had tried to 'grant you the most favourable terms in their power ... a very essential advantage in your favour ... that if you decline it ... the terms offered by the other Party shall be immediately accepted.'[81] When Roxburgh had written on 4 July, he was just recovering from a fever and this had obviously depressed him, for two days later he had revised his limit to Ps 16,000 which Ross felt was sufficiently close to the recommended Ps 18,000, that Roxburgh would accept this higher rental. The precariousness of these negotiations was highlighted the next day, when Ross sent a copy of the letter from his contacts on the Revenue Board, David Haliburton, Charles White and Thomas Cockburn, to Sir Charles Oakeley, the Governor, in which they stated that an effective ultimatum had been sent:

> As it appears however, doubtful from Mr Ross's letter, whether Mr Roxburgh will rent the District on the terms proposed, and as it is of importance at this late season, that some arrangement should be immediately determined upon, we recommend, that the Cowle be sent to the Chief & Council to be in the first instance tendered to Mr Roxburgh, & in case his refusal, that the Proposal of the head of the Inhabitants of 24,000 M[adras]PS per annum for 5 years be accepted without further reference to the Presidency.

To this is added a comment from Ross: 'The offer of the Inhabitants of Corcundah is (as you see in the inclosed Copy of the letter from the Revenue Board) still said to be 24,000, but this is because the Chief & Council still keep back a later Offer of the Inhabitants wherein they propose no more than Ps 22,000.'[82]

By the end of July, Ross wrote to say that when Roxburgh received the cowle, he would have 24 hours to make up his mind, and was persuaded again by Mr Evans to take the lease.[83] By the beginning of August, Roxburgh had decided to go ahead with the lease but still had misgivings about some of its terms which Ross explained were 'the stated forms of the public papers'. Another problem which Roxburgh raised again, having mentioned it earlier, was the need for some form of protection from the 'Hill Plunderers', but the answer was to apply through the Chief of Council at Masulipatam.[84] The politics of the Chief and Council now began to come into the open, for they withheld the cowle from Roxburgh who was advised 'to keep yourself as little connected with, as the most likely means of being independent of either the Chief & Council or the Collector, or their creatures, as possible.' It turned out later that the claimed reason for this delay was 'that the River was then so high that people could not pass it. As the detention was not less than 10 days I distrust the report & suspect that his people have attempted to deceive him'.[85] The letter of 9 August also commended the idea of Roxburgh raising some 'Hill Rangers' and having his friend, Captain Denton, to be in charge of them.[86]

Complications were to continue, for Roxburgh was informed that the customs of Corcondah would not come to him but would be levied and held by the Collector, as he did at the farm leased to Roebuck and Abbott.[87] In the first of these two letters, Ross also gave Roxburgh a slight rap over the knuckles for reacting the way he did, rather than writing direct to Ross for advice. However by the end of the month, after discussions with Mr White and Mr Cockburn, Ross was able to say 'that all the Murtapha Taxes have been ordered to be abolished in the Havily Districts – & that you must not levy them but that you are to Collect every other Duty as may have been usualy levyed by other Zemindars altho you will act with as much moderation as may be consistent with your Interest & the ease of the people'.[88] Once this was cleared, Roxburgh was able to devote time 'in settling the Business of Renting the Farms to the Undertennants, ... & I hope that you have accomplished it to your satisfaction.'[89]

22 *Copy of a letter from Andrew Ross to William Roxburgh, dated 6 March 1793, showing an example of the handwriting of one of Roxburgh's writers, and the condition of one of the better manuscripts.*

The unusualness of the whole arrangement is shown by the letter from the Revenue Board to the Court in London: 'Tho' we are of opinion that your system ought generally be to exclude the employment of europeans, in the farm of your Lands; it may sometimes happen to be of your Interest, to encourage Persons of distinguished talents and industry, tho' not natives of India to engage in undertakings of this Nature. Doctor Roxburgh's abilities have been manifested, in various useful researches and his Character and Disposition qualify him, in a peculiar manner, for promoting objects of Improvement, both in Agriculture and Commerce: we chearfully acquiesced therefore in the proposition of the Board of Revenue, for renting the Corcondah District to that Gentleman.'[90] The final problem was left unresolved, the protection of the District by raising a corps of peons.[91]

INSTALLATION OF A MANAGER

However, in June 1793, Roxburgh had been invited to move to Calcutta on the death of Col. Robert Kyd, and he left the Carnatic for Calcutta where he arrived in November, to be the new Superintendent of the Botanic Garden there. This meant that he had to appoint a manager for the Corcondah property, who was initially Mr Robson but was eventually handed over to Roxburgh's friend Captain Denton. As a result of the continuing problems with the Collector and to collect his family, Roxburgh had to return south briefly in January 1794.[92] The Revenue Board were in the process of holding an inquiry into the behaviour of the Collector, but Ross's reaction was that

> nothing can be so much in favour of your good conduct, as the misconduct of your opponent both here and at home (tho' it may give some trouble in the mean time) & the worse they behave [the] better for you – if you will stand up [for yourself] the Collector should now [be brought] upon to account for the Charges which were brought against [him in] the letter of the Revenue Board – where they speak of the detention of the Cowle and many other things.[93]

The interference of the Collector continued, but turned to Roxburgh's advantage, for he had allowed an abatement in the rent for the farm of Roebuck and Abbott.[94] This is in spite of the fact that when Tudor, Willis, 'his Dubash [man of affairs] & the Gentoo Accountant Caululvarup Nursia Punt came to Caucanaud [Cocinada] at the time when Mr Robson was of ill Health & there they stayed 2 days ... in order to agree for ⅜ more than our Proportion.'[95] The matter was finally resolved in an undated letter which accompanied one of 30 March, in which Robert Gardener, the Chief at Masulapatam, was instructed 'that the District [of Corcondah] shall remain under your immediate authority without the Intervention of the Collector.'[96] The problems of the rent were still causing difficulties, to such an extent that 'in consequence of the depopulated State of the Country, the unexpected difficulties that had arisen to the collection of the Revenue, from dissensions among the Inhabitants and the loss he had suffered from the overflowing of the Guadavery, that he might not be required to pay for the present year, more than he had received from the Country, and that the Rent might hereafter be at or near what the late Zemindar was assessed, being Madras Pagodas 14,692 per Annum.'[97] Roxburgh must have threatened that if this request was not granted, he would relinquish the farm, so the Revenue Board agreed a reduction of 2000 Pagodas in his annual rent.

When the renewal for the lease, after the first five-year term, was approaching, the Court of Directors, confirming the 'established principle, of not allowing Europeans to be concerned in the farming of any of the company's Lands, might be admitted in the particular case of Doctor Roxburgh,' felt that as he had been appointed to Calcutta, 'we much doubt the propriety of granting him an additional term of five years on the expiration of the present lease in 1798, notwithstanding he appears to have selected proper Persons to superintend the Farm during his absence.' However, the Court left it to the discretion of the Revenue Department at Fort St George, to renew the lease in view of the 'Money he may have expended, the improvements he may have made, and the progress of his proposed Manufacture of Sugar and Indigo.'[98] In fact the renewal was in the name of Captain Denton in 1798 but this did not meet with the Court of Director's approval: 'it was by no means our intention to sanction Military Officers engaging in such pursuits, and we therefore direct that any future proposal of this nature from a Military Officer be not attended to; ... personal residence upon the Farm must be an indispensable condition.'[99] It does, however, appear that Roxburgh's tenancy lasted until 1803 at least.

The belated reaction from the Court highlights a number of factors prevalent at the time and pertinent to Roxburgh. The first is the sheer length of time to get a reaction back from England to a problem: in this case it would have been nearly three years from the time of the initial letter of October 1798 to the arrival in Calcutta of the Court's response dated August 1801. Secondly it confirms the point made by Bayly, that 'by buying into revenue-farms, monopolies and the political perquisites which had been the stock in trade of the eighteenth-century kingdoms, company servants and free merchants effectively made the transition between trade and dominion before the authorities in England knew what was afoot.'[100]

ACHIEVEMENTS DURING THE COROMANDEL COAST YEARS

By the time Roxburgh left the Coromandel Coast in 1793, he had laid the foundations of a very successful scientific career. He had proved his worth to the Company by his work at the experimental plantations he had set out at Samulcottah, for which he had been rewarded by the appointment of Company Naturalist on the Carnatic. Internationally, he had been made a Fellow of the Royal Society of Edinburgh in 1791 having been admitted as a Fellow of the Royal College of Physicians of Edinburgh the previous year shortly after gaining his MD from Aberdeen.

He had published a number of articles on a wide range of subjects, from meteorological diaries through Pringle and Banks with the Royal Society, to work on dyes including the important indigo, and botanical descriptions via Sir William Jones and the Asiatic Society. With this increased renown came increased wealth and a growing family, and one of his future energies was devoted to finding jobs for his elder sons and a good marriage for his eldest daughter. The character of the man comes across as a person who was good at close, detailed work, but lacked confidence. Thus, when sending material to Britain, this was split between ships, and his worries about the quality of his drawings is referred to further in Chapter 6.

Before he left Samulcottah, he had come into contact with Alexander Beatson, whom he would meet again when on his final journey to Britain in 1813, when Beatson was Governor of St Helena. This was, therefore, a period in which he developed a sound knowledge of the way in which the East India Company operated, and with the help of Andrew Ross, used this to gain contacts, scientific stature and a sound financial basis. He also built on his earlier contacts and knowledge, having a humanitarian outlook that was exemplified by his attempts to introduce plants to ensure the well-being of the local inhabitants during times of hardship.

Having lost one fortune, he was in the process of ensuring another. The trade with China was lucrative and indigo was another important export from India: Roxburgh was involved with both.

23 Commiphora madagascarensis (Amyris commiphora). *Watercolour by one of Roxburgh's Calcutta artists, c.1796 (Roxburgh drawing 1053). This is a member of the same family as myrrh and frankincense; it is native to Tanzania, but was formerly cultivated in India for making scent.*

CHAPTER 3

Superintendent of the Calcutta Botanic Garden 1793-1813

24 Colonel Robert Kyd (1746–93), who founded the Calcutta Botanical Garden in 1786. Engraving reproduced in King's 'Preface' (1893).

THE DECISION TO MOVE

In 1786 Colonel Robert Kyd (1746-93) proposed the setting up of a Botanic Garden at Calcutta for the development of economically useful plants and the collection of rarities, something that he was particularly interested in though no great botanist, and he spent the next seven years developing the site at Sibpur. On his death, the position of Superintendent was offered as a paid post to Dr William Roxburgh. Before this garden was set up, botanical work in India had been more or less desultory and dependent on interested individuals. Its establishment saw a recognised centre of botanical activity set up in British India.[1]

Kyd had joined the East India Company as a cadet in 1764 and reached the rank of Lieutenant Colonel in 1782, when he became Secretary to the Military Department of Inspection in Bengal. With the settled life that this entailed, he was able to set up a private botanical garden near his house, at Sibpur, outside Calcutta. In June of 1786, he suggested that the Company should establish a 'Botanical Garden not for the purpose of collecting rare plants (although they also have their use) as things of mere curiosity or furnishing articles for the gratification of luxury, but for establishing a stock for disseminating such articles as may prove beneficial to the inhabitants, as well as to the natives of Great Britain.'[2] The reasons behind this were twofold: memories of the famine in Bengal of 1769 were still fresh in the minds of many Company officials; and there was also the desire to break the Dutch East India Company's spice monopoly. A site was selected by the west bank of the River Hooghly at Sibpur, a few miles from Calcutta, and in March 1787 the Company started requesting plants to stock the garden, although the formal approval from the Court of Directors was not given until July. Kyd, however, was primarily interested in the horticultural aspect of the garden and was a collector of rare eastern plants – hence his comment to the Court that they were not just curiosities – rather than viewing the garden as of scientific importance. By the time he died in May 1793, Kyd had achieved a garden with about 350 species and over 4,000 plants.

The first indication that the post of Superintendent of the Botanic Garden was offered to Roxburgh appeared in a letter of June 1793 from Ross in which he states

> I have this moment rec'd your letter of the 8th with the letter to you from Mr Harris of Bengal giving you encouragement to accept the option which will probably be afforded you of going to Bengal to succeed Coll. Kyd in the care of their Physick Garden, which it is very grateful to observe, it is a Credit to you, & justice to your Character But as I am perfectly satisfied, (& now indeed more than before, on conversing with Mr Cockburn the new Member of the Revenue Board) that the application for Corconda will succeed to our wish, & that the addition which you expect to your allowance will also be attended with good success & both, in a few days, it will be most prudent & consistent for you to remain where you are.[3]

25 Factory at Canton. Anonymous watercolour, c.1780-83. This is very much as Roxburgh must have seen Canton when he visited it at the end of 1775 and beginning of 1776.

This crossed with a letter from Roxburgh in which he must have intimated his intention of accepting the offer from Bengal, for in his reply, Ross wrote of 'your willingness to accept of the offer of going to Bengal'.[4] The first official notification came at the beginning of July, with a letter from the Governor General to the Governor at Fort St George:

> The Death of the late Lieutenant colonel Kyd having occasioned a vacancy in the Office of Superintendant of the Company's Botanical Establishment in Bengal, an Establishment which has proved of much General Utility, and is considered by the Hon'ble Court of Directors as very deserving of Encouragement we are induced from the high Opinion we entertain of the abilities of Dr Roxburgh, to request that you will grant him Leave of absence from his Station at your Presidency and permit him to proceed to Bengal for the purpose of undertaking the Charge in question. We have some reason to believe that it will not be Disagreeable to Dr Roxburgh; but as the Period of his Continuance in the Office will be in a great Degree, optional with himself, we hope that the permission we have desired will be granted without prejudice to his Rank and Claims to promotion at Fort St George.[5]

This suggests that Roxburgh had already been approached and given his reply, presumably the letter from Mr Harris alluded to in Ross's letter of 16 June. This plea was reiterated to the Court of Directors in August and in their reply, dated July 1795, nearly two years later, they emphasised the use of Roxburgh at Madras: 'We recommend it to you to consider whether this Gentleman's Abilities may not be more usefully exerted in the several pursuits in which he was so laudably engaged in the Circars, than as Superintendent of the Calcutta Botanical Garden.'[6]

This use of the phrase 'physick garden' by Ross does show his awareness of the fact that the early botanic gardens were based on the physicians' need for herbal remedies, and indeed that many doctors and mission stations as well as hospitals had there own physick gardens: certainly the garden at Tranquebar, for example, was well-known as a centre for plants of botanical interest.

Ross was worried that Roxburgh's departure for Bengal might jeopardise the plans for Corcondah, for at the end of June he was recommending that Roxburgh would see 'the propriety of remaining where you are till the Business is adjusted (as it will soon) & then judge whether you may not visit Bengal', a view repeated a few days later as the cowle drew ever nearer: 'you will now postpone your visit to Bengal.'[7] This, again, emphasises the likelihood that both Ross and Roxburgh would gain from this venture.

LEAVING THE NORTHERN CIRCARS

Not only did Roxburgh have to think about the move to Bengal, but there were also the problems of his successors at both Samulcottah and Corcondah. The Rev. John approached him about the former, suggesting that Dr Benjamin Heyne should take over which would have the extra benefit of stopping him returning to Europe.[8] In fact, Heyne was known to Roxburgh, for they went together on some expeditions, because of Heyne's 'knowledge of Chemistry': amongst the Roxburgh papers is a report by Heyne on iron mines.[9] This together with the appointment of a manager for the farm at Corcondah appear to have been settled by the middle of August:

> you had to leave to go to Bengal to keep your Botanical employ & put the pepper &c Plantations under Mr Heine. But with respect to Mr Robson the Bd of Revenue to be asked if they have any objection to his Managing the Farm – And it will be requested of the Bengal Govt to allow you to return …. The Board of Revenue will no doubt approve of Mr Robson, so that every thing is in good train.[10]

Heyne's position seems to have been approved fairly quickly, for Ross wrote before the end of August 'that the Govr & Council approved of your putting the articles which are under your charge into the Care of Mr Heine & recommended that Mr H should proceed to you without delay.'[11] It was, in fact, Heyne who succeeded Roxburgh as Company Naturalist on the Carnatic and who also succeeded in running the garden at Samulcottah: the two positions appear to have continued to be occupied by the same person until they ceased although there is no evidence that they were contractually linked.

To add to the pressures on Roxburgh, his wife produced Anna Elizabeth some time towards the end of August or the beginning of September: 'I sincerely congratulate yourself & Mrs Roxburgh on her happy delivery'.[12] However, Roxburgh left his family at Samulcottah and arrived in Calcutta at the end of November, when he took over from Dr Fleming who had been the acting Superintendent since Kyd's death.[13] He obviously viewed the Garden with scepticism, referring to it in a letter to Banks as 'a good Indian Orchard, & Nursery but of too great extent.'[14] He immediately set down the parameters that he felt necessary for the garden:

> I presume the chief intention of the Botanical establishment here to be the same as on the Coast Viz[t] the introduction and general distribution of useful Plants, if so permit me to request that I may be informed in what manner I am to carry it into execution … the principal design of the Botanical Establishment in this Country is certainly as he understands it to be, the introduction and general distribution of useful Plants, and that, if he will acquaint the Board in what Soils and Situations the different Species and descriptions of those plants are most likely to thrive, the Secretary will assist him with the Orders that are necessary for cultivating them in the different Countries.[15]

These Public Consultations also

> Agreed that a place of Residence, on the Ground Appertaining to the Botanical Garden be constructed for the Superintendant. Agreed, further, that he be desired to prepare and lay before the Board a Statement of the Establishment that will be necessary for the Garden, and to enable him to conduct the Duties of Superintendant, annexing thereto the Rates of Wages to be given to the several Descriptions of Persons employed.

At a levee in January, he met the Governor General, Sir John Shore, and 'took the liberty of mentioning House rent', which was agreed at 150 rupees per month until his house should be completed.[16] This function of the Botanic Garden does highlight the different expectations of such an

*26 Banyan tree (*Ficus benghalensis*). Lithograph by D. Redman after a drawing made by James Forbes in 1778, from Forbes's* Oriental Memoirs *(page 28, vol. 1, 1813). In the Calcutta Botanic Garden is a specimen famous for its size and dating from the earliest days of the Garden.*

27 The Old Court House, Calcutta. Hand-coloured aquatint from Blagdon's A Brief History of Ancient and Modern India *(plate XVII, 1805). The view that would have greeted Roxburgh on his arrival in Calcutta.*

institution run by a commercial organisation such as the East India Company, and the more scientific basis of even Kew at that time. The modern Botanic Garden has been described as

> essentially a museum of living plants, ... it is a place where plants can be studied by experts for the furtherance of scientific knowledge, and a place where exhibits are arranged for the education and recreation of non-experts. The maintenance of a collection of living plants involves the assistance from horticulturalists, so that Botanic Gardens are places where botanists and horticulturalists cooperate, to their mutual advantage.

As will be shown, there were elements of such co-operation at Calcutta from a very early stage.[17] Another aspect of plant hunting beginning to emerge in the east was the search for exotic plants that the wealthy at home could grow in the hot houses.

It must be remembered that with Hope's training while he was in Edinburgh, he had been trained to view the whole environment as contributing to the development and healthy growth of plants as well as determining which plants would grow naturally. Thus, he was always looking at the soil and climate, the importance of the latter having been emphasised by his connections with Sir John Pringle and the Royal Society. Equally, the importance of a wide circle of correspondents was deeply ingrained from his time in Edinburgh, and he had, indeed, already started to develop such a circle, which would expand dramatically over the ensuing twenty years he was in Calcutta.

THE HOUSE

At the levee mentioned above, Roxburgh also asked Sir John Shore about leave to collect his family from the Coast and 'settle some concerns there' as he had had time to ascertain the nature of his new job, which he found very much in tune with his own wishes: 'for I have much at heart the task lately assined [*sic*] me by that Government, the introduction of useful Plants, which I had begun before I left the Coast, and now find myself with my present Appointment, much more able to effect so praise worthy an undertaking.' Gone now was the doubt as to whether he would return to Britain and practice medicine, or stay in India as a botanist and natural historian, that plagued him four or so years before. By the middle of the month, he was back in the south, and met up not only with his family but also had an opportunity to enjoy the company of his friends. Amongst the latter were the Kindersleys, who were cousins of Dr James Edward Smith of the Linnean Society, with whom 'we have never been half so merry as when you & Mrs K & the Committee were here [Cocinada], we can scarce say we have had a dance since.'[18]

28 Sir Charles Grey's house and garden, Calcutta, sketched by Marianne Jane James, 1828. Sir Charles Grey was one of the Court judges and this view is particularly interesting for it shows his house of two storeys compared with Roxburgh's three-storeyed mansion, both on the River Hooghly.

Sadly no plans for the proposed house exist, but the first ideas were sent in June although neither the plans nor any estimate of cost were included, but were requested.[19] These were sent at the end of July, 'together with Estimates of three Native Builders, amounting to from Sicca Rupees 25,044 to 28,438'. This was above what he had expected, so proposed that a 'reduction in the long Veranda' would make a saving of 'between four and five thousand Rupees.' This was very much more than the Governor General in Council had in mind and capped the cost 'at the utmost Sicca Rupees 15,000 including every Charge.'[20] Revised plans to fit within this budget were submitted about a week later and were accepted. It is these plans which were transformed into the present building, and give an indication of the reason for the third storey: 'Being solicitous for a third Story, will not, I hope, be deemed improper, because it is my wish to reside continuously in the Botanic Garden even During the Rains.'[21] The extra storey also was able to accommodate his large family, although the top floor was for his herbarium collection. However, as can be seen from Fig. 29, this is no mean house, more a mansion with considerable social pretensions, and must reflect both Roxburgh's own views of his position and the tacit agreement of the Governor General in Council. It certainly seems to equate with the houses for such senior members of Calcutta society as Sir Charles Grey (Fig. 28).

29 Garden Reach. Hand-coloured aquatint by Robert Howell, after J. B. Fraser, 1824–8. This shows the three-storey house built for Roxburgh, and its position overlooking the River Hooghly.

30 A recent photograph of Roxburgh's house.

31 Interior of Roxburgh's house, showing the main door on the first floor.

By September, Roxburgh was asking for reimbursement for the costs so far on the house, which must have been 8,000 rupees, for in May, he was asking for the balance of 7,000 rupees.[22] However, in November, Roxburgh wrote that 'the accompanying account of the money I have already laid out on the House erecting for me in the Botanic Garden, the expence has exceeded considerably the estimates delivered to me by the Master Bricklayer and Carpenter at the beginning of the work' and asked for the difference of 5000 rupees, a request that was turned down as the 15,000 rupees the 'Government consider as a very liberal allowance.' The costs submitted by Roxburgh are given in Table 1 of Appendix 9. However, the house had been expected to be completed 'soon' at the end of December 1794: builders' timings have never been reliable![23] Equally, it looks very much as though Roxburgh planned from the outset to build a larger house than the Company gave permission for, starting with his initial estimate of 25,000 rupees. When this was reduced to 15,000 rupees, he still went ahead to build to a cost of 20,000 rupees: again, how often do people increase the size of their plans as these are being developed, but in this case it does look as though he cut his plans only slightly as a result of the Company's decision, and paid the rest out of his own pocket – once more suggesting that he had already amassed quite a fortune.

The house was repainted and repaired about every three years, although the first costs have not been found, but in October 1803 Roxburgh wrote, 'It is now, nearly three years since that were last repaired, white washed, & painted.' At the same time, the 'two Octagons & the three European Gardeners Houses in the Botanic Garden' were refurbished. When this had been done previously it had cost 'Sicca Rupees 1206..4..3' and so an allowance of 1,500 rupees was given.[24] These repairs cost Rupees 1441..8 (see Table 2 of Appendix 9),[25] and in October 1806 William Roxburgh junior, as Acting Superintendent, estimated that the repairs would cost 1300 rupees when they in fact cost rupees 1227..10.[26] By the time we get to the repairs in 1809, they were rupees 1317..8 (see Table 3 of Appendix 9).[27] What is of interest is the large number of doors and window frames that were planned in the erection of the house, 65, against

32 Interior of the old building at the Tollygunge Club, Calcutta, showing how the interior of Roxburgh's house may have looked.

only 53 that were ever redecorated: by the colour schemes, 23 doors were external and 30 internal. The final set of costs for repair, in 1812, were rupees 1264..8 (see Table 4 of Appendix 9).[28]

CHRISTOPHER SMITH, NURSERYMAN

To assist him with his botanical work at Calcutta, Roxburgh had a number of European assistants and gardeners, including for some time his eldest son, William junior, considered in more detail in Chapter 4. Amongst these were the painters, of whom he must have inherited some as well as bringing his own (see Chapter 6 for a fuller study of this), for in January 1794, he wrote 'it is as yet scarce in my power to fix upon any standing establishment of Garden painters, &ca for the Botanical Department, I therefore beg leave to have it continued as heretofore, till I am better acquainted with the abilities and usual usages of those I may find it necessary to employ.'[29] The painters referred to were the watercolourists who produced the collection of drawings that are now held in Calcutta and Kew. The problem with the gardeners and nurserymen was a different issue and will be discussed at this point as they give insights into Roxburgh's character.

The first nurseryman to be appointed after Roxburgh arrived, caused by the death of Mr Hughes, was Christopher Smith who had been assistant on the *Providence* during its voyages of 1791-93 in search of bread fruit trees in the company of Captain Bligh.[30] He came on the recommendation of Banks and had been elected a Fellow of the Linnean Society before departing from England. His appointment was confirmed in February 1794, at a starting salary of 2,000 rupees for the first year, 2,400 rupees for the second and thereafter 2,800 rupees, compared with the 1,500 rupees per month that Roxburgh was receiving as Superintendent plus his medical salary as Head Surgeon on the Madras Establishment. To help him on his start in Calcutta, he was also given a loan of £100 which would be repaid out of his salary and a free passage to Calcutta.[31] Problems emerged almost immediately, for in January 1795 another nurseryman had arrived at Calcutta before Smith, Thomas Douglas, who had been employed by Mr

Light at Pulo Penang before the latter's death. Roxburgh described him as 'a Man of abilities in his Line, which induces me to request permission to Employ him in the Hon'ble Company's Botanical Garden, till Mr Smith arrives', which was approved.[32]

Smith arrived in April, having spent time at Madeira and the Cape collecting plants, on 'instructions received from Sir Joseph Banks', which incurred the relatively high cost of £33 5s.[33] His arrival at the Botanic Garden highlighted the parlous state of the accommodation provided for the European gardeners, for there was nowhere for Smith to live and 'the two Bungalows in which the two European overseers reside, are in a very ruinous condition.'[34] From the letters written by Smith and about him, he appears to have come to India to make his fortune, and matters relating to money crop up right from the start, for in June 1795, he complained that the £100 advanced to set him up was being deducted from his salary before he could afford it: 'As I am not a rich man, it will distress me to have that Sum deducted at present. Therefore I humbly request that you will be pleased to postpone the stoppage till the first of January next',[35] an arrangement that was agreed.

By the end of the year, Roxburgh had sent Smith to 'Mallacca & the Spice Islands, (for by this time we imagine they are in our possession) for nutmeg, clove, &c &c plants and seeds', for which he was being paid an extra 200 rupees a month allowance, thus doubling his salary.[36] Once more Smith set out with a loan, this time from Roxburgh for 500 rupees, 'to enable him to proceed without waiting to trouble Government for a supply of money to fit him out',[37] a practice repeated while Roxburgh was at the Cape. Smith seems to have used this opportunity to invest his unused salary, for in January 1798 his agents were approaching the Secretary to the Government for his arrears, amounting to nearly 8,700 rupees, which was agreed, giving a good indication as to how Europeans when away from Calcutta could begin to generate some wealth; a quick calculation shows that this sum is in fact more than his salary since he arrived in Calcutta – again problems with Smith and money.[38] This was confirmed when Smith wrote to Banks in 1799, when he stated 'when I left Bengal in October last, I was worth £2,000 sterling, which sum is in the hands of Messrs Campbell Clark & Co. Wine Merchants & a House of Agency in Calcutta; at 12 per cent per annum', having said that 'you have made a poor Man rich'.[39]

33 *Detail of* Melicope tetrandra. *Watercolour by one of Roxburgh's Calcutta artists, c.1800 (Roxburgh drawing 1411). This small tree, was described by Roxburgh from the island of Penang, and must have been drawn from a specimen growing in the Calcutta Garden.*

Roxburgh must have begun to get impatient with the success of Smith's collecting and his general behaviour, as Smith had written in October 1797, 'by the advice of the Commercial Resident at Amboyna, he is to remain there, collecting, & sending plants to the Hon'ble Company's settlements, untill he had orders from hence to the contrary.' Added to this 'the Boy Samson, who you directed to be sent to Mr Smith, absolutely refuses to return.'[40] When Roxburgh found that 'nearly three fourths [of the plants] have perished during the voyage', he discovered that this was more from the state in which they were planted and packed, rather than any negligence on the voyage. In addition, many of the plants that Smith had sent to Pulo Penang in 1796 were still in their boxes, so Roxburgh suggested that they should be distributed to 'the European Gentlemen & Chinese, settled there, who have lands to cultivate them on'. Smith did indeed seem to be using this tour to make as much money as possible, for in June he sent a copy of a letter in which the Deputy Governor at Fort Marlborough and Captain Hugh Moore of the ship *Phoenix* had agreed for Moore to transport nutmeg and clove plants from Amboyna to Fort Marlborough, in the ratio of 2:1, whereby Moore was to receive $2,000 (these figures were for Spanish dollars, equivalent to about 4s. 2d., now about 21p) for delivery, rising to $3,500 for the safe delivery of 300 plants with an additional $7 per plant for the next 50 plants: the ship set off with 1,004 nutmeg and 581 clove plants.

While Roxburgh was at the Cape in 1798, Dr John Fleming took over as Acting Superintendent of the Botanic Garden, and he seems to have been thoroughly impressed by Smith, writing

> on Mr Smith's diligence and activity in executing the Commission with which he was entrusted, too much praise cannot be bestowed. How well he employed his time during his residence in the Moluccas will appear from a perusal of two Lists which I herewith enclose, One of plants collected by him and sent to the Cape, the West Indies, Europe, and to different parts of India, the other of the Moluccan plants now in the Botanic Garden.[41]

Whether it was carelessness or duplicity, Smith's accounts which Fleming passed on contained 'several errors in the Calculations and additions', and currency conversions were based on figures supplied by Smith, giving much ground for additional potential 'errors'.[42]

By the end of the year, Fleming was suggesting that on the strength of Smith's success at the Moluccas, he should 'be directed to proceed to Amboyna, and the other Eastern Islands for the same purpose as that on which he has sent on his former deputation, I mean to collect useful Plants and Seeds to be dispatched to this and over other settlements as often as he met with an opportunity.'[43] A month later, Fleming sent in two bills for Smith's 'deputation to Pinang and the Eastern Islands', one for 800 rupees, for cash advanced 'to enable him to proceed without delay' and the other is detailed in Table 3 below.[44] This is an interesting set of figures for previously Smith had been charged 35 rupees per chest for plants, suggesting either that he had been overcharged before or these chests were in fact packages.

67 Chests		
10 Baskets		
32 Pots		
15 Casks Water		
124 Packages @ 10 Rs	1240	
Mr Smith's Passage with Servants, &ca	260	
	SᵃRˢ 1500	

Table 3 Costs of plant collecting, by Christopher Smith: 'To Freight of the following per The Diana to Pinang, dated 18 October 1798'.

Smith appears to have retained a direct line of communication with Banks, his patron, for in June 1799 he wrote a long letter. He admitted in it that many of the 64,052 plants he had sent from the Moluccas had not reached their destinations alive, 'in consequence of a long voyage, detention at the Cape, & St Helena for convoys, and in particular as there was no person of practical knowledge on board to take care of them.'[45] As Roxburgh always ensured that he got one of the officers, either the captain, surgeon or purser, to look after them, it seems surprising that Smith had not learnt to do the same after his experience on the *Providence*. In May 1798, he arrived in Madras with 37,089 plants of which he left 2,066 there, sailing in June for Calcutta with the remaining 35,023; this section of the letter continues with the information that 'I have been successful in bringing one Nutmeg tree 14 feet in height, bearing four nutmegs, from the Island of Banda to Calcutta, one of which arrive at Maturity and was presented to the Rt Hon'ble the Earl of Mornington.' By the omission of any mention of the other plants, and the fact that that paragraph started with the statement 'notwithstanding a circuitous voyage', it would seem reasonable to assume that the rest may have perished; confirmed by the previous comments of Roxburgh's, that three quarters of the plants did not survive.

He then recounted his progress in getting 10,000 spice plants from the Calcutta Botanic Garden, under the approval of Fleming, for Prince of Wales Island where he had procured 33 acres for their cultivation, which would become a 'national benefit'. This was in contrast to the Calcutta Botanic Garden which 'it is supposed by some Gentlemen that [it] will fail, the expenses are enormous; and very little of any importance produced from it. (This will ever be the case so long as the <u>head</u> is deficient in practicale knowledge;) the annual expenses exceeds £4000 Sterling.' In view of the fact that 10,000 spice plants had been shipped from the Calcutta Botanic Garden, this seems more than unjustified. At the same time, Roxburgh was writing to Banks, 'I am concerned at the fate of Smith's Amboyna plants, tho from all I have ever seen of his labours, I had great reason to fear what has happened. He is exceeding industrious, but wants what is necessary to make his industry useful.'[46] As Roxburgh was not liable to denigrate people unnecessarily and even here he was giving due praise where it was deserved, it does suggest that there was a deep underlying friction between these two men. This view is supported by the comment of Dr J. E. Smith, quoted above. This predicament is especially true when the future of these spice plantations is considered (see Chapter 11). Although Smith was Superintendent of the Company's Spice Plantations at Prince of Wales Island, he was still also technically Nurseryman at Calcutta, and his pay was tied to the latter position.

Roxburgh's view of Smith's horticultural expertise is further reflected in a letter of May 1800, in which a consignment of

> 38 boxes of Gomootoo Plants (Saguerus Rumphii) and one of Black Peppers, have been brought for the Botanic Garden, eleven more have gone to pieces on board ship during the passage from the insufficiency of the boxes they were planted in. No letter accompanied them, but am informed they were sent by Mr Smith the Nurseryman who sent the collection mentioned in my letter of the 19th of March last, and am sorry to say that this extensive consignment is in no better state than the former; which makes me again take the liberty of suggesting that those large collections of Plants brought so far by sea, at a very great expence, are not so likely to answer the purpose, as by sending fresh well ripened seeds.[47]

This lack of attention to detail cropped up again when Roxburgh reported that 'during my absence at the Cape, some thousands [of spice plants] were brought into the Garden from Amboyna by Mr Smith the Nurseryman, a great part thereof were again carried away by the same person to Prince of Wales's Island, but the particulars of these transactions I cannot ascertain.'[48] Smith's economy with the truth is also shown in a letter by Roxburgh in 1804, where he states that 'above five sixths of the number

34 *An Indian woman from the time of Roxburgh. Etching from Lord Valentia's* Voyages and Travels to India *(London, 1802–6).*

[of clove and nutmeg plants] shipped had perished'.⁴⁹ On a final note on Smith, he is assumed to have died at Penang in 1807.⁵⁰

As Roxburgh was responsible for Smith to some extent, for it must be remembered that he was also sending details of his work to Banks, the Superintendent was worried by the profligacy of Smith and also his lack of attention to detail. Both these practices leading the Company into extra expenses, a fact that the Directors were always aware of but from their distance from the point of cost usually unable to do much about. The so-called excellent work that Smith has been credited with must therefore be severely questioned when viewed from Roxburgh's and thus the Calcutta end, rather than the highly acclaimed view that Smith sent back to England.

THOMAS DOUGLAS AND PETER GOOD, GARDENERS

As mentioned above, Thomas Douglas came to Calcutta in 1795 and, when Smith went to the Moluccas, he was re-employed on a temporary basis, and Roxburgh referred to him as 'the Acting Nurseryman, was employed under the Marine Board to superintend the Experiments on hemp &ca at Reshera.' Roxburgh was needing extra help at the Garden and Douglas was not prepared to give up the work he was doing to help out, so Roxburgh asked that he be allowed to employ Samuel Wheeler, which was authorised.⁵¹ Like Smith, Douglas appeared to push for the maximum returns, by demanding that he should be paid as both the Nurseryman and the Superintendent of the agricultural experiments at Rishera, something that was not countenanced; this work at Rishera, on hemp, for Roxburgh, is considered further in Chapter 10. However, when Douglas returned to Calcutta later that year, he and Wheeler were both retained on the staff.⁵² The benefit that Smith got from his patronage by Banks can be seen from the fact that, although he was appointed on an increased salary, when Douglas, acting as Smith's stand-in, asked for the same increments to be paid, was told 'that the Governor General in Council does not deem it to be expedient to comply with his request'.⁵³ Once more like Smith, Douglas tried again to apply for these arrears, while Roxburgh was in England, with the same result as before, although by this time Wheeler had died.⁵⁴ Nothing daunted, Douglas tried yet again later that year, but this time with letters from Roxburgh which he had been able to trace, and so was granted part of the arrears.⁵⁵ This letter also gives an insight to Douglas's employment record, for he was at Rishera from the middle of April 1800 to the beginning of April 1801, where he was performing the experiments on hemp for Roxburgh. The other point of interest in this series of letters is the time taken for the Council to act upon Douglas's own letters: when Roxburgh or Fleming wrote to the Secretary, the matter was usually dealt with in the next few days, in the case of Douglas, the matter often took a month – an indication of the prioritising of people's claims on the Council's time.

When Lord Minto arrived as Governor General, Douglas tried once more to get the remaining arrears paid, and persuaded Francis Buchanan, who had been Acting Superintendent of the Botanic Garden at the time, to support him. From this set of correspondence, it appears that Douglas in fact arrived at Reshera in September 1799 on the death of Mr Sinclair. Buchanan's letter is a carefully worded document, giving both sides of the argument, possibly as a result of his advice given earlier not being taken, and he ended with the statement, 'Should you think that this explanation is not satisfactory I can only advise you to request that Lord Minto would have the goodness to take Mr Darell's letter into consideration, and determine what ought to be the fair construction put on its contents.'⁵⁶

The end of Douglas's time in India was presaged in a letter he wrote in November 1808. On the excuse of taking plants from Calcutta to the Cape, St Helena and England, he stated,

> I would hope the expences that may be incurred for my passage to England will be
> defrayed by Government – and as there are many Kinds of Trees, Shrubs, and Herbs,

in Britain, that would be useful in India, it is also proposed that a Native, or a Chinese Gardner, may be sent to England where I would assist him in selecting the requisite sorts of Plants and seeds for him to bring to the Honorable Company's Botanic Garden.[57]

Roxburgh's recommendation was fairly damning, both for the way that Douglas had gone about this application and its content:

no application has been made to me, for plants for the Cape of Good Hope, nor do I conceive any of the plants of this Country will live there, without the aid of artificial heat ... To St Helena many may be sent with every prospect of advantage to that Island. For His Majesties Botanic Garden at Kew, ... I have been in the habit of sending to that Garden for above twenty years.

As far as plants being returned to India, Roxburgh went on to say that Banks and Aiton 'send whatever I apply for, under the care of some friend or acquaintance, amongst the Commanders of the Companies Ships, with much greater chance of success, than if under the care of a Native of India.' Not surprisingly, Douglas was told that if he was 'desirous of taking any plant or seeds at his own expence, he [Roxburgh] is authorized to select and furnish him with such plants and seeds, as he may consider to be most useful and acceptable.'[58] Fifteen years earlier, Banks had applied a different approach, for he employed Peter Good to accompany Smith to India before returning with a consignment of plants and seeds. The draft instructions are very specific as to his responsibilities and are worth repeating:

The first duty of every Person who embarks on board ship whatever his Station may be, is to conduct himself respectfully towards the officers of that ship & to obey as far as he is Able such orders as they may choose to give to him, they will not call upon every man who is a passenger to assist in the management of the ship unless there is real occasion for his help & of this they who are sailors are better judges than any Land men can be. Whenever therefore you are called upon to assist you must do it readily, cheerfully & willingly.

In Case you sail in the same ship with Mr Smith your business in the outward bound Passage will be to obey his directions in taking care of the Plants Committed to his charge by doing this you will not only do good service to the East India Company at whose expence you are victuald during the voyage but you will also Learn the art of taking care of Plants at sea in which he has much experience.

When you arrive at Calcutta you are to continue your assistance to him in the Companies Garden of which he is appointed Gardiner & to obey such directions as the Superintendent of the Said Garden or Mr Smith shall give you during your stay in India.

The Principal business for which you are sent out is to take charge of such Trees Shrubs & Plants as shall be sent home for his Majesties use in the Ship on board which you are placed & of them you will have the sole management & direction during the homeward bound voyage.

You will therefore employ yourself whenever you have it in your Power & your Principals the Superintendent & the Gardiner of the Companies Botanic Garden will chiefly employ you in collecting such wild Plants or the Seed of them as are not to be found in Kew Gardens & in Planting them & Such others of the Companies Garden as are to be sent to England in Proper pots Boxes or tubs fit to be placed in the plant Cabbin, & in taking Care of them after they are planted & in attending them which you must not fail to do in person when they are removed from the Garden to the ship & also in taking care of them when on board the ship & you will specialy remember that during the whole time they are on board it will be highly improper for you ever to sleep out of the ship.[59]

Three paragraphs follow, the first stipulating that only plants intended for the King are to be placed in the plant cabin under Good's control; the second set out that he is 'not on any account or consideration to part with any Plant or the Seed of any Plant or any cutting Graft Slip Sucker Scion or offset of any Plant Tree Shrub or Bulb'; and the last paragraph instructed him to inform the 'Principal Gardiner at Kew' as soon as

he landed at the first English port of his arrival and the state of the plants.

With the parallel facts that Smith was sent out under the patronage of Banks and that he had another of Banks's protégé's under his direction, it is not surprising that there were conflicts between Smith and Roxburgh. The very strong ownership of the plants that Good brought home with him gives an impression of the value bestowed on them by Banks. It is also interesting to note that Roxburgh had already started sending plants back from Calcutta by this stage, and sheds some light on the developing relationship between him and Banks. This latter point is of particular interest when it is born in mind that Roxburgh was beginning to publish his research articles through the Asiatic Society under Sir William Jones rather than the Royal Society under Sir Joseph Banks, as will be discussed further in Chapters 6 and 7.

Roxburgh was sent a copy of the instructions which reached him in March 1795 and a month later the plants were ready to embark onto the *Royal Admiral*, Good having incurred costs of under 100 rupees.[60] Unfortunately Roxburgh was not well during Good's short stay in Bengal 'otherwise the collection would have been much more extensive,' but it did nevertheless contain 82 plants with a number of each. However, ever aware of the potential economic importance of new plants, he sent bags of seeds of *Agrostis linearis* (now *Cynodon dactylon*) with Good, one for St Helena, one for Kew and the third for Banks.[61] As a final note on Peter Good, he accompanied Robert Brown on his journey to New South Wales in 1801 as the gardener when Ferdinand Bauer, the famous botanical artist and younger brother of Franz Bauer, was the botanic draughtsman; Good died on this trip in 1803.[62]

One other European gardener is mentioned by Roxburgh, Allen Bowie, and it is worth highlighting the differences between him and Smith. Bowie is described by Roxburgh as 'the European Gardiner', as was Smith, but Bowie 'has been long in the Botanic Garden'. However the crucial difference between them was their salaries: Smith, as mentioned above, started on 2,000 rupees a year rising to 2,400 rupees, whilst Bowie was on 'only fifty Rupees per month' – 600 rupees per annum, a quarter of what Banks had arranged for Smith.[63] Roxburgh was always meticulous and very

35 Nymphaea rubra. *Ink and watercolour drawing by an anonymous Calcutta artist, based on a drawing made for Roxburgh at the Calcutta Botanic Garden, c.1794 (Roxburgh drawing 657). The fact that only the botanically important floral details are coloured suggests that this was made for an unknown botanist, rather than the full coloured copies made for various aristocrats and amateurs.*

fair about getting the garden expenses paid, and this discrepancy could well have been yet another factor in the possible difficulties between him and Smith, however experienced he might have been after his voyages on the *Endeavour*.

GARDEN EXPANSION AND LABOURERS

Almost as soon as Roxburgh arrived at Calcutta, he was expanding the garden even though he had complained the previous month that it was too big, by arranging for the purchase of land 'contiguous to the Botanical Garden purchased from the Executors of the late Col. Kyd.'[64] Six years later, the garden had 'been much enlarged for the introduction of numerous new Plants from every quarter of the world, By various and some of them extensive experiments in Husbandry' and as a result Roxburgh was allowed to take on 25 additional labourers for the 'ensuing season'.[65] By 1805 yet more land was required, for the planting out of teak trees and sago palms, which again was permitted, as it 'can be effected without any additional Increase in the present Establishment of the Botanical Garden'.[66] This piece of land extended from the Botanic Garden to the land belonging to Sir John Royds, and measured 160 biggahs. However, it needed a large labour force to prepare it, so young Roxburgh (his father was in Britain at the time) asked for the use of as many convicts 'as can be spared, to bring into order the additional ground for Teak and Sago trees'.[67] The ground was liable to flood during the rains, so a request to build two flood gates was also granted, at a cost of not more than 600 rupees. The cost in fact came to just under that figure (see Table 4). Although the use of the convicts was initially granted, they could not be spared as they were heavily involved on the roads 'and other essential work ... in the vicinity of Calcutta'.[68] However, young Roxburgh appeared to have got his hands on a number of convicts, for his father, on his return, gave a detailed description of the status quo in November:

> I find the number of Gardeners and Labourers (one hundred of all descriptions) are only able to perform the work of the Botanic Garden alone, and the two additional pieces of ground, which have been lately added for Teak and Sago-palm Plantations chiefly, must be neglected, if more men are not allowed.
>
> The first of these two pieces was completely planted during 1804. The Second, by far the largest, extends from the east side of the first, along the side of the river, 1,000 yards, to the west side of the ground, at present occupied by Sir John Royds; has not been brought into order, nor any part thereof planted. The working men of from 180, to 200 Convicts, (amounting to about 150 on an average) have been employed since the 19th of January last, making a Bund, or Bank along the river side, to keep out the tide, and the freshes from the river; and it will require fully as much more time to complete the bank. For the above stated reasons, I trust Fifty more labourers will be allowed, to enable me to get the first plantation brought into order, and the second piece of ground levelled, and planted as quickly as possible. At present the Teak plants in the nurseries, intended for this plantation, are fully large for moving, and if not transplanted soon, they will be greatly two [*sic*] large, consequently lost, most of the Sago-palm plants are already too large, many of them will of course be lost in transplanting, and the longer it is delayed, the greater will the loss be. A loss which cannot be replaced without sending to Bencoolen for more seed, which will require at least three years before the new plants will be ready to transplant, those now in the nurseries, are from five to eight years old.[69]

The hundred gardeners came from all over the Far East, some were Indians, others Malays and some Chinese, according to various letters written over the twenty years that Roxburgh was in Calcutta, and there is even reference to a Portuguese gardener, who was paid 12 rupees per month.[70] This use of convicts as a source of available and cheap labour was a common practice to the end of the British rule in India. A brief study of Roxburgh's work on teak will be considered in Chapter 8 and his attitude towards developing practices that were labour intensive and thus providing employment for the natives, is further developed in a study of his work on the introduction of sugar in Chapter 12.

40,000	Bricks at 6 Rupees per 1000	240.. ..
900	Maunds Saorky at 16 Rupees per 100 Maunds	144.. ..
250	Do Chunam at 57 Rupees per Do	142..8..
8	Bricklayers for 1 Month and 15 Days at 5 Rupees	60.. ..
	Jagry Earthen pots &c	11.. ..
	Sicca Rupees	597..8..

Table 4 Expenses incurred in building two Flood Gates for the additional Ground for Teak and Sago Plantations attached to the Honorable Company's Botanic Garden near Calcutta.[71]

GARDEN IMPROVEMENTS

As soon as Roxburgh arrived at the garden, he started work on improving it, first by building a 'wall and railing, along the River side of the Botanic Garden ... as far up (beginning at the lower octagon) as the Bridge immediately about the Large Banyan Tree, which is nearly half the extent of the Garden along the Riverside.'[72] There were in fact two octagons in the garden at this time. These seem to have been common introductions at this period, the one at Culzean in Ayrshire being completed in 1814 and Franz Bauer depicting a hexagonal tea pavilion in 1776, but, with the Company's stringent controls on expenses, it would be more likely that these had been built for special plants.[73] These two at Calcutta were in such a parlous state when Roxburgh arrived that he described them as 'in so open & ruinous a state, that I thought it necessary to have them repaired, & Windows put in the Westernmost ones, before the rains set in ... at my own expence ... the roof of the Easternmost Octagon requires to be renewed, as the temporary it lately had, was only sufficient to make it last during the season.'[74] The cost of these repairs was 464 rupees.[75] In the previous December he had paid 'Ramtonoo Dutt's bill for making a Terrace of an Octagon ... Passed in Council the 6th September '93 500' rupees.[76] This reference to putting in windows seems to confirm their use for special plants.

Two years later during a gale, 'a Bungalow in the Botanic Garden in which one of the European Gardiners resided, was blown down upon him by which one of his legs was fractured. At the same time, and by the age of the bungalow of the other European is so much injured as to render it unsafe to live in, nor is there any house for the Nursery man.' Roxburgh continued by asking for 3,000 rupees to build bungalows for the nurseryman and the two gardeners, which was granted.[77] The relative costs of housing is interesting, on the basis that each of the bungalows cost 1,000 rupees which was five times the monthly salary of the European gardeners, compared with Roxburgh's house which was more like ten times his salary. When he sent in his expenses in June, there was still extra work being done: 'the number of coolies have been greater by 30 for some time past than when I took charge of the Garden, in order to clean up and enlarge the Tanks, cut drains, &ca during the dry Season, these may be discharged in the course of another month.'[78] By September, 'the additional labourers are now discharged & the expence reduced as much as can well be', then he continued with a description of the reason for much of the extra work, 'considering the great extent of the Garden & encrease of labour from the great numbers of useful trees and other plants lately introduced, such as teak (which thrives astonishingly well), Bourbon Cotton, India of various kind Cochineal Nursery, Cinnamon, Sappan wood, American Babool, Anou (A valuable species of Sago Palm, sent from Dr Chas Campbell from Sumatra, it also yields a fibrous substance of which Cables, Cordage &c are made.); Dammer Poo (Sent from Pullo-pinang by Mr Smith the Nursery Man.) Ingrafted fruit trees of various kinds from the West Indies, America & Europe – to these may be added the Pimenta or all-spice of Jamaica and Olive of Marseilles.'[79]

Year	Month	Roxburgh's salary	rent	Workmen's wages	Draftsman	Other
1793	May			640..14	100..	
	June			696.. 6	100..	
	July			643. 2	100..	
	August			642..12	100..	
	September			648..12	100..	560.. 4
	October			676..13	100..	
	November					
1794	March	1500..	150..	818.. 8		
	April	1500..	150..	902..12		
	May	1500..	150..	853.. 8		
	October	1500..	150..	959.. 5		
	November	1500..	150..	1038.. 2		
	December	1500..	150..	1064.. 2		
1795	January	1500..	150..	1053.. 6		
	February	1500..	150..	1067..12		
	March	1500..	150..	1388.. 2..9		452..

Table 5 Expenses incurred by the Calcutta Botanic Garden during the beginning of Roxburgh's tenure as Superintendent. The figures in the 'Other' column refer for September 1793 to the terrace for the Octagon plus a bill for Sundry materials; the figure for March 1795 is made up of 121 rupees for food and boat hire for the shawl goats mentioned in Chapter 7, and 331 rupees as expenses incurred by Christopher Smith, previously mentioned. Not included are the costs of building Roxburgh's house, as these were paid later. Sadly, later documents do not specify what the expenses were, just that they had been submitted.[80]

Even assuming that the cost of the draftsman is included in the wages for the workmen after the first six months, which was when Roxburgh arrived at Calcutta, the monthly costs rose significantly in the missing four months of June to September 1794 (from 850 rupees to 1000 rupees) and there was another large increase in March 1795. It is unfortunate that further accounts have not yet been found to follow this through for more than Roxburgh's initial 18 months. It does highlight, however, the very much more active part taken by Roxburgh in the development of the Botanic Garden than Kyd in its initial years, reflecting their very different views of the role of the Botanic Garden.

There are two contemporary accounts which give some indication on the content, layout and general regard for the Garden at the time.

I visited the Botanic Garden, which is under the care of Dr Roxburgh. It affords a wonderful display of the vegetable world, infinitely surpassing any thing I have ever before beheld. It is laid out in a very good style, and its vast extent renders the confinement of beds totally unnecessary; yet, I think, it is a pity that a small compartment is not allotted to a scientific arrangement. The finest object in the garden is a noble specimen of the Ficus Bengalensis, on whose branches are nourished a variety of specimens of parasitical plants, Epidendrons, Linodendrons, and Filices. The water, also, is beautiful, being more covered with red, blue, and white Nymphaeas. Utility seems more to have been attended to than science. Thousands of plants of the Teak tree, the Loquat, the grafted Mango, and other valuable fruit and timber trees have from this place been disseminated over our Oriental territories; and at present it is a complete centre, where the productions of every clime are assembled, to be distributed to every spot where they have any chance of being beneficial. The nutmeg was in considerable perfection; but the Mangusteen, though often brought, has never survived its transplantation one year. The chief novelties are from Napaul and Chittagong. Most of the West Indian plants are making their way here, and will probably thrive well. It is by far too hot for European vegetables, and of course many even of our pot herbs are in the list of their desiderata.[81]

This from the pen of the man, Lord Valentia, who was so critical of John Roxburgh at the Cape. It shows, however, that he could give praise where he felt it due, but was critical of areas with which he disagreed. There is also a paradox in the *raison d'être* of the Garden, in that the Court of Directors were running a profitable trading company, and only secondarily (and not always willingly) a scientific institution. Roxburgh, by using the economic arguments at the same time as following his own personal interests, was an excellent exemplar of the ability of such men of science to tread the narrow balance.

The care and time that Roxburgh devoted to his visitors, is shown by the diary entry of Maria Graham, whose husband (Captain Graham, brother of Robert who later became Regius Keeper of the Royal Botanic Garden Edinburgh) accompanied Valentia on his tour of the Garden seven years before.

> I came here just now in order to go to the botanical garden, where I went yesterday with my friend Dr Fleming, who introduced me to Dr Roxburgh and his family, with whom we breakfasted. Before breakfast we walked round the garden, and I was delighted with the order and neatness of every part, as well as with the great collection of plants from every quarter of the globe. The first that attracted my attention was a banian tree, whose branches Dr Roxburgh has clothed with the numerous parasite plants of the climate, which adorn its rough bark with the gayest colours and most elegant forms. In another part of the garden the giant mimosa spreads its long arms over a wider surface than any tree, except the banian, that I remember to have seen. The Adansonia [the Baobab], whose monstrous warty trunk, of soft useless wood, is crowned with a few ragged branches and palmate leaves, seems to have been placed here as a contrast to the beautiful plants that surround it …. I will only mention one other tree, the Norfolk island pine [*Araucaria excelsa*], which reminds me in every one of its habits of the firs of Northern Europe, but that it seems to grow higher and lighter, which may be the effect of the heat and moisture of this climate.
>
> The botanical garden is beautifully situated on the banks of the Hoogly, and gives the name of Garden-reach to a bend of the river. Above the garden there is an extensive plantation of teak, which is not a native of this part of India, but which thrives well here; at the end of the plantation are the house and gardens of Sir John Royds, laid out with admirable taste, and containing many specimens of curious plants. After having visited the garden, Dr Roxburgh obligingly allowed me to see his native artists at work, drawing some of the most rare of his botanical treasures; they are the most beautiful and correct delineations of flowers I ever saw.[82]

THE CAPE OF GOOD HOPE

Roxburgh's health, never strong, suffered in the humid heat of Calcutta, and by the end of 1796, he was laying the foundations for a trip to the Cape of Good Hope. He reached this new British acquisition in 1798, and a fuller study of his botanical work and importance there is given in Chapter 14. The important points that need emphasis here relate as much to the developments of his ideas as to his family.

36 *Blettermanhuis, Stellenbosch, built about 1789, showing the type of house that Roxburgh might have lived in at the Cape. This house was built by Hendrik Bletterman, the last magistrate of Stellenbosch to be appointed by the Dutch East India Company.*

He recalled his son William junior from Edinburgh University to start preparing him as his successor as he could not arrange for him to get a Writership with the Company, at Calcutta, through Banks. This is discussed further in Chapter 4. However, the fact that William junior, and daughter Mary, joined him at the Cape where he had John working for him, does suggest that he was very well aware of the importance of this additional British colony and its botanical wealth. His suggestion, developed further in Chapter 14, that he might get involved in a new botanic garden here also suggests that he may have reached some sort of realisation of his own career development in Calcutta: new challenges were needed and the Cape was one opportunity. Other than producing a list of Cape plants, that has not been traced, this was not to be, and he returned to Calcutta after 18 months in the better climate of the Cape, with his health temporarily improved. While he was at the Cape, he travelled enormous distances (see Fig. 127, page 214), suggesting that his claim of ill health was not as dire as he made out – supported by the idea of him possibly developing a Botanic Garden there. Travel was either on horse-back or more usually in ox-wagon, described by Lord Valentia: 'These waggons are the only machines adapted to the roughness of the roads, as they have every advantage of strength, and difficulty of being overset.'[83]

SALARY MATTERS

When Roxburgh and his family returned to Calcutta at the end of 1799, he tried to get the arrears of pay awarded to him, a problem that was quickly remedied.[84] However, it raises two important points that are appropriately considered at this point. The first is that he must have lived the 18 months at the Cape off his own wealth, indicating once more that he had (i.e. after his losses following the hurricane of 1787) accumulated at least a moderate fortune. The second point concerns his salary, an aspect not yet considered, together with his promotion within the Medical Service.

Roxburgh was appointed Assistant Surgeon at the Madras General Hospital in May 1776 which equated to the rank of Lieutenant, whose pay in 1796 (earlier figures are not readily available) was 224 rupees per month when on garrison or cantonment duty, rising to 284 rupees when in the field.[85] His promotion to Surgeon, on 27 November 1780[86] when he was at Nagore, would have increased these figures to that of a Captain, 321 rupees and 411 rupees respectively. With his move to Samulcottah, and certainly by the time he was made Company Botanist, he was no longer on the active list, and when he was recommended for appointment by seniority to the Third European Regiment, the Government could not spare his services, but continued his employment as Superintendent of the pepper plantations. For the rest of his time in India, he remained, slightly anomalously after 1793, on the Madras Establishment, but this caused only a slight delay in the payment of his salary when he moved to Calcutta although there were other complications (see Chapter 5).[87]

On his appointment as Company Botanist (at the time the titles 'Company Botanist' and 'Company Naturalist' were used interchangeably), he was awarded the same extra allowance as that given to Patrick Russell, of 60 rupees per month on top of his military pay which would have been in the region of 320 rupees per month, but for some reason he was paid only 40 rupees. Andrew Ross 'spent some hours at different times in the Offices of the Sec^y, the Accountant, and Paymaster', who demanded 'an authenticated Copy of the Orders of the Board'.[88] The formal application to redress this error was made by Ross in February 1790 who wrote four days later to give 'you the satisfaction of knowing that I have applied (with success) for the removal of the difficulty which arose to your salary as Professor of Natural History & Botany'.[89]

It appears to be, once more, Andrew Ross who took up the question of Roxburgh's salary, for at the beginning of January 1793 he wrote that he had taken 'occasion to speak to Sir Charles Oakeley of the inadequacy of your Allowance, to your long & meritorious Service to the Company & to the Public & to your constant expences'.[90] The intimation was that if Roxburgh would 'make an Application to the [Medical] Board' due attention would be paid to it. Two months later, Ross was giving an urgency to this and explaining what the application should contain: 'to make an Application for an addition to your monthly [salary] as Botanist & what you profit in the natural History way & therein state the length of time the labour, the attention, the expence – the risk to your health – & other inconveniences to which you have been exposed.'[91] This was to be put forward in April but Ross was still waiting for an opportune moment in May.[92] By July, when he had had a chance to discuss it with his friend, the Madras Government Secretary White, it was deemed that it ought to be held back until the question of the cowle for Corcondah was resolved.[93]

His appointment to be Superintendent of the Botanic Garden at Calcutta was supposed to be 'without prejudice to my rank & claim of promotion at Fort St George, … and even to retain my appointment on that Establishment.'[94] Roxburgh wrote back a fortnight later that 'the Salary of a Head Surgeon, which is the Rank I hold on the Coast Establishment, is Sicca Pagodas 2500 Annually.' The reply he received was slightly devious: 'he be desired to state whether the Salary he alludes to as receivable by a Head Surgeon at Madras, be attached to the Rank merely, or to the office annexed to it, and, if to the latter, whether there be any person now under that Presidency in

37 Memorial to Roxburgh in the Calcutta Botanic Garden. Photograph taken in 2005.

37 Memorial to Roxburgh in the Calcutta Botanic Garden. Photograph taken in 2005.

Receipt of the Salary which Mr Roxburgh claims.'[95] This elicited one of the few rather abrupt letters that Roxburgh wrote:

> I cannot take upon myself to say whether the Salary stated in my Letter of the 26th Ultimo be attached to the Rank merely of a Head Surgeon, or to the Office annexed to it, as I am the only one on that Establishment, whose standing in the Service gave him that rank, that has been absent on other duty since the institution of Head Surgeon took place, and it was between two & three months after my acceptance of the Office I have now the honour to hold under the Supreme Government, that I attained to that rank by the death of a senior Surgeon Mr Gordon.[96]

This must have been a delaying tactic or the ingenuousness of the Calcutta Council, for it was general practice that there were a defined number of appointments at each level, and promotion was purely based on length of service rather than any other criteria. It must be borne in mind that he was also receiving the much larger salary of 1,500 rupees per month as Superintendent.

When he left Bengal on leave in 1805, he once more had problems with pay, but this furlough pay was allowed at the rank of Head Surgeon as he would have forgone his salary as Superintendent which would have been paid to Dr Fleming, his deputy while away. This was granted in January 1806 together with the retiring pension attached to that rank.[97] The Madras Government was insistent, however, that this pay should come from the Bengal Presidency, and indeed, Madras was also insistent that Roxburgh could no longer 'have the same claim to employment under this Presidency as those Medical Officers who have been engaged in the performance of active duties on the Public Service.'[98] The whole problem was that Roxburgh was by this time the longest serving member of the Madras Medical Service but, because he had been in Bengal for so long with no active service, the Madras Government stuck to the rules prohibiting him from having a salary higher than that of Head Surgeon.

The final problems on his pay and pension came when he finally returned to England in 1814, when he first requested for the arrears of his pay while on the journey home, and then for a salary and pension linked to the rank attached to being a member of the Medical Board which by seniority he would have reached had he remained on the active list.[99] After he died, his executor, Alexander Boswell, wrote a number of times to get his pension paid, ultimately unsuccessfully.[100]

SUCCESSION TO ROXBURGH

This leads on to the matter of the succession to Roxburgh as Superintendent. As early as 1806, after Roxburgh had survived the exceptional length of 30 years in India, Francis Buchanan was jockeying to be heir, for he wrote to Dr J. E. Smith: 'Roxburgh is pushing hard to get his son [William junior] appointed his successor and I am endeavouring to secure the succession myself. Could you obtain from Lady Hume either her interest for me or at least her being quiet; it might be of service.'[101] What is of interest in this, is that Lady Hume should be of such importance to Buchanan, for she was also a correspondent of Roxburgh's, supplying him with plants, and the year before a parcel of plants for Calcutta had been despatched by her through Lee and Kennedy, the London nurserymen.[102] Smith was obviously sympathetic to this approach, for Buchanan, who was in London at the time, wrote more fully later that month:

> I am much obliged to you for the trouble you have taken in applying to Lady Hume and shall state to you fully the case. Young Roxburgh was some years ago appointed Assistant to his father which place he now holds and of which I shall not attempt to deprive him. Afterwards he was recommended by the Court of Directors to succeed his father should he be found <u>qualified</u>. On his father leaving India the son would have accordingly been appointed Superintendent of the Botanical Garden; but it was considered that a part of the duty of that office should be to describe the vegetable productions and to communicate these descriptions to the learned of Europe; and as young Roxburgh was not qualified for any such thing the appointment was deferred until the pleasure of the Directors should be known. What I wish therefore is to be appointed successor to Old Roxburgh whenever he chooses to retire. He of course wishes to get his son appointed and says that science is not necessary; that a cultivator alone is required. If Lady Hume will have the goodness to speak in my favour provided the appointment is still considered as open, she will do me essential service and she can do Mr Roxburgh no injustice unless she be of opinion that science is not a proper requisite for the office.[103]

38 *Portrait of Lady Amelia Hume (1751–1809) by Sir Joshua Reynolds; she was a correspondent and patron of Roxburgh.*

This letter highlights a number of differences between Roxburgh and Buchanan. Roxburgh had pushed Buchanan forward shortly after he came to India, writing to Banks that Buchanan was 'a worthy, valuable man, & no doubt the best Botanist in India'.[104] Banks was, however, not impressed by this largesse, stating that 'I admire his Talents both as a Botanist & a Geographer & hope sometime to become acquainted with him. I cannot however subscribe to his being the best Botanist in India tho his descriptions & Drawings of Birmah Plants are in my hands, because I think that you are a better one.'[105] This closeness between Buchanan and Roxburgh was also shown in a letter of 1797, writing, 'I have had leisure to bestow some time on the study of Botany and my discoveries I regularly communicate to Roxburgh with whom I am very intimate.'[106] This patronage of Buchanan by Roxburgh is further stressed by Charles Allen, who wrote: 'Once again Dr Buchanan had cause to be thankful for the patronage of his good friend Dr Roxburgh, for he was now [1804] invited by Lord Wellesley to become his personal physician, and at the same time to take charge of his Institution for Promoting the Natural History of India.'[107] Secondly, Buchanan appears to have used the success of Roxburgh's drawings to promote his own as early as 1796, for a Public Letter of that year states that 'from the satisfaction which you expressed at the receipt of Doctor Roxburgh's drawings and from the liberal encouragement which you have invariably given to the pursuits of a scientific nature, we entertain no doubt that the drawings of Doctor Buchanan will also be highly acceptable to your Hon'ble Court.'[108] However, it must be remembered that drawings were the best method for recording data, so there was an inevitability in Buchanan producing drawings and sending them to the Court of Directors. Behind Buchanan's drive to succeed Roxburgh as Superintendent was his strong belief that the post should go to

39 St John's Church, Calcutta from the south-west.

a qualified botanist, an epithet that he felt did not apply to William Roxburgh junior. It was generally accepted that the training for botanists at this period was by attending the relevant courses at medical school, as William Roxburgh senior had done and indeed Buchanan himself also with Hope. As William junior had only attended at most part of the course of Materia Medica,[109] it would seem that Buchanan's fears were well founded.

This was not the first time that Buchanan had benefited from Roxburgh's generosity, for the previous year, through Roxburgh's patronage, Buchanan was offered the job of surgeon on a Company embassy to the previously closed court of Ava. Although politically a disaster, for Roxburgh it was an enormous success, for Buchanan returned with a magnificent haul of specimens, drawings and botanical notes.[110] When Buchanan was in Rangoon, Roxburgh had sent seeds that Buchanan had forwarded to him, to Dr Smith at the Linnean Society in Buchanan's name.[111] This helpfulness was repeated on a number of occasions, such as when Roxburgh took on Buchanan's engagements at the Asiatic Society in April 1797 and putting him forward to Banks.[112] It is also interesting to note that Buchanan approached Smith for patronage rather than Banks, raising once more the issue of the relative importance of the Linnean Society against the Royal Society at the beginning of the nineteenth century. It is also possible that Buchanan turned to Smith from Banks after the latter published his descriptions and drawings under his own name.

Lady Hume first appeared as a correspondent in 1802, when Roxburgh was planning to send her a box of roots, to share with Banks, the box being so 'large that each may have abundance'.[113] It is possible that it is amongst these roots that Roxburgh included a *Roxburghia*, for he wrote later that 'I believe it was me that sent Lady Hume the roots, & probably before the plant had blossomed with me in Bengal. The original is a native of the Coast of Coromandel. What I sent Lady H. was from Chittagong.'[114] This correspondence with the Humes continued right to the end of Roxburgh's life, for in a letter to Lambert in 1814 he wrote:

> When you see Sir A. Hume, pray put him in mind of my request from Chelsea just before we left it. He can, I know help me much if he pleases, with many of the plants still wanted in India, & now is the best time for his Gardeners to gather such seeds as ripen in his Garden, & also with bulbous & tuberous roots.[115]

Although there is no evidence either way, it is interesting to speculate whether Buchanan's approach to Smith was caused by Roxburgh canvassing for his son to be assigned as his heir while he also was in London. Buchanan failed at that point to get himself nominated as Roxburgh's successor but, following the death of William Roxburgh junior in 1810, he must have once more set the ball rolling, for in a letter from Dr J. Fleming, Head of the Medical Board in Calcutta and who had been Acting Superintendent when Roxburgh went to the Cape, are the comments: 'I wish particularly to be informed what your views are in respect of the Botanic Garden in the event of Roxburgh's death or resignation', and added 'when I mentioned R's Death or resignation, I believe I might have omitted the latter alternative, for I am convinced that he has given up all intention of ever leaving this Country.'[116] However, by the time that Roxburgh left India, Buchanan must have had some long term change of heart, for although he did run the Botanic Garden during 1814-15,[117] Nathaniel Wallich wrote to him, 'if I succeed to the Botanic Garden I am the most happy man in India. Your interference has had great weight in my favor with his Excellency. He told me himself that you had recommended me for the garden.'[118] It is probable that by this date Buchanan had inherited the estate at Leny, near Callander in Scotland, to which he returned in 1815.[119]

Nathaniel Wallich, a Dane, had been appointed a surgeon at Serampore in 1807 but became a prisoner of war after the settlement was annexed in 1808. He was released at the behest of Roxburgh in June 1809 to be employed by the East India Company at the Calcutta Botanic Garden.[120] Roxburgh looked to him for answers to a variety of natural history problems: 'a few days ago, I received your letter in answer to mine of the second, which pleased Dr Fleming much, and he sent off immediately your opinion of the Insect to Mr Burt, the gentleman who made the discovery up at Mutra'.[121] In his letter of January 1810, Roxburgh wanted Wallich to go to the mountains of Chittagong plant hunting, but this had not been accepted, possibly because Wallich did not consider his travelling allowance adequate, and John Roxburgh was sent instead. Wallich succeeded as Superintendent when Buchanan left until James Hare was appointed in April 1816 and then again on Hare's departure in August 1817 till Wallich finally left India in 1846 but with breaks when he went to London (1829-33) to distribute the massive East India Company herbarium, and when he had to go to the Cape for his health in 1842.[122]

SUMMARY OF ROXBURGH'S LAST TWENTY YEARS IN INDIA

By the time Roxburgh came to leave Calcutta in 1813, he had raised its Botanic Garden from a small, comparatively unimportant garden to the pivotal scientific garden belonging to the Company in the east. It had grown in size and stature, and although still relying on the by now out-dated Linnean sexual system of classification for some of its lay-out, the simplicity of this system for non-specialist botanists did make it easier for these men to use and understand how to identify plants. It contained plants from all parts of the globe, donated and collected by a very wide variety of people with whom he had corresponded over the years, considered further in Chapter 7.

Scientifically, he was seen as the doyen of Indian botany, having codified at least the floras of Bengal, the Carnatic and the northern areas of the Ganges plain, in the manuscript version of what became his *Flora Indica*. He had plans to develop this flora into a publishable reference work himself, with numerous engravings, but this project never came to fruition, with his manuscripts eventually being published by his colleague William Carey (see Chapters 6 and 7).

A reflection of his importance is seen in the size of mansion that Roxburgh was allowed to build as the Superintendent's residence. Unusually for the time, this was a three storey building which is still an important part of the Calcutta Botanic Garden. His botanical expertise was also developed in his stay at the Cape of Good Hope, which is studied in more detail in Chapter 14. His dealings with his own gardeners and plant collectors is shown by the work of Christopher Smith, sent out by Banks on a comparatively high salary and who was already a Fellow of the Linnean Society, giving an indication of his social standing: in view of these two facts, it was not surprising that relations between Smith and Roxburgh were at times difficult. Roxburgh's working methods are also shown in how he cared for Thomas Douglas.

CHAPTER 4

Family

INTRODUCTION

It is worth at this point interposing a brief resumé of Roxburgh's own family, although a List of his Family appears in Appendix 7. A study of Roxburgh's family exemplifies many of the late 18th- and early 19th-century attitudes. When Roxburgh arrived in India in 1776, it was common for this basically masculine European society to take 'bibis' and children from these liaisons were common. This practice was particularly common among the younger and more junior members of society, partially from lack of resources to support a wife.[1] The changing view was reflected when first in 1786 an order was passed banning the Anglo-Indian orphans of British soldiers from travelling to England, reinforced by a further order of 1791 that stated that no one with an Indian parent could be employed in the civil, military or marine branches of the Company.[2] Thus, in common with many Britons in India, Roxburgh arranged for John to be educated in India while the rest of his children were sent home for their education.

40 Letter written to his father by John Roxburgh, aged about twelve years, dated 28 October 1789 and written from Tranquebar.

Equally, in line with 18th-century *mores*, Roxburgh looked after the careers of his male children, arranging for John and William junior to work as botanists, while the next two boys, George and Robert, got cadetships in the Bengal army. Similarly, the two elder daughters, Mary and Anna Elizabeth, had been found suitable husbands, both in the Company service. Another aspect of his family life was the much shorter life expectancy of the wives: hence Roxburgh had three over the nearly forty years that he survived in India, itself no mean feat, as we shall see.

JOHN ROXBURGH (c.1777–1823)

As a man of probably 25 years old, arriving at Madras, Roxburgh would have been older than many of the married men stationed there. Writers and Cadets could be as young as thirteen, and many died soon after arriving; add to this that, with short life expectancies, men married young, often in their early twenties. Additionally, there was a general acceptance of what has been referred to as 'the role of colonial comfort women was so well established that many commentators would refer to them, especially to those appointed to servicemen of rank, as "wives".'[3] What we do know about Roxburgh's oldest son, John, in spite of articles which go from the extreme position that he 'had no existence'[4] to an answering article proving that he did exist but giving fairly brief and erroneous details about him,[5]

41 Drawing of caterpillars dated 16 June 1793. Although it is annotated as being by 'James Roxburgh', it must in fact be by John (then aged 16) as James was not born until 1802. This is confirmed by the fact that it is bound in a book along with the letter shown in Fig. 40 signed by John Roxburgh. An explanation for the 'James' could be that he was often referred to as Jack and whoever annotated this put the wrong Christian name to it.

is quite considerable. He was born comparatively shortly after his father arrived at Madras, sometime in 1777,[6] as Roxburgh himself put it, 'a natural son produced by the sallies of youth before marriage'.[7] He was educated at the Moravian Missionary at Tranquebar, by the Rev. C. John, with whom Roxburgh had some understanding, that John would be supplied with clothes and sundries, and extras given in the form of 'no money at all but only cloth, Chints, red & white Handkies, chieves Marchies & Gingam for my Children' plus 400 Pagodas per annum.[8] The first letter to mention him is when he must have been about 12 years old, and shows that his education was based on traditional lines, 'Jack shall become a compleat Natural Philosopher & for this Reason I will acquaint him with the German & Latin Language as well as with the English.'[9] At this stage, he appeared to be making good progress in reading even if his start at writing seems to have been somewhat delayed.[10] He already had an interest in natural history, for 'he is growing Eugenia jambas in his garden, and has collected curious spiders'[11] and the previously quoted letter remarked that the only problem with being 'continually employed in our yard in digging the ground, sowing & planting like a gardener, which is only attended with that inconvenience, that he needs to shift twice a day all his clothes.' From this, I have a picture of a typical naughty boy, loving to get his hands dirty and paying more attention than his teachers want to the insects and plants than his studies: memories of my own childhood in India! The idea from a very early stage of this education was to train him to help his father, for the next letter stated: 'Your Jack you shall soon have as soon as he can assist you by writing and drawing.'[12] An example of John's skills are shown in the accompanying Fig. 40.[13]

John's lack of progress became evident the following year: 'he comes on but slowly … . However if you leave him some years longer under my care, I hope he still will turn out well and become a useful member of society, especially when he comes to the years of discretion.'[14] This letter also gives an indication of his stature, 'I don't know for certain his age & I would wish you would mention it to me. According to his size he seems to me & to every one a Boy of Ten Years,' this when he was nearer sixteen! However, this did not stop him from being a problem: 'he is the greatest Play fellow full of tricks amongst all my Children & that he causes me more Trouble than all the others together …. However his reading [is] so poorly as not one Child reads which has been instructed only a few Months.'[15] This is corroborated by the letter which he wrote to his father also in 1789: the quality of the writing is poor for his age.[16]

The next we hear of John is in 1799, when Roxburgh took him to the Cape and left him there to collect plant specimens.[17] The instructions that he was given by his father were in such detail that they do suggest that Roxburgh was aware of the guidance that John needed: among the instructions were the comments that John was

> to collect seeds of all sorts particularly of the most beautiful & most useful plants. Also the bulbous & tuberous roots of all plants and specimens of all such plants as he thinks of Erica which he still wants specimens of will accompany this, 3 or 4 specimens of each sort to be sent (to Dr R −) and one or two to Sir Joseph Banks, Bart, Soho Square, London. A portion of each sort of seeds and bulbs you will set aside for Dr R & this portion must be divided into two, to be sent by different vessels for fear of accidents. Seeds of European plants growing in the Gardens about the Town as also bulbs of European flowers, will be very acceptable in India, tho not in Europe.[18]

Two years later, he was still at the Cape, partially on his own account and partially collecting for the Company, when he wrote to Banks with a collection of Ericas and Proteas.[19] His success may be judged from the fact that he was sending material to J. E. Smith at the Linnean Society and to Aylmer Bourke Lambert.[20] It appears as though his own botanical skills were not great at recognising the plants, for in a letter to Lambert he stated,

> I would send the species of <u>Protea</u> which you wish to have. I am very sorry I had them not at hand, as soon as I get them I will transmit the specimens with the greatest plesure by the first conveyance; I will be obliged to you in your next letter for their Numbers, I can refer to the list which is left with me of the sorts sent to you & Sir Joseph Banks by Dr Roxburgh.[21]

This diffidence is confirmed in a letter from Viscount Valentia:

> I called [at Cape Town] the next day on young Roxburgh, who showd me yours & Lambert's letters. He is inclined to collect, but perfectly ignorant of what is rare & valuable in England. I gave him as much information as I could. I have desired him to collect for me & forward the things to Europe; whether he will or no, is more than I can say. He had nothing valuable.[22]

As John Roxburgh had not travelled to Britain, his lack of awareness of what the plant collectors there craved is not surprising, particularly if he was not particularly well-read, as seems the case. However, he did send plants to Lambert who was sent 'all his collection particularly <u>Erica</u> & <u>Protea</u> which I hourly expect'.[23]

As far as his father was concerned, John's time at the Cape was not wasted, for

> John Roxburgh at the Cape is not my Brother, but a N****l Son, who I left there merely to collect for me, & to try his own luck in doing the same for others, if he thought proper. He was educated in India, which will apologise for any little inaccuracies. He has sent me a great many additional specimens, so that I must have near 200 sp of Erica, & about 100 of Protea.[24]

Equally, a study of *Hortus Bengalensis* produces the fact that John was responsible for three introductions from the Cape to the Calcutta garden: *Ornithogalum capensis* in 1802, and *Crotalaria argentea* and *C. capensis* in 1804. There is one introduction credited to John in *Hortus Bengalensis* which raises a number of interesting points: *Cheiranthus*

Zizyphus vulgaris Nº 12.

42 *Watercolour of a wood sample of* Ziziphus jujuba (Z. vulgaris) *painted by John Roxburgh, and submitted as part of his documentation to be reinstated as Nurseryman to the Calcutta Botanic Garden in 1820.*

mutabilis was introduced by John 'from Madeira', but no date given for this. There is no reference in any of the literature of John having travelled north west of the Cape or visited Europe, which such a reference would seem to suggest: a satisfactory explanation for this introduction would therefore be that John introduced it from the Cape after it had arrived there from Madeira.

Roxburgh took John back to Calcutta in 1804, for at the end of that year there is a letter which states:

> The great progress you have made in drawing, particularly Botanical Drawing, induces me to think that it will prove greatly to the advancement of the Art of the Native painters now employed, or may hereafter find necessary to employ, are put under your charge. I have therefore appointed you *Head* Painter in the Botanic Garden with the usual allowance of one hundred Sicca Rupees (Sa Rs 100) per month from the first of August last, out of which you have to pay the whole wages, or other expenses incurred by all the native Painters employed by you in & for the use of the Botanic Garden.[25]

However, if the Indian head painter died in 1814, having worked for Roxburgh for 20 years, it suggests that this position of John's was very short-lived, as he stated that he had been 'appointed to the office of head Nurseryman to succeed the late Mr Wheeler' in 1804, giving him 'an aggregate allowance of 301 S Rs per Month' (see Chapter 6 for more details on Roxburgh's painters and drawings).[26] It was the latter job that appears to have occupied him for the majority of the time, for in 1810 Roxburgh wrote,

> I will cheerfully spare the Nursery man, Mr John Roxburgh, from this Garden for the purpose of making Botanical researches into these Countries [Chittagong and the eastern frontier, now in Bangladesh] during the ensuing season, and that he will be perfectly satisfied with the same extra allowance, (vizt 200 Rs per month) which was to have been given to Dr Wallich to defray his travelling expences.[27]

His success on these plant hunting trips is evidenced by the number of introductions he was responsible for in 1810 and 1811, approximately 80 species from Chittagong and Silhet. This plant-collecting would have suited him far more that being left to his own devices at the Cape, for he could send effectively whatever he found and his father and brother William junior would have been able to identify these introductions and decide on their value. His expertise as an artist does survive, in his painting of *Ziziphus vulgaris* (see Fig. 42).

After William Roxburgh left Calcutta in 1813, John stayed on as Overseer until a difference of opinion with Nathaniel Wallich, the new Superintendent, caused them to separate their ways. One explanation was put forward by Desmond, that this may have sprung from a difference of opinion between Roxburgh and Wallich in 1809 which led to the latter's departure from the garden in that year, but this seems unlikely in view of the letters between Roxburgh and Wallich which indicate a strong sympathy between the two men. John was reinstated in April 1816 by James Hare when he became Superintendent, but was again dismissed in March 1819, after Wallich reappeared as Superintendent in August 1817. There is a Memorial sent by John to Joseph Dart, Secretary to the Governor General in Council, to which are attached a number of copies of letters between Wallich and John Roxburgh regarding the latter's dismissals. The first occasion seems to have been because Wallich accused John of striking one of the former's servants, but after Wallich left and Dr James Hare took over, John asserted that 'the man, who was a private servant of Mr W − & not belonging to the Garden was in the habit of carrying lies to his master ... . Mr Roxburgh denies being guilty of personal rudeness to Mr Wallich.'[28] Wallich never seemed to have given reasons for this second removal, although John did indicate that it may have been due to leaving the Garden during working hours, so that he could court and marry Catherine Young, an accusation he denied. As Desmond stated, it does seem as though Wallich was being harsh, which suggests that there was a basic clash of personalities between the two men.[29] What he did in the following years before his death, is not known.[30]

John married Susanna Benedict in Calcutta in December 1806 and she died, possibly in childbirth, in 1818, as their last son was christened six months afterwards.[31] They had four children, all born in Calcutta, although as we have seen he collected for his father in Chittagong and the eastern frontier in 1810-11. After Susanna died, John married Catherine Young by whom he had another son. After John's death, his son John Peter died aged 16 in 1831 a ward of the Free School.[32] This predicament of the poverty of half-castes is reflected in the statement of William Huggins, writing in 1824:

> I shall now direct your attention to a class of men at present numerous, and increasing very fast in India, I mean Indo-Britons, or as they are often called, country born and half-caste men; ... formerly they could hold commissions in the regular service, or civil appointments; and, as far as I am informed, did not labour under disabilities of any kind; however, their situation is now materially different; they are incapable of holding commissions in the regular service, or leading appointments; are admitted only to insignificant employment and stand on the same footing as other natives.[33]

As early as the late 1780s, there was growing concern about the numbers of these descendants of usually young British men, whose number was thought to be more than 11,000 in the British coastal settlements by 1788.[34]

In spite of his poverty when he died, he has a plain but substantial grave in South Park Street Cemetery which I was able to identify when visiting this graveyard.

43 *Memorial to John Roxburgh in the South Park Street Cemetery, Calcutta.*

WILLIAM ROXBURGH JUNIOR (c.1780-1810)

When Patrick Russell referred to the death of Roxburgh's wife in 1788, this was, according to Sir George King, Mary Bonté, whose father may have been of Swiss or French extraction and the Governor of Penang, but no record of this marriage has been found.[35] They had two children, William, born about 1780, and Mary, born about 1784, but again there is no record of either of these births or baptisms among the records held at the British Library. Mary's story, as far as it impinges on her father is slight, except that she married Henry Stone of the Company's Civil Service by whom she had three children, and she died shortly before her father and was buried in Greyfriars Churchyard in Edinburgh.[36]

Young William Roxburgh was educated in Edinburgh for nine years from the age of about nine, the same age as George Roxburgh was sent home in 1798. Writing in January 1799, Banks had said of young William, that he was 'much pleas'd with his open & honest countenance. Mr Ormes his uncle & the Gentleman to whom his education was entrusted give him the best of character'.[37] It has not been possible to find out who this gentleman was nor is there any record of a Mr Orme(s) living in Edinburgh at this time, in the contemporary *Post Office Directories*. What this might mean is that the Mr Orme was Robert Orme, the historian of India who had worked for the East India Company from 1743 until 1759. If this is the case, it raises some interesting further questions about Roxburgh's own origins. Roxburgh senior tried to get Banks to arrange for young William to get a Writership in the Company in 1798, but without success, and instead asked for him to be appointed as Assistant to his father 'in the Botanical line'.[38] His education was described in the same letter as being 'so conducted, as to enable him soon to become a good naturalist', and he had, indeed, been in Edinburgh for nine years for this purpose.[39] Roxburgh had obviously started manoeuvring much earlier, for in 1796 Banks had intimated that he could not start asking the Court of Directors until the sale of the fasciculi of *Plants of the Coast of Coromandel* had paid for themselves.[40]

Father and son were reunited at the Cape at the end of June 1799 as William and his sister Mary had sailed together,[41] and all returned to Bengal on the same ship.[42] This would suggest that William did not complete the year's course of Materia Medica for which he matriculated at Edinburgh University.[43] Father and son, presumably with family, arrived in Calcutta by 16 November when Roxburgh reported back for duty, but William was kept idle until Roxburgh heard from Banks about his future.[44] However, in April 1800, the Court of Directors agreed that

> William Roxburgh ... as peculiarly well qualified, from his education and natural talents in the botanical line, to assist Dr Roxburgh (his father) in his botanical researches at your Presidency, we have permitted him to proceed thither for that purpose, and whilst so employed, you will grant him an allowance, not exceeding 250 rupees per month.[45]

This had reached Calcutta by at least September for Francis Buchanan wrote that 'I received yesterday yours of 26th September, ... I congratulate you on Williams appointment although it certainly would have been better to have got him a writer, yet the garden will be a handsome provision for him and the opportunity he will have under your tuition he will soon become proficient.'[46]

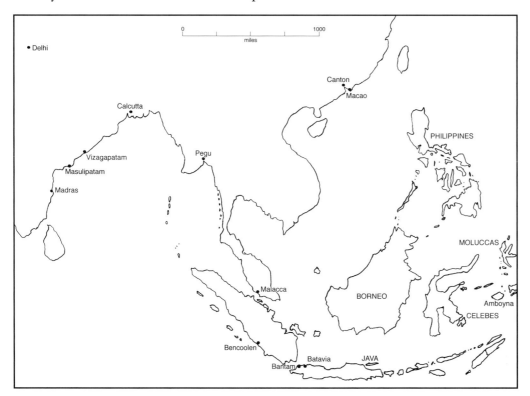

44 *Map of the Far East, showing the settlements belonging to the British East India Company and those of other European nations, and some of the places that Roxburgh and his correspondents visited (adapted from Prakesh, 1998).*

By July 1802, young William was in the Moluccas, 'in search of seeds & plants for this Garden, will, I hope, put us in possession of Rumphius's Agallchum tree.'[47] From there, he went on to Prince of Wales Island where he sent a large consignment, 'amongst them two new Artocarpus's, two new Arecas, and many other very interesting seeds, plants and specimens. From Amboyna he returns to Prince of Wales Island, & from there to Ceylon, & we still want many of the plants of that Island in this Garden.'[48] On this plant hunting trip to the east, young William had taken a gardener from Calcutta with him, a Mr Beale, who had to return due to ill health in March 1803 and another, Mr Mollies, who returned in October accompanying 100 boxes of spice plants. This is taken from a detailed set of accounts of his sojourn in the East Indies from June 1802 until February 1803.[49] Before paying this total sum of 1,486 Spanish dollars, Roxburgh was asked to verify that these were reasonable which was expedited a fortnight later.[50] The success of these collecting trips can be gathered from such comments as 'The supply of young plants landed at Bencoolen in May 1803 by Mr Roxburgh, Viz. 22,323 Nutmegs and 7003 cloves, fortunately places the plantations on that Island, on a most respectable footing.'[51] More on the spice plantations will be considered in Chapter 11, but young William was instrumental in these early years in assisting their development to a considerable degree.[52]

As far as searching for new species for his father's Botanic Garden is concerned, the first plants came in 1800, from Rajmahal in Orissa. But the following year he had a successful visit to Chittagong in East Bengal (now Bangladesh), sending back 11 new species, plus four more from Rajmahal. Over the next few years, when he was also collecting spice plants, he was responsible for the introduction of a further 20 species, mostly from the east: Sumatra, Malaya, Moluccas, and Prince of Wales Island.[53]

Already it was becoming accepted that young William would take over the running of the Calcutta garden from his father: 'providing William Roxburgh junior conducts himself suitably, he will take over from his father when the post becomes vacant, due to death, removal or resignation of Dr Roxburgh.'[54] This manoeuvring was in preparation for Roxburgh to go on home furlough, which was granted in March 1805.[55] Over the next nearly two years, there are a number of letters signed by young William as Acting Superintendent, the first in June 1805, the last in April 1807.[56]

Young William returned to Bencoolen in 1808, to continue his plant hunting activities. However, this was not an easy journey, for 'Poor fellow, he has been taken by a French Privateer, when on his way to Bencoolen, to take care of the companies spice plantation. He was however sent on shore near that place, with loss of all he had with him.'[57] He obviously survived this hazard of eastern travel, for the next reference to him is, 'I however hope soon to get some specimens, and tidings from my son regarding the Camphor tree, all I have yet learned is that it grows far inland on the north end of the Island' of Bencoolen.[58] By this time he was referred to as the Superintendent of the Company's spice plantations in Sumatra.[59] This problem with pirates and the capture of shipping was a common problem in the islands of the Far East, particularly by the French.

Young William died aged about 30 and was buried at Padang in September 1810.[60] However, the family tradition for 'natural' offspring continued, for he left a pregnant local girl whose name is difficult to decipher but appears to be Sukeowa. Their son was born after his father died and was baptised William in February 1812, but must have been born not later than June 1811.[61] It is possible that his executors, one of whom was his brother-in-law, Henry Stone, may have taken the infant in, because the executors of young William approached the Secretary to the Public Department regarding William's salary which they claimed was less than what had been agreed when the post of Superintendent of the Spice Plantations had been created in 1803.[62]

MRS MARY ROXBURGH (NÉE HUTTEMANN)

After William Roxburgh's first wife died in 1788, he did not remain on his own for long, as he married Mary Huttemann in June 1789.[63] She was probably the daughter of the German missionary John Henry Huttemann who was buried at Negapatam in 1787, and the brother of George Samuel Huttemann (1769-1843) who was Head of the Free School in Calcutta that was attended by John Peter Roxburgh (see above).[64] They had eight children of whom the eldest, George, died shortly after his father,[65] and two (Bruce and James) were instrumental in the publication of their father's *Flora Indica* in 1832. The eldest daughter, Anne, was sent home with a Mr Boswell, probably Bruce, in 1801: 'the box of specimens I find must be sent on Board as baggage belonging to a very charming young Lady who calls me Father, Anne Roxburgh. She is under the care of Mr Boswell & going with him to Edinburgh for her education.'[66]

This second Mary was a great help to Roxburgh, for it appears that she was dealing in some way with silk, as she had sent some to Dr Anderson in Madras in 1793.[67] The success of this is seen two days later, when Ross wrote that he had 'delivered the Skains of Silk to Dr Anderson & he thinks that Mrs Roxburgh has much merit in producing such as samples' and that by getting someone instructed in the production of silk 'Mrs Roxburgh will make more money than will be required for the Education of all the Children that she has already brought you & may still bring.'[68] From one of Buchanan's letters to Roxburgh, four years later, it appears that her assistance was more than trading but also keeping accounts: 'In my hurry I forgot to get from Mrs Roxburgh an account of what you have advanced to the [probably Asiatic] Society on my account.'[69] This practice of wives assisting their husbands seems to have been quite a common occurrence, for Humphry Repton's wife 'provided the "hidden investment" characteristic of small businesses at the time, bearing and raising of future partners, the provision of physical and moral support systems as well as various secretarial tasks.'[70] After 16 years of married life, Mary died in 1804 which may have precipitated Roxburgh setting out for England in January 1805.[71]

MARY ROXBURGH (NÉE BOSWELL)

Roxburgh arrived in England on 12 July, taking a house first in Southampton Row then moving at the end of the month to 2 Charles Street, Brompton (possibly the modern Charles II Place, off the King's Road), 'close by the Botanic Garden' (now known as the Chelsea Physick Garden).[72] Although there is little extant evidence, Roxburgh must have kept up his connections with the Boswell family, for Buchanan wrote in 1798 'I dined yesterday with Boswell who is very well and in good spirits and I am happy to hear from him that you are so much pleased with the Cape.'[73] This is more than likely to have been Bruce Boswell (1747-1807), the second son of Dr John Boswell with whom Roxburgh stayed while in Edinburgh and would probably have met then. In 1801, Roxburgh was acting as agent for him in Calcutta, depositing a sum of 20,000 rupees with the Sub-Treasurer as an interest-bearing Government loan.[74] With these Company connections, he was also probably the Bruce Boswell who was captain of the East Indiaman *Chesterfield* and Mary's uncle, which would have allowed Roxburgh to keep in touch with the family as well.[75] These continuing connections could also explain the fact that the two sons of John Boswell, with whom Roxburgh stayed in Edinburgh, have names which are repeated in two of Roxburgh's sons, Robert and Bruce. These two boys, aged six and five respectively, had been sent home at the end of 1802 for their education, probably to Edinburgh.[76] This must be put in context, for William Roxburgh junior was sent home, at about nine years old, to his uncle, Mr Ormes, and this move for Robert and Bruce probably indicated a growing closeness with the Boswell family. Indeed the swiftness with which the marriage of Roxburgh to Mary Boswell took place in November 1805 must reflect this, for there would hardly have been time for Roxburgh to arrange anything before leaving Calcutta after his second wife's death two months before his departure.

45 *Mary Boswell's model of the west front of Edinburgh University. A number of these models, showing Robert Adam's (never-completed) design for the new University building, were made – probably as a handicraft project in an unidentified finishing school for young ladies in the early 1790s. The University has two versions, and Mary Boswell's was deposited with the Royal College of Physicians of Edinburgh.*

In November 1805, Roxburgh married his third wife, yet another Mary.[77] This was Dr John Boswell's second grand-daughter by his second son, Robert (1746-1804). Mary took on a family which by contemporary standards was big enough to cause comment, for Roxburgh's friend Lambert referred to it as 'his <u>Family very large</u>!!!! & he finds some difficulty to find a House to hold them all!!!!'[78] The eldest of these, George, probably came home with his father and was awarded a Cadetship in the Bengal Cavalry in January 1806 and with the seven other children of Mary II's, plus servants, would indeed have required a fair sized residence. The Court of Directors were getting impatient for a decision on a date for Roxburgh's return, for in September 1806 he wrote to William Ramsay, Secretary to East India House, that it was his 'intention to resume his situation; ... at as early a period as my health will permit'.[79] He was obviously looking after his health at this point, for the letter was written from the fashionable spa city of Bath. When Roxburgh was given permission to return to Calcutta in early 1807, he was accompanied by his new wife, daughter Anne and a native servant but permission was refused for a European man servant; thus he must have left behind Robert, Bruce, Elizabeth, Sophia, James and Henry, presumably all for their education.[80] By this time the family was living at 36 Bernard Street and Roxburgh was still writing to Dr C. Taylor, Secretary to the Society for the Encouragement of Arts, Manufactures and Commerce, at the Adelphi, in July.[81]

One of Mary's skills is shown by a model in the Royal College of Physicians of Edinburgh's Library of the east front of the new plan for Edinburgh University. This is taken from the design of Robert Adam, dated 1789, when Mary would have been in her mid-twenties. This would suggest that she was the beneficiary of a more enlightened education, which was earlier proposed by the Rev. Dr James Fordyce in his *Sermons to Young Women* that was published in 1765, in which he recommended that his listeners should not merely dance but apply themselves to history, biography, geography, astronomy, moral and natural philosophy, as well as 'severer studies to every prudent length'. Mary must have been fairly strong, for she gave birth on the voyage out to India, another daughter, Sibella, and both mother and daughter were reported to be in good health on arrival in Calcutta, where Mary 'is delighted with this Garden, & will gather and preserve specimens for yourself which will in your Eye, render them doubly valuable, will it not Mrs Lambert?'[82] Mary had two more children, Mary Anne and William, born in 1812, after his half-brother had died. This youngest William was one of the early pupils at Edinburgh Academy, in 1827-29, attending Mr Ferguson's class, gaining his MD from Edinburgh University in 1835 and eventually retiring to

Ipswich in 1886.[83] He gave a medicine chest to the Royal College of Physicians Edinburgh in 1877, which is inscribed 'This medicine chest belonged to my great-grandfather, Dr John Boswell.' The presumed descent was from Dr John Boswell to his eldest son Robert on whose death in 1804 it was passed to his daughter Mary. When she married William Roxburgh the next year it may then have been used by her husband and was then kept by Mary until her son William graduated in 1835.

From this, it can be seen why confusion has arisen in the past over who was who, with four William Roxburghs and not only his three wives being Mary, but then confounding the confusion by having two daughters of the same name as well.

REVIEW OF WILLIAM ROXBURGH'S FAMILY

By the time Roxburgh died, he left a family well provided for, reflecting his successful life in financial terms. His family also reflects the attitudes of the time, with an increasing concern over children born of mixed origins and their reduced career paths, shown by John Roxburgh, against those of European parentage who were sent home for their education and who could be employed in all departments of the Company, seen in his legitimate sons. The success of Roxburgh's provision for his family is also shown by the 'gentrification' of his son Bruce who was granted arms and crest in 1853, using his father's botanical success as well as the Indian connections.[84]

Much confusion has been caused by the repetition of names, with Roxburgh calling two of his sons William: William junior acting as Acting Superintendent while his father was in Britain in 1805-07 and who died in 1810 and caused additional confusion by leaving a son who was also christened William; followed by Roxburgh's youngest son, William, born in 1812. Further problems were caused by the fact that all three of Roxburgh's wives were Mary as was his eldest daughter.

This study of Roxburgh's family also highlights the short life expectancy of Europeans in India at the time, with two wives and two of his children predeceasing him. What was unusual, and possibly reflects his own medical knowledge, was that none of his children died before they were grown up and none of his wives died in childbirth. The emphasis placed on family tradition is shown by the fact that it was his two sons, Bruce and James, who arranged for the publication of Roxburgh's *Flora Indica* in 1832, and his youngest son, William, who bequeathed his grandfather's medical equipment to the Royal College of Physicians of Edinburgh, after becoming a physician himself.

46 Medicine chest belonging to Dr John Boswell. This was possibly also used by Roxburgh, and was given by his wife Mary (née Boswell) to their youngest son William who presented it to the Royal College of Physicians of Edinburgh.

47 Roxburgh as an old man. Drawing by W. J. Hooker, c.1814. The drawing or specimen shown in the folio volume is, appropriately, Roxburghia gloriosoides (see Fig. 129).

CHAPTER 5

After Calcutta 1813-1815

THE VOYAGE FROM BENGAL

At the beginning of 1813, at the age of nearly 62 and after 37 years in India with just the two breaks, at the Cape 15 years before and in England eight years previously, Roxburgh finally felt it necessary to return to Britain. He requested leave because 'my health has suffered so much during the last twelve months that it is deemed necessary for me to take a voyage to a Colder Climate for its reestablishment. I have therefore to request permission of His Lordship in Council to proceed to the Cape of Good Hope, or St Helena, in failure of a conveyance to the former, and eventually to England, should it be necessary.'[1] However, he did not consider that this was a permanent departure, for he went on 'it is however my full intention to return from the former place by the Mauritius, with the view of bringing from thence and Boourbon [*sic*] many plant and seeds which have not yet found their way to this Garden.' The move to the Cape for health reasons was also becoming common, especially since this was in British hands. Roxburgh had arranged for his friend Henry Colebrooke to take charge during his absence, all of which was approved. A month later he had organised berths on the East Indiaman *Castle Huntly* for 'me, Mrs Mary Roxburgh and three Children, vizt Sibbella, Mary Ann and William Roxburgh on board of the Ship as Passengers to St Helena with two Native Servants named Gunmi and Alliera for whom the usual deposits have been made',[2] which was again agreed. The need for deposits was to guarantee the return of the natives who would thus not become a burden on anyone in Britain, and it is interesting that Roxburgh had achieved a level of affluence to afford these two servants, probably an ayah (nanny) for the children and a bearer.

It is worth recording what had happened to Roxburgh's other children. William junior had died in 1810 and Mary had married Henry Stone in 1803, and so do not need to be considered. Of the children of the second marriage, George had received a Cadetship in the Bengal Cavalry in 1806 and survived for another two years in the east, before being killed by lightning on Java in 1815;[3] Anna Elizabeth had married Robert Henry Tulloch at Serampore in 1811;[4] Robert and Bruce had also got Cadetships in the Bengal Army, the latter transferring from the infantry to the cavalry in 1814 with the nomination of Joseph Cotton, one of the Directors and a friend of Roxburgh's.[5] That leaves Elizabeth, Sophia, James and Henry of whom there is no mention, and who were all too young to be self-sufficient, so one can only assume that they had been left behind on Roxburgh's previous visit to England in 1807 when Elizabeth would have been eight and Henry three.

By the beginning of March they were near Ceylon. Roxburgh was impressed by the *Castle Huntly*, 'a very fine 1200 ton Teak ship, built at Calcutta within the last 18 months' but less impressed by the journey down the Bay of Bengal in a fleet of nine ships. He had hoped to have long enough at Colombo to visit the 'new Botanic

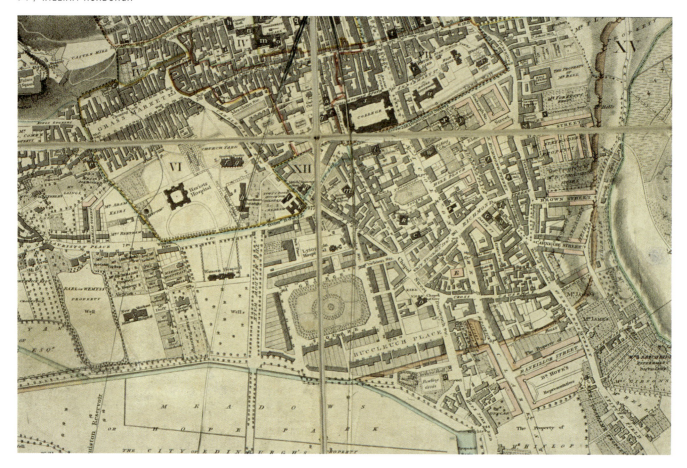

Garden establishing' there particularly as he had some seed from Bengal for it.[6] By the middle of June he had been on shore at St Helena 'one week, and have already derived benefit by the change' and intended to spend 'some time' there, aiming to 'proceed by the next Fleet, or by the first China ships of next season.'[7] However, writing at the end of August to Dr Taylor, he was 'still alive, you see, and that is all. I am now reduced to a perfect skeleton, and weak in proportion.'[8] The letter to Banks is important, for it gives a glimpse of the vicissitudes of sailing plus the connections and familiarity that Roxburgh had with at least some of the hierarchy of the Company: one ship, the *Marquis of Wellesley*, grounded going down the Bay of Bengal and had to put in to Bombay for repairs.

Also mentioned in this letter from St Helena was one set of packages to be delivered to either Banks himself, Mr Amos, Roxburgh's agent in London, or Mr Davis, a director of the Company. The fact that Amos was included as one of the recipients suggests that this may have been some goods that were to be sold and also some plant material. Roxburgh was also friendly with Captain Joseph Cotton, another director of the Company, through his work on hemp, both having published on this through the Royal Society of Arts, and who sponsored Bruce Roxburgh's transfer to the cavalry. The letter finishes 'to please myself & Colonel Beatson, I shall begin upon the plants of this Island as soon as the Fleet has sailed' which brings us back to Beatson, whom Roxburgh had come across twenty years earlier, when he was helping in the surveying of the Circars, to consider canalisation of part of the Godavary with which Beatson was also involved which will be dealt with in detail in Chapter 13. Beatson, it should be added, was Governor of St Helena from 1805 to 1813. Roxburgh's work on the flora of St Helena is considered in Chapter 6.

48 Kirkwood's map of Edinburgh (1817), showing Park Place (on the join) to the north of George Square. The site is now beneath the McEwan Hall and the late 19th-century Medical School of Edinburgh University.

IN BRITAIN

Roxburgh, once in England, again requested that he should receive his salary while on St Helena, but the Court had stringent rules which it applied: 'if any Servant of the Company shall quit the Presidency to which he belongs, other than in their actual Service; his Salary and allowances shall not be paid during his absence; and in the event of his not returning or coming to Europe, they shall be deemed to have ceased from the day of his actual quitting such Presidency.'[9] The fact that he was able to live on the island for nine months, and get a loan of £500 from its government, once more indicates the security of the financial position he had reached by this time.[10]

Roxburgh arrived in England by the middle of May 1814[11] but in a very sickly state, for he was described by Smith: 'I fear poor Roxburgh is dying at Chelsea.'[12] This is confirmed by the very shaky hand of a letter he wrote in the next month in which he said that

> the state of my health, ever since my arrival in London, has been so bad as to deprive me of the pleasure I expected to derive from the society of my scientific friends .... Since the effects of the mercury began to subside, I have been somewhat better, but still too weak to venture out of my room .... I fear we must give up all thoughts of Scotland for the present, & remain for the winter in or near London, or cross the Channel, & take up our abode in France during the winter.[13]

This was written from the house at 16 Cheyne Walk that he had taken, possibly as it would have had a view across the Thames, reminding Roxburgh of his house in the Botanic Garden at Garden Reach. It is also literally a five-minute walk from the Chelsea Physick Garden. In spite of his ill health, he still asked the Court to send the drawings and descriptions of Indian plants to Sir Joseph Banks at Soho Square, so that he might be able 'to complete the last half of the 3rd volume of Indian plants, being read.'[14]

In spite of these comments, Roxburgh was writing from Edinburgh in the beginning of October, that 'we have now been here about two weeks, & still so weak as to be obliged to stay where I am during the winter, without I gain sufficient strength to allow me to go to London by land, to go by sea in winter is out of the question.'[15] Even as ill as he was still, he could not stop working, for he continued that he had just seen John Lunan's *Flora Jamaicensis* (published in 1814), and had made out 'a list of some of the most useful trees &c still wanted in India. I send you annexed a copy of the list, & beg you will do what you can to procure me their seeds, to send to Bengal .... Do my good friend what you can to help me to be useful to our India Empire.' This aim was repeated to Banks, 'whether I am ever able to return to India, or not, my anxious wish, to the last hour of my existence, will be to add all in power to the vegetable riches of our possessions there.'[16]

It was unfortunate that Roxburgh returned to Edinburgh in one of the coldest winters for some time, confirmed by his recording a temperature of 20°F mentioned in his letter to Banks, and in December stating that 'the winter has been & still is very severe, too hard for an old Indian constitution like mine even if I had no complaint'.[17] He was lucky, though, that he was not a year later, for in April 1815 the Tambora volcano in Indonesia erupted with such force that Sir Stamford Raffles in Batavia ordered two warships out because the explosion that he heard was presumed to be enemy action; the eruption was of such force that it reduced the height of the volcano by 5,000 feet, spewing enough ash into the atmosphere to reduce the global temperature by 3°C, and causing the 'year of no summer'. However, this cold did not stop him visiting the Edinburgh Botanic Garden, then still at Leith Walk:

> when I arrived here I was able for once only to go the length of the Botanic Garden, where I met a very exclusive collection of Tropical plants; many of them my old Indian friends, but they are miserably lodged, & every way hampered, tho the Gardener, Mr [William] McNabb, seemed to spare no pains to make the most of such accommodation as the Garden affords.

49 Park Place, showing the Reid School of Music (Edinburgh University) on the left. Wood engraving from William Ballingall's Edinburgh Past and Present *(facing page 64, Edinburgh, 1877).*

He continued: 'when I first studied Botany under Dr Hope in this Garden 42 years ago, it was in its infancy, & then sufficiently large, tho the soil was always bad, vizt a poor coarse sand.'

He bought the house in Edinburgh he was living in, at 4 Park Place, from John Brown, an architect.[18] He referred to it in his letter to Banks, 'we have fortunately got an excellent well built house, in a warm situation on the south side of Edinburgh, & rather without than within it.' It is interesting that he should look for a house on the south side of Edinburgh rather than the New Town: the early Georgian area around George Square was still very much a prestigious residential area, Sir Walter Scott living in George Square itself, whilst the New Town was only slowly filling up. George Square had only been completed in 1785 after James Brown, a probable relation of John Brown, had started developing the Ross House policies north of the Meadows in 1766. Although he was working on both his Flora of India and a Flora of St Helena,[19] Roxburgh's plans for further publishing and botanical research were overtaken by events, for his letter to Smith was his last extant record, as he died on 18 February 1815, and was buried in his father-in-law's tomb in Greyfriars Kirkyard.

FROM POOR MAN TO RICH MAN

A study of Roxburgh's will gives an insight into the importance of his family connections and the importance of the Boswells.[20] The executors were: Alexander Boswell, Writer to the Signet in Edinburgh, presumably Roxburgh's brother-in-law who would have organised Roxburgh's burial in the Boswell tomb; Mary Roxburgh, Roxburgh's widow, and Alexander's elder sister; George Stone, a barrister of Lombard Street in London and possibly the brother of Henry Stone; William Egerton, Alexander and Mary's brother-in-law; Henry Stone, Roxburgh's son-in-law and widower of his eldest daughter Mary who had died the year before and is also buried in the Boswell tomb; and Robert Tulloch, husband to Anne Roxburgh, the eldest surviving daughter – the last three men were all in the East India Company's Civil Service. This will was made in June 1814 while Roxburgh was still living in Chelsea but was proved in Canterbury in March 1815 for some reason. The witnesses, apart from James Amos, Roxburgh's long standing agent, were all family: Alexander Boswell, John Anderson and his wife Elizabeth who was Mary's sister.

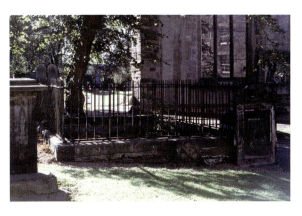

50 The Boswell family tomb, Greyfriars Kirkyard, Edinburgh.

His bequests raise the problems of Roxburgh's origins, for after the first to his 'dearly beloved wife Mary Roxburgh' who inherited the 'furniture, plate, Horses & Carriage', the next three bequests were to the families of his step-brothers and step-sister: Thomas Parkhill of Ayr, an annuity of £50 and his widow £20; Isabella Humphrys of Symington an annuity of £50 plus £100 to her heir on her death; and William Parkhill of Glasgow, son of his late half-brother John Parkhill, £200. On the basis that investments were yielding 10 per cent, the capital portion of the estate would have been £1,400. The house in Edinburgh went jointly to Mary and her sister-in-law Isabella, the widow of her brother Robert.

Next was care of Roxburgh's botanical drawings and manuscripts, 'from being lost to the public', which were to be in the care of: Thomas Colebrooke, a member of the Supreme Council of Bengal; Samuel Dennis, a Director of the Company; Henry Colebrooke, of the Bengal Establishment and who took over running of the Botanic Garden in 1813; Robert Brown, Sir Joseph Banks's librarian and curator from 1810 to 1820; and Dr James Edward Smith, President of the Linnean Society. All these men were of considerable stature on the world's botanical and/or Indian scientific scenes.

The residue of the estate was to be invested in 'Government or other Public Stock' for dividends to be paid: ¼ to wife Mary for life if she remained a widow; the sons George, Robert, Bruce, James, Henry and William each to receive £3,000 when they reached 25 years of age, with the interest for the first three to go to them meanwhile, and for the younger three, Mary to have the interest for their maintenance and education; Robert Tulloch, son-in-law, an annuity to the highest sum paid by the Calcutta Civil Fund; and his four daughters plus the four children of his deceased daughter Mary, each to have the same proportions when they reached 21 or married. On the basis that this means that each of his children was to get the same amount, equivalent to £3,000, this suggests that this portion, ¾ of Roxburgh's estate, was valued at approximately £36,000, making a total of about £48,000, excluding the value of his drawings and descriptions, and excluding the bequests to his half-siblings and their families, nor does it include the value of his house in Edinburgh, the contents and his 'horses and carriage'. This sum which must in total have exceeded £50,000 was a considerable fortune at the time and is an indication of how far Roxburgh had moved in the nearly four decades that he spent in India.

The views from Britain of how the employees of the Company returned home were used by a number of contemporary and later writers. Thus Samuel Foote, who wrote the play *The Nabob* in 1778, described the returning Sir Matthew Mite and his fellow nabobs as 'these gentlemen, who from the caprice of Fortune, and a strange train of events, have acquired immediate wealth, and rose to uncontroled power abroad, find it difficult to descend from their dignity, and admit of any equal at home.' Towards the end of the play, young Thomas, the hero, stated: 'For, however praiseworthy the spirit of adventure may be, whoever keeps his post, and does his duty at home, will be found to render his country best service at last.' Similar attitudes towards the ease with which wealth was generated in the east were also given by Jane Austen.

To 'cut a figure in London society and perhaps enter politics', £100,000 was deemed necessary, and comparatively few Scots achieved this level of wealth. A more modest fortune was £20,000 which was sometimes referred to as barely a 'competency'.[21] To give this wealth of Roxburgh some further contexts, Charles Allen referred to 'the enormous sum of £15,000 [paid] to the widow of the late Colonel Colin Mackenzie to secure the bulk of his antiquarian collection'.[22] Also, to give an idea of what could be purchased with sums of this order of magnitude, Henham Hall was built for Sir John Rous in Norfolk in the 1790s, designed by William Wyatt. This classic mansion, of three storeys and seven bays, cost £21,370 to build.[23] On the basis of this, it may be said that Roxburgh came back to Edinburgh very comfortably set up financially, even if his health had suffered in the process.

However, to have lasted nearly forty years in India was in itself unusual. James Forbes, writing in 1813, recounted 'of the nineteen youths with whom I thus commenced my juvenile career [in 1766], seventeen died in India many years before my departure [in 1784]; one only besides myself then survived; with him I formed an early friendship, which continued without interruption to his death, for he also has since fallen sacrifice to the climate, and I have been for nearly ten years the only survivor':[24] virtually eighty per cent died well within the first fifteen years. A walk round the cemetery of St John's Church in Calcutta with its large number of graves of the young who so quickly succumbed to the diseases of the tropics, is a depressing experience, that so many of the bright young men and women lasted so short a time.

This will also gives an interesting insight into Roxburgh's views about the relevant ages of inheritance. All of the male members came into their inheritances on reaching the age of 25, whilst the daughters inherited on reaching 'the age of Twenty one Years or on the Days of their respective Marriage with the consent of their Guardians'.

There is a codicil which refers to a 'Considerable property in Nagapatam' which has not been traced, and which should be added to the value of his total estate: it would seem to be unlikely to be some real estate which had been rented from the East India Company as there would have been some record in their archives, in which case it would appear that this refers to some sort of stock. Roxburgh, in fact, died a man of some fortune and position.

SUMMARY TO PART 1

This review of William Roxburgh's life has not looked in any detail at his botanical or other scientific work, which will be explored below, but has given a framework within which this can now be viewed. Coming from an unknown background, he was nevertheless given an advantageous start, with the patronage and connections of the professional leaders of the Edinburgh Enlightenment of the 1770s, probably through some link with the Boswell family. Once at Edinburgh, he benefited from at least one year at the Medical School studying surgery and anatomy, as well as botany. With introductions through some of the people he stayed with, met and studied under, he was able to make contact with the leaders of the British scientific community in London, particularly Sir John Pringle, a Scot who, as President of the Royal Society, was able to forward the careers of up-and-coming young men from north of the Border.

Following a fairly common route, he spent two journeys as a Surgeon's Mate before getting a posting to Madras as Assistant Surgeon. With his strong botanical and scientific education, he was also lucky to meet up with Johann Gerhard König, a pupil of Linnaeus, who was beginning to describe the Indian flora in the Linnaean manner. The death of König, the loss of all Roxburgh's own papers in the flood of 1787 followed by the opportunity of becoming Company Naturalist, gave him a chance to review his own position as a descriptive botanist, while at the same time experimenting with the introduction of economically useful plants. This period in the Carnatic was also an opportunity to do some important field work, both botanically and surveying, introducing him to the vicissitudes of travelling into the interior as well as, for example, to Ceylon. Backed by the rationalist and humanitarian Edinburgh education, the famines of India had a profound effect on his attitudes, so that he was always aware of the need to improve the crops, farming opportunities and employment prospects for the natives as well as the economic and financial constraints imposed by the Company.

Being a product of the mid-18th century, he would have been aware of the potential for wealth creation for the residents in India before he arrived: a draw for many an impoverished Scot. The loss of his first small fortune at Coringa did not stop him from

51 Costus speciosus. *Living specimens of this ginger-relative were sent by Roxburgh to Kew on no fewer than ten occasions (more than that of any other plant). Hand-coloured drawing by one of Roxburgh's Indian artists, c.1792 (Roxburgh drawing 215).*

Costus speciosus

starting again, first by renting a farm for sugar and indigo at Corcondah, and then, from frequent references to his agent, Mr Amos, he must have done a fair amount of private trading. This meant that at his death he had acquired an estate worth approximately £50,000.

Two other areas need to be considered at this stage, which have been alluded to in the preceding pages: his health and his achievements. The former is mentioned in nearly forty of his letters but only very rarely is anything specific given as to his actual ailments. The clearest indication was given in a letter from Sir William Jones to Banks in 1791, 'Roxburgh will do much on the Coast, if he can be relieved from his terrible head-aches'.[25] Life expectancy was sufficiently short for those living in India, that health was always a concern, and is reflected in the time and effort that Roxburgh gave to finding a cure for malaria, working for many years on *Swietenia febrifuga* as an alternative to Peruvian Bark, a natural source of quinine, which was very expensive in India because of the South American monopoly. Prior to his move to Bengal, there is only one reference to his health, in 1790, writing to Patrick Russell: 'I have made up my mind to a longer stay in India than I ever intended. I must therefore endeavor to do something while my health permits.'[26] This could, however, have been an allusion to the likelihood of his health breaking down. Other than a bout of 'sickness' shortly before he left the Coromandel Coast, what seems to have caused much of his hardship was the climate of Bengal from which he suffered almost as soon as he arrived there: 'I am now from home on account of a bad state of health which I have been more or less afflicted with ever since I left the Coast but more so for these last two months.'[27]

As mentioned before, three years after arriving in Calcutta Roxburgh started to approach the Government for permission to leave on account of his health, writing: 'My constitution has suffered so much during these last twelve months that I am under the necessity of soliciting the Hon'ble the Governor General in Council permission to proceed on a voyage to sea for the recovery of my health.'[28] However, a fortnight later he wrote to Banks: 'my health is much as when I last mentioned it, however I find I can not well remove from hence for the present, so must take my chance a little longer.'[29] This continual mention of his health was very typical of his time, but, writing to Smith in 1799, he complained of 'headaches, noise in the ears & Giddiness' which fit the symptoms of tinnitus very closely.[30] Again, after his return to Calcutta, in 1801, he wrote that 'if my headaches would leave me I should be one of the healthiest men in India',[31] which must give credence to the fact that his health was genuinely weak, exacerbated by tinnitus, for which there is no real cure even now; it can lead to depression and is more common among older people, all of which could apply to Roxburgh. Being a malady attached to the ear, it would also explain why Roxburgh appears to have been a poor sailor.

Having said this, there are also mentions of his suffering from fever on a number of occasions, but with his interest in finding an alternative to Peruvian bark for malaria, this could explain why he managed to survive for so long and also why most of his family also survived.

The idea of using ill-health as a reason to go to the Cape and develop the garden there has already been mentioned, and from the moment that this idea was first floated to the date of departure does suggest that he may have devoted some time and effort to preparing the ground for the position of Superintendent at the Cape. Again, he used the fact that his health suffered on the trip to the Cape to explain why he planned to stay longer than originally thought.[32] However, he did indicate at the end of his stay at the Cape, that 'my health is upon the whole better, if I could get my head well, all would be well, but I fear that will never be'.[33]

As has been mentioned before, he was a man of immense energy, it being claimed that he travelled 'over 40 miles in a morning'. The same obituary gives another example: 'while being conveyed in a his palanquin between Calcutta and Madras, in the midst of one of the extensive forests that overhang each side of the road, he

52 *Commillah, the residence of Joseph Buller, showing a palanquin, of the type that Roxburgh must have travelled in. Hand-coloured aquatint from Blagdon's* A Brief History of Ancient and Modern India *(Plate 41, 1805).*

suddenly leaped from it, to the utter astonishment of his bearers, ran to the spot where he had marked a particular plant for which he had long searched in vain, and bore it back in triumph.'[34] However, the veracity of these stories does not always stand up to careful scrutiny. In another of his obituaries appears the story of Roxburgh being informed by Colonel Hardwicke that 'a vegetable butter called *Fulleva* or *Fullevara*, by the Hindoos was discovered to be the produce of a tree growing in the Almora mountains. No sooner was this fact, relative to an article used for various economical purposes by the natives, disclosed to Dr Roxburgh, than he took a journey thither, expressly for the purpose of ascertaining the precise tree, and discovering the process by which its butter was procured. It proved to be a new species of *Bassia*, nearly allied to Parke's African butter tree.'[35] In Roxburgh's article on the *Bassia butyracea* is the true account: 'On Captain Hardwicke's departure for England, in the beginning of 1803, he gave me a small quantity of the above mentioned substance, observing, that the only account he could give me of it was, that it was reported to him to be a vegetable product from Almorah, or its neighbourhood, where it is called Fulwah of Phulwarah. In consequence of this information, I applied to Mr Gott, (who is stationed in the vicinity of that country,) to make the necessary inquiries; and from him I procured an abundance of well preserved specimens, at various times, in leaf, flower and fruit.'[36]

Having said that, the distance covered by some of the later plant hunters was quite extraordinary. There are, for instance, examples of Ernest Wilson (1876-1930) covering 33 miles in a day, and George Forrest (1873-1932) covering 1,000 miles in a month: although they may have lived a century later, their means of transport had not changed – by foot and horse, although Roxburgh may have used a palanquin at times, as well. Nearer to Roxburgh's time, David Douglas (1799-1834) wrote in 1826 while plant hunting in the Rocky Mountains of the United States, 'as the extreme distance [between his headquarters and what he referred to as the highest establishment] does not exceed more than 800 miles, frequent journeys can be made to and from each in the course of the season.'[37] Some idea of Roxburgh's journeys is best gleaned from his time at the Cape, which is dealt with later (Chapter 14).

Having raised doubts about some of the stories relating to him, Roxburgh was certainly an energetic person, for on the journey to England in 1805, when the opportunity for taking a lot of exercise was limited, he 'got pretty stout during the voyage home, & have in general been pretty well since my arrival'.[38] Equally, as will be shown in Chapter 14, he covered large areas in South Africa. The British weather did not suit him, for in December 1806 he wrote to Smith: 'last spring when you were in London I was confined to the house by illness, and all the summer I have been on the opposite side of the Kingdom, endeavouring to recover my health. This climate does not suit my Indian constitution. I am therefore preparing to return, & mean to leave this spring.'[39] He and his new wife seem to have delighted in being in Bengal, for in September 1809 he wrote to Banks: 'Mrs Roxburgh as well as myself are better here than in England, we must both have been made for a warmer climate than the Latitude of 56-60 North.'[40]

This continual awareness of his health appeared again in the first letter to Wallich, in which he complained of a sore throat.[41] The next mention of his health was on his way to St Helena in March 1813, when he 'had suffered much during the whole of last year, from a severe & obstinate dyspepsia, which I have been long advised to go to sea for, as the only chance of recovery.'[42] These inconsistencies in his attitude to his own health are highlighted by two letters, two days apart: 'I have been on shore one week, and have already derived benefit by the change' followed by 'I was compelled to leave Bengal on account of my health, which sincerely repent for I have got daily worse during the whole of the long passage of 4 months to this Island, where I must remain for I have not strength to go on or return just now.'[43] There is one allusion which might hold the key to this conundrum, for he wrote to Mackenzie that 'Botany will help to drive away sickness and the <u>Blue Devil</u>.'[44] This could be depression, similar to the 'Black Dog' from which Sir Winston Churchill suffered, but in the case of servants of the Company, brought on by boredom, stressed by Sir William Jones as a cause of illness, as will be discussed further in Chapter 7. For an active man such as Roxburgh, even he may have had his moments when ennui took control, hence the apparent swings which have been highlighted here.

One aspect that has not yet been considered is the scientific and botanical experiments that Roxburgh performed, from work on dyes to his plantations of spices and teak. These will be developed in Parts 2 and 3.

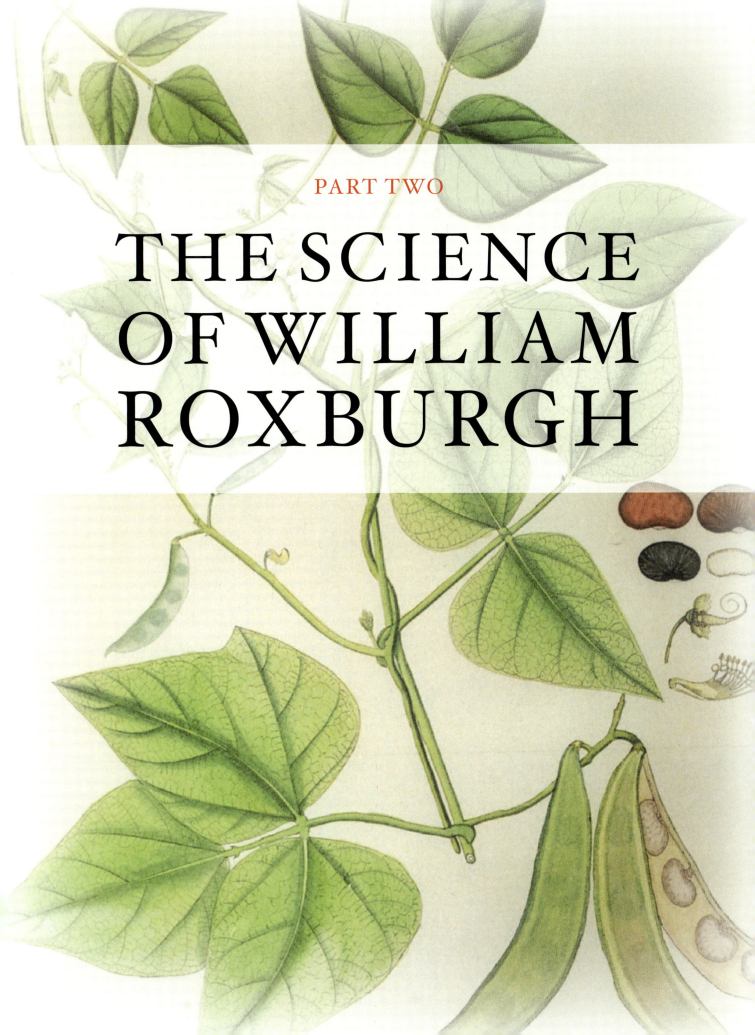

PART TWO

THE SCIENCE OF WILLIAM ROXBURGH

CHAPTER 6

The Botanist

INTRODUCTION

Reference has been made in Part 1 to William Roxburgh's scientific importance. The next two chapters, making up Part 2, treat these topics in more detail. As he is known more for his botanical contributions, these are considered first, in this Chapter, with his wider scientific interests being covered in Chapter 7.

For a botanist working in territory that had not been investigated in any detail from a European point of view, and this particularly applied in India where there had been a long indigenous tradition of herbal knowledge, the first duty was to collect and describe the plants. Following close on this was the dissemination of these plants, especially to Europe. However, in the East, there were a number of practical problems associated with this: for living plants to reach Europe, they had a long and slow sea journey, which involved crossing the equator twice. Not surprisingly, few plants arrived in London still alive, so much thought had to be given on ways for the information on these new discoveries to reach Britain. One way round this problem was to send non-living material, either in the form of drawings and descriptions, or as pressed herbarium specimens: Roxburgh used both methods. There was also the demand for living specimens and Roxburgh devoted much time and effort in trying to find a suitable method.

Over the years, Roxburgh did succeed in introducing a large number of plants into Britain, as well as introducing a wide range into India. On the back of all this work he produced a large number of publications, both in London and India, covering taxonomic descriptions as well as treatises on the production and uses of numerous economically valuable plants: he was, after all, an employee of a commercial trading company.

TRAVEL

With the training that Roxburgh had at Edinburgh, it is certain that he started collecting his own specimens very soon after arriving in India (he lost his first collection in the hurricane of 1787), and would have immediately come across the two major problems: travel and deterioration of specimens. After the first wave of plant introductions from the Cape during the late 17th and early 18th centuries, many hundreds of new North American species reached English gardens between 1730 and 1790.[1] However, with the expansion of interest towards the east, new plants were being discovered but, because the tropical exotics being imported could not grow out of doors in Europe, the initial interest had to be in the contents of natural history cabinets rather than the 'botanical' gardens which adjoined many of the mansion houses which were being built.[2] Added to this, the East India Company was always aware of the potential benefits of discovering economically useful plants, and had, as mentioned in Chapter 3, employed König as the first Company Naturalist, to search

53 *Ox cart decorated for a wedding. Although this photograph was taken in the Guntur District in the 1950s, by the author's mother, the scene and form of transport can have changed little since Roxburgh's time.*

for and collect such material. Roxburgh was lucky in that he started his career in India close to König and so could draw on the latter's interest and accompany him on some of his plant-collecting expeditions. They first met in 1778-79 before König went on his journey to Siam and Malacca, and after his return in 1782 much of their botanical work was done together.[3]

Travel, however, was slow, difficult and dangerous. It usually took place over land by boat on rivers and in palanquins. As the boatmen would not prepare their food while on water, the boats had to be moored early in the evening to give them time to set up camp before they started cooking.[4] Travel by either method was therefore slow and arduous. On land,

> Officers of Government, and men of rank are covered in a palankeen, or more properly a palkee, an Asiatic luxury, as yet unknown in Europe. It is composed of a shell, or frame, about six feet long, and half as broad, fixed to a long bamboo, forming a bold curve at the centre, which there rises about four feet from the frame. Over the bamboo is spread a canopy of cloth, or velvet, the length of the shell, adorned with fringes and tassels of gold, silver, or silk; and the frame contains a bed and pillows, covered with silk, and so disposed that you may either sit up or recline, as is most agreeable. The palankeen is carried by four men, who with relays, travel at a great pace.[5]

This was confirmed by Viscount Valentia, writing 20 years later, adding that

> bearers of our palanquins had been ordered at the different towns, to be placed at stages of about ten miles from each other, so that we had every reason to hope we should proceed without difficulty from one residence to another, intending to travel during the night and halt in the day …. For each palanquin were required eight bearers, which formed a complete change; we had also three mussal or link boys, and three men to carry our luggage …. At half after seven in the evening, having taken leave of our friends, we partly undressed ourselves, and well wrapped up in bedgowns went to bed in our palanquins, and proceeded on our journey. The motion, though incessant, was by no means violent.'[6]

Roxburgh obviously travelled widely, from such comments as the fact that 'he had set out on a excursion to the Island of Ceylon'.[7] There were dangers all around, from the vagaries of sea travel which could mean ships foundering or being taken by French privateers, as happened to young William Roxburgh in early 1808, and being attacked by tigers which Roxburgh's plant collectors escaped on at least one occasion let alone 'the still more savage (human) inhabitants.'[8] The ever cautious Roxburgh therefore took additional precautions. He sent some 'alyum Paddy', seeds and letters, copies of which were sent on the *Marquis of Wellesley* whose Captain, Le Blanc, Roxburgh referred to: 'I have always found him particularly obliging'; a second set was despatched with a Dr Anderson, 'a passenger on the Hindoostan, one of the ships of this Fleet'; and a third set with Captain Paterson, 'commander of the Castle Huntly'. The *Marquis of Wellesley* 'got on the ground going down the Bay of Bengal' and so had to sail to Bombay for repairs, but the other two appear to have arrived without further problems. This care of spreading things round so that at least some items would arrive, was carried to the extent that Roxburgh on this occasion also split what was being carried on the *Castle Huntly*.

It is not surprising, therefore, that the costs of transporting boxes of plants was comparatively expensive: in 1805, the cost of transporting two boxes each way between Calcutta and Fort Marlborough cost 64 rupees; whilst in 1799 the cost had been four rupees per package between Banda and Calcutta.[9] Another example of the necessary costs of plant collecting was given by Roxburgh in 1811 (Appendix 10, Table 1) and Appendix 10, Table 2, gives an example of the costs of collecting spices for the plantations which William Roxburgh junior managed on Prince of Wales Island during 1802 to 1803. The cost of presents and payment to the local chief and his servants in 1811, 43½ rupees, amounted to almost 20 per cent of the total costs. Some idea of the daily cost of the various people who accompanied Christopher Smith can be gleaned from comparing the rates for the pulawah (headman) of 2 annas 6 pies (equivalent to about 1p per day), to the dandies (labourers) who received nearly 1 anna 3½ pies per day (equivalent to about ½p per day), while the peon (guard) and mangay (overseer) each received about 1 anna 5½ pies per day (equivalent to slightly over ½p).

The second Table in Appendix 10 gives an indication of how these trips were funded in the field: by advances from the local Company Commandants and Commissioners. What is of interest is that young William Roxburgh was paying his eight labourers about the same rate as Smith was paying to his overseers (1 anna 5 pies per day). It is also possible to get some idea of the cost of procuring the plants: William Roxburgh handed over 21,483 nutmeg plants and 6,910 clove plants at a total cost of rupees 9595..10..10

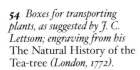

54 Boxes for transporting plants, as suggested by J. C. Lettsom; engraving from his The Natural History of the Tea-tree *(London, 1772).*

which is the 4,505 Spanish dollars given in paragraphs 28 to 32: an equivalent to £1,079 10s. 2½d. (£1,079.51).

The problem associated with this transport of living plant material was that much would arrive at its destination no longer alive. A particular example of this was the consignment which accompanied Roxburgh's letter of 15 December 1796: Banks wrote on his copy of the letter 'Boxes per Berrington thrown overboard. Boxes per Thetis nothing alive sent to Kew. Seeds came safe. Dec[r] 23 1797.' The boxes on the *Thetis* contained 83 plants consisting of 53 species whilst the *Berrington*'s boxes had 82 plants from 47 species, many species being the same on the two ships.[10] There is also a sad vignette of Banks driving post haste to Kew on a summer day in 1790 to see new Indian plants – probably shipped by Roxburgh – to find of the scores of plants sent only four stumps were left and they were apparently dead.[11] This does highlight a major concern of the plant hunters of this period, particularly from the east where the journey was long and the interest of the captains not necessarily committed to the survival of these plants unless there was

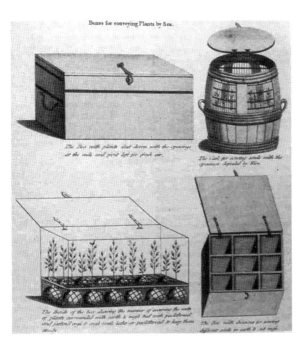

a surgeon on board who would take over the duty of their care. In his early days in India, in the late 1770s, Roxburgh had tried a successful method, 'as far back as twenty nine or thirty years ago I sent to Sir John Pringle, the seeds of various plants preserved in the hardened mucilage of gum arabic, which he wrote me reached him in very excellent conditions, particularly the seeds of several species of Mimosa.' Roxburgh then continued to describe exactly how he had prepared this covering.[12] On another occasion he felt that the presence of his daughter Mary 'may have been some help towards their being taken care of'.[13]

Although sending plants in their dormant, seed, stage was used for decades, the problem was that the recipient did not always know the appropriate conditions for germination. Later he found a more effective method of transporting live plants, in a box that he had specially made, where his comments attached to the diagram speak volumes of the vicissitudes which the plants had to undergo: 'a grating to keep of Goats &c and to give air, & light to the plants ... the 4 feet raises the box above the deck, to secure if from salt water.'[14] The goats were the fresh meat that the ships carried on board.

This was a definite advance on the boxes promoted by John Ellis in 1770 but similar to the one he recommended five years later.[15] Writing in the 1770s, Ellis had published a treatise on transporting plants with diagrams to show how it could be achieved, and added that 'more of the same seed may also be sown after the ship had passed the Tropic of Cancer'. The problem of predators on board was also dealt with, for he continued that 'if very small bits of glass are mixed with the earth or thrown plentifully over its surface in the boxes it may prevent mice or rats from burrowing in it and destroying the tender roots of the plants and growing seeds.'[16] Writing in 1800, Roxburgh gave directions for taking care of growing plants at sea which was later published by the Royal Society for the Encouragement of Arts. Here he gave a summary of generally accepted practice: plants should be protected against the salt sea spray, given light and air, watered in moderation and their growth controlled as they passed into warmer climes; baskets with roots and succulent plants could be hung over the stern or from projections above the wheel house; insect pests were to be ruthlessly eliminated; seeds were to be kept in a cool dry place, preferably below deck, and he was still proposing that they should be covered with the mucilage from gum arabic; and roots should be packed in dry sand.[17] William Salisbury had proposed a successful method of transporting live plants, by packing them in sphagnum moss (*Sphagnum palustre*), which worked well for plants sent to St Vincent and Sierra Leone. Roxburgh, though, did not find this a satisfactory technique, writing 'Mr Salisbury's Sphagnum palustre has cost the Honorable East India Company a great deal of money, without anything to show for it .... Three times, I think it is, that Mr Salisbury has sent large collections to this garden, and constantly without a single one alive.' In a subsequent letter, published at the same time, Roxburgh continued: 'The voyage is too long, and too many climates to pass through for your moss to answer.'[18]

However, the improving methods of transporting the plants could easily be destroyed, as happened often, when they were left either still on board ship or at the Company's warehouse: Banks gave a nice excuse for this: 'I am well aware that Plants are always left too long on Board the Ships that bring them home. As soon as a Ship get into the River the natural joy & high spirits which are the result of revisiting ones native country, sets all order & discipline at defiance.'[19] The problem of transporting live plants from the east was not overcome until Dr Nathaniel Ward invented his Wardian case and first tested it in 1833. Meanwhile, for Roxburgh and his contemporaries, the answer was to send seed and dormant tubers.

55 *Roxburgh's box for transporting plants, which highlights two problems that he faced when moving plants by sea: salt water could get underneath and be soaked up into the soil, thus killing the plants, hence Roxburgh's suggestion that they should have feet to keep the base off the deck; and the growing plants could be eaten by the live-stock, such as goats, that were kept on board for fresh meat, so Roxburgh suggested the metal grill. Sketch included by Roxburgh in a letter to William Mackenzie dated 22 June 1811.*

POSTAGE

There is one other area of the costs of running the Garden which can be glimpsed, through a series of three documents of 1808: the considerable cost of postage. Roxburgh started by stating:

> I am sorry to find it necessary to observe to you, that I have not seen in the recent regulations for the General Post Office, the Botanic Garden amongst the excemptions of postage. I have therefore to request you will have the goodness to inform the Right Honorable the Governor General in Council, that without this indulgence which is coeval with the Garden, is continued to it, the loss of the Hon'ble Companies Botanic Garden, and to science in general will be very real. Few days pass wherein I do not either receive, or send away parcels of seeds, specimens of plants, or Botanical drawings, all of which must cease if the indulgence is not continued. I therefore hope this establishment will be added to the list of excemptions of postage.[20]

The size of this correspondence can be gleaned from one particularly well documented set of letters, which Francis Buchanan sent on his trip to Nepal as part of Captain Knox's embassy: between 3 January 1802 and 4 April 1803, he sent 117 letters and packets of seeds, an average of about one every four days.[21] Similarly a glance through Appendix 5, the List of Correspondents, gives an idea of the number of people with whom he was in touch, with one or two being particularly frequent correspondents, such as Dr Berry who was on the Madras Establishment, and William Hamilton from America. The initial response to this was to allow only those 'letters and parcels bona fide on the public service, and bearing your signature and designation, will be exempt from the payment of postage.'[22]

The Post Master General did not seem to accept this decision, for in January 1809, Roxburgh was complaining that 'postage on all such [bona fide public service] letters and parcels, as above stated have been refused at the Post Office until the postage was paid.' Roxburgh continued that 'the late Post Office regulations, has nearly annihilated every benefit this Garden used to derive from the excemption of postage, which it formerly enjoyed'.[23] This effect was particularly severe for his mail from 'the southward of Ganjam, where the postage must now be paid before any letters or parcels can be received at the several post offices'. Luckily the Regulations were over-ruled, and Roxburgh was allowed to revert to his free postage, and his large correspondence and movement of plants, seeds and drawings was able to continue unimpeded: obviously for the Company, a financial decision in that the gains from plant introductions and dissemination as well as the good will, outweighed the cost of postage.

PLANT INTRODUCTIONS

The combination of better methods of packing and the reduced cost of postage of plant material had its effects. The success of the improved method of transporting plants can be gleaned from the Record of Accessions at Kew, and is further supported by Roxburgh's correspondence with Banks: writing in January 1796, Roxburgh wrote: 'You mention Kew Garden having been enlarged & express a wish for as many plants as I can send. You may be sure it will give me much pleasure to have it in my power to contribute towards replenishing it. I have now ready five chests of plants to go.' These were sent on the *Francis* and *Hillsborough* which sailed at the end of the month.

Roxburgh sent plants from India to England almost every year from 1784 whenever he was in the country except those years he was either at the Cape when he still sent plants but from there. The accession records start in 1793 when only about five live plants arrived; in 1797 this had increased to nine and the following year this had grown to 38 plants. In 1800, about 63 species were received, a number of which had already been sent, in 1803 the number was 57, and in 1804 the number was 99 plus a number of seeds. Thereafter, Roxburgh appears to have sent seeds with quantities given in Table 6. The problem with these lists is that frequently it has been impossible to

Year sown	No. of species	Total for year	Recipient	Page reference
Nd but 1805	192		Kew	70-76
1805	256		Banks	97-105
nd but 1805	99	547	Kew	106-108
nd but 1806	74		Kew	153-155
1806	386	460	Kew	211-219
1807	157		Kew	286-291
1807	197	354	Kew	291-297
1808	46		Col. Herret	339-340
1808	774		Kew	340-345
1808	79		Kew	345-347
1808	150		Kew	348-352
1808	163	1212	Banks	358-362
1811	51		Banks	50-51
1811	11		Kew	51
1811	144		Kew	60-62
1811	21		African Inst	65
1811	86		Kew	75
1811	173	486	Banks	79-81
1812	59		Kew	100
1812	42		Kew	110-111
1812	71	172	Banks	111-112
1814	89		Kew	167-168
1814	123	212	Kew	172-173

Table 6 The number of species of plants that William Roxburgh sent to Kew via Sir Joseph Banks. The details refer to *Records of Accession at Kew*. Although there are a large number of duplicates (see Table 7 for number of species), when these numbers are compared with the Summary of Plant Introductions in Appendix VII of Carter, it does suggest that Roxburgh's contribution has been absorbed into the introductions accredited to Banks and that Carter grossly underestimated the number of these introductions.[24]

identify the species that Roxburgh sent, often because he either gave only a generic name but also he used local names that have not been identified.

From a study of these lists, using only those plants specified as coming from Roxburgh: for instance, about 110 were given in the acquisition records as coming from Christopher Smith from the Calcutta Botanic Garden in 1795 via the offices of Peter Good, who was responsible for a collection of 19 species from the Cape in 1800 that were forwarded to Edinburgh. However, many of these species appear to have been sent by Roxburgh at other times, although there is no record at Kew of Roxburgh sending any plants from the Cape himself, although, from his correspondence, he did send large numbers, as will be seen in Chapter 14. Equally, a number of the plants, such as *Brunfelsia americana*, *Carolinaea insignis* and *Pitcairnia angustifolia* would appear to be plants that Roxburgh had received from places as far afield as America, and a number from further east, including China.

Year	Number sent
1793	1
1796	96
1797	9
1798	36
1800	50
1803	43
1805	342
1806	303
1807	450
1808	213
1809	160
1810	55
1811	240
1812	146
1813	183

Table 7 Number of plants sent each year by Roxburgh to Kew.

The total list of these identified species comes to 1034 that the Kew Acquisition Records state that Roxburgh was responsible for sending to the King's Botanic Gardens: the names were checked against Roxburgh's *Flora Indica* and, where they do not appear in that, they were checked against the index to Hooker's seven-volume *Flora of British India*. Of this list, three were given either different colour varieties or geographical varieties: *Abrus precatorius* with black and red varieties plus one where the colour was not specified, *Canna indica* (in fact from South America) with yellow and red forms, and *Carissa carandas* from Bengal and China; there are also a few which were probably incorrectly transcribed on their arrival, such as *Epygea cordata* and *Exygea cordata*, and for a further few no name can be found, for plants named, for instance, as *Casulia axillaris*. There are, nevertheless, more than a thousand species that Roxburgh sent to Kew, and for a considerable proportion these must be assumed to have been their first introductions; this is an area still needing further research, to see if there is any information on whether (or which) of these species sent as seeds germinated. It is interesting to note that virtually half were sent only once, but a number were sent frequently: *Nyctanthes arbor-tristis* and *Sophora tomentosa* were each sent nine times, and *Costus speciosus* and *Leea macrophylla* ten times each.

It is also worth stressing that the three years when the greatest number of species reached Kew, were those in which Roxburgh was himself in Britain (1805-07). It must therefore be assumed that these seeds were brought by Roxburgh as well as others being sent by William Roxburgh junior under his father's instructions.

Roxburgh did not only send plants to London, for there are a number of other important recipients. Although often sent to Sir Joseph Banks, Roxburgh frequently used him as a 'post box' to forward them to other people and organisations. For instance, there are references requesting him to send plants on to Edinburgh on a number of occasions as well as to Henry Dundas.[25] He also sent plants to America and the Cape, further east to Port Jackson (Australia), and, on his route home, to St Helena. A list of at least some of the people to whom he sent plants during his years in India and his short stay at the Cape is discussed in more detail below (Chapter 10).

The following people were specifically mentioned in Roxburgh's correspondence as recipients of his plant material, either as living plants or seeds: Sir Joseph Banks, the Court of Directors and King's Botanic Garden Kew; Aylmer Bourke Lambert; Dr James Edward Smith and the Linnean Society; Professor Daniel Rutherford and the Royal Botanic Garden Edinburgh; Dr Patrick Russell and The Royal Society; Dr Wright and The Royal Society of Edinburgh; The Royal College of Physicians Edinburgh; Sir John Sinclair and the Agricultural Board; John Vaughan and the American Philosophical Society, Philadelphia; Benjamin Barton and the American Philosophical Society, Philadelphia; Mr Molesworth; Sir John Murray; Henry Dundas; Lady Hume; Lord Valentia; James Vere; Charles Greville; The African Institution, London; and Sir George Young, both at St Vincent and the Cape of Good Hope. Three additional recipients were mentioned in a letter to Banks in January 1809: Carl Thurnberg, Carl Willdenow, and Nicholas Jacquin. Overall a very international list.

There is also a letter from William Moorcroft to Roxburgh[26] which requested Roxburgh to send varieties of Mountain rice to: Sir Joseph Banks, Lords Somerville, Egremont, Winchelsea and Heathfield, the Duke of Bedford, Sir John Sebright, Sir F. Dashwood, Mr Marshal, Mr A[rthur] Young, Mr Rogers, Mr Parker of Mundew near Watford, Herts, Mr Nowel, Mr Thomas Scarisbrick of Scarisbrick Hall, Lancs, Mr Whitbread, Mr Dymole and Mr Fane of Tetsworth.

Another example of the sheer quantity of plants that were being distributed by Roxburgh at this time was given in January 1794 (see Table 8), when he listed 2,071 plants contained in '38 chests & other packages of growing plants sent to various parts of Bengal, the Carnatic, Circars, England, St Helena, West Indies, &ca between the 15th of November 1793 & 15th of January 1794.'[27] In addition, 'considerable quantities of seeds have also been transmitted to various parts, such as those of Teak, of the various sorts of Hemp cultivated in Bengal, of Virginia Tobacco, of Conelian Indigo, coffee of Arabia, &ca.' These new plants were greeted with enormous excitement, and three occasions show this very clearly: Banks wrote in 1796: 'The Plants you have sent home by the Royal Admiral arrived in the finest order possible. The Ship unfortunately for us put into Bristol from whence they were brought to London in a waggon not without some loss which was vexatious but unavoidable. Next to the collection which Capt Bligh brought home it is the largest addition Kew Gardens have ever received at one time.'[28] This list does not seem to equate with anything in the Kew Acquisitions Records.

There is, again, a list of 166 plant species that Roxburgh sent under the care of Sir John Murray, addressed to Banks, in December 1796, but there is no record of what happened to what would have been thought a well-looked-after consignment. Then in March 1801, Lambert, often a very exuberant correspondent, wrote:

> Sir J. B. has begun to work on the plants of Dr Roxburgh sent over by Mr Brown & last week gave me my first share of them. It is certainly one of the finest collections that ever came from India. I have only got as yet as far as part of Triandria which was as much as one man could conveniently carry.[29]

This excitement was exceeded six months later, when Lambert wrote:

> I have just received a long Letter from our good Friend Dr Roxburgh with one of the finest collection of seeds I have ever had from him among which is Smithia, do you want any of it? Milne who is just by here has already sown many of them, who will raise them if any one can – Also a chest containing I believe one of the largest collections

56 Carissa carandas. *Hand-coloured engraving from Roxburgh's* Plants of the Coast of the Coromandel *(plate 77, London, 1798), based on a drawing by one of Roxburgh's Indian artists, c.1790 (Roxburgh drawing 8).*

57 Flemingia macrophylla (Hedysarum trinervium *but also referred to by Roxburgh as Flemingia congesta). The genus was named by Roxburgh after his friend, Dr John Fleming, who was a member of the Madras Medical Board. Copy of a drawing made, c.1802, by one of Roxburgh's Indian artists (Roxburgh drawing 1278), and probably sent to the East India Company in London, given later to Kew and passed on to Edinburgh as a duplicate.*

of specimens that ever came from India at one time …. I opened the Chest & was five hours taking out layer after Layer & did not I can assure you get a quarter through the <u>Cargo</u>. I hope you will come as soon as you can & make some stay with us & give me your assistance & take what you want as I could not go on any farther.[30]

Sadly, the letter did not specify further the species that caused all this excitement.

The traffic in plants, needless to say, was not one way, and an insight to the problems of the increasing number of incoming plants and different species can be gleaned from numerous references but, in particular, in 1800 Roxburgh was complaining about 'the Labour in the Botanic Garden, in consequence of its having been much enlarged for the introduction of numerous new Plants from every quarter of the world.'[31] The range of the origins of these plants can be gleaned from a study of *Hortus Bengalensis*, and include, for example, the West Indies, Brazil, Peru, Virginia, the Canary Islands, Europe, Arabia, St Helena, the Cape of Good Hope, Ceylon, the Andaman Islands, Tibet, Nepal, Bhutan, Sumatra, New South Wales and China. Another example of the

Plants of Bengal, and Malay

Fruit trees, such as Mango, Jack, Jumboes of three kinds of ingrafted oranges	260
Of peach, some of which were ingrafted	250
Of China fruit Trees, such as the Loquat (*Mespilus Japonica*) Leechee (*Sapindus edulis*) Wampes, Mulberry Ennaslun &ca	639
Of Timber Trees, such as Teak, Toon, Fever Bark (*Swietenia febrifuga*) Sissoo, Ebony, &ca	50
Of Tallow trees	35
Of Sumatra board leaved Indigo (*Asclepias tinctoria*), of Nerium Indigo, of Annotto &ca	450
Of the Nicobar Bread fruit or Mellore	21
Of Marha Coffee	6
Of Ornamental shrubs &ca	360
Total	2071

Table 8 Numbers of plants sent from Calcutta between
November 1793 and January 1794.

number of species that were being introduced is the growth of species in the Botanic
Garden during Roxburgh's 20 years as Superintendent, from about 350 in 1793 to 3,500
in 1813.[32] Equally, Lambert twice commented to Smith about Roxburgh sending 'one of
the finest collections [of seeds] that ever came from India'.[33] Not all the introductions
were intentional, for a number of plants recorded in *Hortus Bengalensis* are described
as 'accidental', suggesting that they germinated in the soil in which selected plants
were transported.

THE BOTANICAL DRAWINGS

The need for botanical drawings at this period was determined by two factors. The
first was that it was the only way to show people in Europe what the living plants
looked like. The reason for this was that, as has been shown above, often the plants
would rot on their way home unless carefully packaged, and certainly the chances of
getting living plants to survive the journey were slight. Secondly, there was always a
major problem with herbarium specimens in India being destroyed by decay and ants,
so that watercolour paintings were an acceptable substitute for herbarium specimens,
on which descriptions of new species could be based, the drawings thereby becoming
'types'. Added to this, there was a growing interest in Europe of books of the emerging
new flora as fresh areas of the globe were opened up with keen naturalists often being
on the voyages of discovery, with Banks on the *Endeavour* being a particularly famous
example. Such knowledge, indeed, was deemed to be part of a liberal education. The
engravings for these books were based on the paintings done either by the collectors
themselves or, as was especially the case in India, by local artists.

As the quality of these drawings was crucial to the accuracy of the published
etchings, it was not surprising that Roxburgh was kept informed about reactions to
current publications. Thus amongst the Roxburgh manuscript collection are various
reviews of such books. For example, of Smith's *Plantarum Icones*, the *Monthly Review*
stated 'how valuable a publication must this appear to be when it shall have to display
a body of original figures, of great exactness, of the first authority, and of the choicest
rarity.'[34] Another advantage of these drawings was that 'paintings were needed to show
the specimen in its living shape and with its true colours. Drawings, in fact, served not
only as a sentimental reminder of excited reaction to new sights, but were a definite
contribution to science for they could be sent or taken to England to help identify
and classify specimens. They were by far the most accurate means of recording natural
history specimens and the easiest to store and transmit.'[35]

One problem with the accuracy of these drawings, however, was that Roxburgh,
like so many of his contemporaries, used Indian artists who had been trained under

Mughal and other indigenous traditions. As the old Indian empire declined these artists sought employment and patronage from the new rulers, who required a different style of presentation and emphasis. In the Mughal tradition, the main subject of a painting was made to dominate by increasing its size, or reducing that of the surrounding items; equally, the concept of directional light was not part of common usage. This is seen in some of the drawings produced for Roxburgh by his native artists, which can have a flatness produced from use of blocks of colour, and the problems of light is shown by shadows not always being consistent.

Having said that, the detailed drawing that was essential to the Mughal miniaturists was of such value that 'William Tennant, a Company chaplain, said of such Indian artists that "the laborious exactness with which they imitate every feather of a bird, or the smallest fibre on the leaf of a plant, renders them valuable assistants in this department." They had to learn to paint flowers in a completely alien naturalistic idiom.'[36] Earlier, Desmond wrote that 'a certain stiffness and lack of suppleness in their [Indian artists'] drawings suggests their unease at working in this unfamiliar style. This formality does, however, endow the flower drawings with a naïve decorative appeal.'[37] This appreciation of the Mughal style was echoed by Richard Mabey, who expanded the relationship:

> There was an accomplished tradition of flower painting in Mughal culture, chiefly of delicate miniatures built up by layer upon layer of brilliant body colour. The paintings were finished by the use of fine brushes, which were drawn across the paint to add texture or surface detail. In this way it was possible to suggest the lustre of petals or the leatheriness of leaves.[38]

The collection, description and drawing of plants had occupied Roxburgh right from the start, and once he recovered from the trauma of the hurricane he started again. Holttum claimed that Roxburgh learnt much from König, including the excellent drawings by Indian artists that Roxburgh trained to see the significant detail. However, this perspicacity was more likely to have been a legacy of Hope's teaching, who used drawings extensively during his lectures. Having said that, there are cases where slight inaccuracies have produced taxonomic problems, such as the increased angularity of the stems of *Caralluma adscendens*, as White and Sloane pointed out.[39]

Roxburgh was sending copies of the drawings and descriptions to the Rev. John, for the latter frequently mentioned them in his letters, but it is not evident whether these were sent for sale or comment.[40] Added to this, the idea had been mooted that König would produce an illustrated flora of Indian plants but first with his death and then the fact that Russell was more interested in snakes than plants meant that the idea went on hold.

However, the first record of drawings being sent to England does not appear until 1790 in a letter from Andrew Ross, and to Patrick Russell not until August of that year.[41] The first 19 drawings and descriptions to be sent to Banks were despatched on the *Houghton* in September 1790 with a further 23 three months later, making the 42 mentioned in the list of September.[42] In the letter to Banks, Roxburgh asked 'candidly give me your opinion of them, which would render them more perfect in future. I have in all about 700 nearly completed, 200 of which are grasses.' In view of the fact that Roxburgh had lost all his drawings in the flood of 1787, and there appears no evidence that he had drawings kept elsewhere, this is an amazing rate for the Indian artists to have been working, or is a reflection on the number that Roxburgh employed. Certainly later in Calcutta he must have had a number, for, as mentioned in Chapter 4, he appointed John Roxburgh as Head Painter in 1804. With the letter of 1787 also went a collection of 19 living plants and the seeds of a further six plants. Over the following years, Banks selected 300 of these drawings and descriptions which appeared as twelve fasciculi bound into three volumes, appearing as Roxburgh's first great oeuvre, *Plants of the Coast of Coromandel*.[43]

58 Zingiber montanum (Z. cassumunar). *Hand-coloured engraving, based on a drawing by one of Roxburgh's Calcutta artists (Roxburgh had this plant drawn twice, drawings 501, c.1794, and 1507, c.1808). Published by Roxburgh in a paper on gingers (and their relatives) in* Asiatick Researches (*vol. 11, plate 5, 1810*).

These 300 hundred drawings were made by Indian artists whom Roxburgh trained. Much has been written about the problems of getting these Indian artists to paint in the European tradition, and the changing effects of this instruction can be seen in the development of the Roxburgh drawings.[44] The care with which Roxburgh took over the drawings can be gleaned from the fact that they were shown to a number of people by Andrew Ross in Madras, exemplified by his comment in 1792:

> the Box of Drawings – the contents of which I shewed to Lord Cornwallis – Coll Alr Ross [Councillor Alexander Ross, previously Mayor] – & others of the Family. To the Sons & the [?]Vakeels of Tippoo Sultoun – thro' their Attending Officer Capt Doveton – To Sir Charles & Lady Oakeley – Mr Home the [portrait] Painter (who is a judge & gave them great commendations) and to several others – who are pleased with such subjects – & in particular the two Doctors Anderson & Dr Berry the last of whom kept them many days in his hands on purpose to study them & he was very thankful for having such an opportunity.[45]

Of Lady Oakeley, Ross was to say the following year, 'very clever & accomplished as she is, in that, & every thing else, & has always expressed the highest approbation of your drawings,' nevertheless the sight of the drawings in Dr James Smith's book (*English Botany*), 'will be a better guide for you.'[46]

Volume	Fasciculus	Drawings	Date published
I	1	1 – 25	May 1795
	2	26 – 50	Nov 1795
	3	51 – 75	Aug 1796
	4	76 – 100	Jan-Mar 1798
II	5	101 – 125	May 1799
	6	126 – 150	May 1800
	7	151 – 175	Apr 1802
	8	176 – 200	May 1805
III	9	201 – 225	Jul 1811
	10	226 – 250	May 1815
	11	251 – 275	Mar 1820
	12	276 – 300	Mar 1820

Table 9 The dates of publication of the various parts of William Roxburgh's *Plants of the Coast of Coromandel*.

Banks, however, was less easily satisfied, for Roxburgh wrote to him testily:

> It would have given me much satisfaction if you had mentioned what the defects were that my drawings and descriptions had. I would then, probably, have been able to rectify them in those that are still to finish, but as you have left me in the dark, I must jog on until I hear from you on the subject, which I sincerely beg you will point out by the first conveyance.[47]

Roxburgh's sensitivity to criticism is also evident in this letter, for he stated, 'I am far from thinking that I am possessed of abilities to make them perfect yet I hope I am possessed of foresight sufficient not to expose my ignorance to the world.' Two years later, Roxburgh had still not heard back from Banks on the nature of the criticisms: 'Some two years ago, Sir J. Banks wrote me that my drawings labourd under the same defects Dr Russells did, but to this hour I know not what those defects are, so mine must continue to be wanting.'[48] By 1795 there is a greater objectivity to the drawings, for Roxburgh wrote that 'there is a degree of stiffness runs through the whole of my drawings, particularly the first, which now begins to wear off, so that I think you will be better pleased with those I am about to send than with any of the former.'[49] When Banks did eventually respond to give comments on the drawings, he was complimentary: 'You are right in saying that your Draughtsmen improve. Many of the last Drawing are charming, the Grapes in particular valuable in the highest degree to Botanists as the Fructification is delineated with accuracy & on a large scale.'[50] Some of these improvements must have originated from Roxburgh having seen such publications as Curtis's *Flora Londinensis* and his *Botanical Magazine* which were both in his library, and he had dealings with A. B. Lambert who later published a lavishly illustrated book on pines. As mentioned above, Roxburgh had also seen Smith's *English Botany*.[51] A stylistic study of the three major sets of drawings still needs to be done, from those at Kew, Calcutta and Edinburgh, as the style of those in Roxburgh's *Plants of the Coast of Coromandel* were done under the eye of Banks in London.

Roxburgh had these drawings done at his own expense and there is no evidence of payment by the Company for artists until he reached Calcutta and they remained in his possession: 'The 900, now completed, are a present to the Company, for they have no claim whatever to them, as they are entirely done at my own expense before I left the Coast.'[52] However, he may have inherited a draughtsman when he moved to Calcutta. Shortly after Roxburgh left Calcutta in 1814, there was a request for a pension for 'the Widow of the head painter of the Botanic Garden lately deceased. He was employed in that capacity for a period of 20 years and had been previously in similar employment for nearly five years under Dr Roxburgh in drawing plants of the Coast of Coromandel.'[53] This uncertainty regarding the ownership of the drawings is further highlighted by a series of letters in 1818 regarding both copies of these drawings and the originals which ended in 1820 with the decision by the East India Company that they did belong to the Company.[54] Their final deposition with the Company was probably in part due to the intervention of Henry Colebrooke who looked after the Calcutta Botanic Garden when Roxburgh left for St Helena.

59 Curcuma angustifolia. *Hand-coloured engraving, based on a drawing by one of Roxburgh's Calcutta artists, c.1808 (Roxburgh drawing 1511). Probably drawn from a specimen growing in the Calcutta Botanic Garden, to which it was introduced by Henry Colebrooke. Published in* Asiatick Researches (*vol. 11, plate 3, 1810*).

The Drawings in question were by Doctor Roxburgh considered as his undoubted property; and accordingly, when he was embarking for St Helena, they had been put up by him with his belongings to accompany him to England; and it was at my particular and express request, that the Drawings were relanded, and, as a personal favour left in confidence with me, under an engagement on my part to convey them to him in England, or to leave them to expect his return in India, as he should desire by a communication to be made to me from St Helena. But as he had not communicated his wish upon this point when I quitted Bengal, I left the Drawings with Mr John Roxburgh, to be delivered to you, to be dealt with as Doctor Roxburgh (who was yet living) might direct .... I distinctly remember, that Doctor Roxburgh affirmed, that a great number of the earlier drawings were made at his own expence, when he was receiving no allowance connected with such matters ... the Artist who was alone employed by Doctor Roxburgh upon the original drawings, died when I was in charge of the Botanic Garden.[55]

The present whereabouts of the Roxburgh drawings is given in Appendix 3, the sets in Calcutta and Kew being both almost complete, the former being the set that Roxburgh retained while that at Kew being the set sent to the East India Company via Banks. Detailed studies of these two sets have been made by Robert Sealy and M. Sanjappa.[56]

The full list of the despatch of the drawings is given in Table 10 and Appendix 3 gives the institutions that hold copies of the drawings.

Drawing Nos	Date sent	Ship	Comments
1 – 19	Sept 1790	Houghton	
1 – 18	Apr 1795	Red Admiral	
20 – 42	Apr 1791		
101 – 200	Dec 1791	Phoenix	
201 – 300	Feb 1792	?Northumberland	
	Jul 1792	Doulton	
301 – 400	Aug 1792		
401 – 500	Jun 1794		Delayed after move to Bengal
501 – 905	Dec 1794		Received Aug 1798
1 – 640	Dec 1795		Copies
906 – 1009	Dec 1795		
1010 – 1100	Dec 1796	Berrington	
1101 – 1200	Dec 1797	Rose	Received Jan 1799
1201 – 1500	Nov 1802		After return from Cape
1501 – 1758	Feb 1808		After return from England
1759 – 1922	Jan 1809	Bengal	Lost at sea
1923 – 2000	Jan 1810		
1759 – 1832	Feb 1811		Copies to replace losses
2001 – 2100	Feb 1811		
2101 – 2150	Jan 1812		
2151 – 2411	Jan 1813		
2412 – 2572	Oct 1814		Sent by H. Colebrooke

Table 10 The dates on which drawings were sent to England. Dates are those on which evidence is given for the despatch, and were taken from correspondence from William Roxburgh to Dr J. E. Smith and Sir Joseph Banks as well as the Public Consultations. The date for the second hundred may be May 1792, for a letter of that date has recently been discovered in the Royal Botanic Garden Edinburgh Library in which it states that 'the ~~third~~ [*sic*] second Hundred Botanical Drawings'.

60 *William Carey (1761–1834). Stipple engraving after a portrait by Robert Home, published in Eustace Carey's* Memoir of William Carey, M.D. *(London, 1836).*

PUBLICATIONS

A full list of Roxburgh's publications appears in Appendix 1, from which it can be seen that his early articles were on meteorology (see Chapter 7), dyes (see Chapter 9) and then more descriptive works once he reached Calcutta. A fuller study of those relating to economic botany will be found in Part 3.

Some time in 1812, Roxburgh must have given William Carey at Serampore his list of plants growing in the Botanic Garden, for in October Carey wrote to the Governor General, Lord Minto, 'I am about to print a catalogue of the plants in the Honorable Company's Botanical Garden at Calcutta from the manuscript of Dr Roxburgh which he kindly communicated to me for that purpose. I have no expectation of any gain from publishing this work, and have undertaken it solely from a wish to promote the science of Botany by furnishing the lovers thereof with the most compleat list of the botanical productions of Asia that can be procured from the very best authority now existing. As some considerable expence must be incurred in the printing of this work, which will make about three hundred pages in Octavo, I shall consider it as a great favour if Government will encourage the publication by subscribing for fifty copies of it at eight Rupees a copy.'[57] Eighteen months later, Carey took 'the liberty of sending to you the fifty copies of the Catalogue of Plants in the Honorable Company's Botanical Garden, which were subscribed by Government. The number of columns it was necessary to get into the page, has occasioned us employing a very small type, likewise increasing the Length of the page. This has occasioned the Number of pages to fall short of the number calculated. I mention this to obviate any enquiry which may be made on that head.'[58] This confirms the view of Robinson, that both parts of *Hortus Bengalensis* were published in 1814 even though the second is dated 1813.[59] Roxburgh's problem was lack of access to literature and authenticated specimens, thus knowing what had already been described under what name. This led him to misapply many names: to apply them to species other than that intended by their original authors (ie the types of the species). Taxonomists have a very clear set of guidelines about the naming of species: when more than one person has described and given different names to a species, the correct name is given to the first description that was published. However, when a person such as Roxburgh described a species and gave copies of the description to a number of people (Roxburgh had a number of copies of his Indian flora copied before it was published), nevertheless, the first *published* description takes precedence. This problem of priority of names was compounded by the fact that *Flora Indica*, in which many species were first described in the manuscript, was not published until well after Roxburgh's death, by which time many names were in comparatively common use. As Mabberley put it: 'The problems of nomenclature associated with the delayed publication of William Roxburgh's posthumous *Flora Indica* are well known to all who work on the botany of Asia.'[60] Thus Robinson mentioned that 185 had already been used for other species, and Karegeannes gave a good example of how the confusion was caused: '*Begonia apter* Roxburgh, was a new name when listed in the catalogue of 1814 [*Hortus Bengalensis*] but, unfortunately, by the time the description was published in 1832 [in *Flora Indica*] another species had been published under the name, and his had to be renamed later.'[61]

This confusion was further confounded by the fact that Roxburgh was himself aware of this weakness, frequently running into problems by giving a plant a name that had already been used. This was not uncommon, for Sir Joseph Hooker wrote: 'Wallich has eight names for *Pteris aquilina*, and I do think he has two names for three-quarters of the species in the early part of his catalogue, besides Don's, Royle's, Edgeworth's, Roxburgh's and often De Candolle's.'[62] This is shown, even at generic level where a number of the genera which Roxburgh named had to be renamed later (see Appendix 8).

In only one article written by Roxburgh is there material that indicates some of the processes through which these may have gone before publication. In 1810, Roxburgh published through the *Asiatick Researches*, an article on the Scitamineae,[63] in which he paid respect to the work done on these genera by first König, and then Retzius and most recently William Roscoe: 'To these authorities, I gladly add my own experience and suffrage.' Having quoted Sir William Jones's article in which he, in turn mentioned Retzius's comment that 'the genera in this order will never be determined with absolute certainty, until all the scitamineous plants of India shall be perfectly described', Roxburgh then continued by emphasising the importance of the 'living, or recent state, that the flowers can be well understood; particularly the nice structure of the anther, which is here of more importance in determining the genera, than in any other order.' He then proceeded to describe the ten genera, naming the species in each with their descriptions, all written in English. On a number of occasions, he included parenthetical remarks which showed that his confidence in giving these plants their places was not always great: thus 'all the species (known to me,)' in *Hedychium coronarium* and 'which renders it so difficult to distinguish the species, that without the aid of colour, I should despair of making their specific characters discriminative' in *Curcuma*.

The problem here was that Roxburgh was often dealing with plants second hand, in that he worked on those that were sent to him by collectors rather than collecting and describing them from first hand field work. The number of people from whom he was in correspondence for the acquisition of the plants is enlightening:

Phrynium virgatum	Dr Anderson	Madras
Curcuma aeruginosa	Dr William Carey	Pegu
Curcuma comosa	Mr Felix Carey	Pegu
Curcuma leucorhiza	Mr Glass	Bhaglepur, Bihar
Curcuma angustifolia	Henry Colebrooke Esq	Nagpur

Curcuma viridiflora	Dr Charles Campbell	Sumatra
Curcuma sp.	Colonel Hardwicke	Duab
Amomum angustifolium	Captain Tennent	Mauritius
Zingiber rubens	Dr Francis Buchanan	Rungpur
Alpinia mutica	Mr William Roxburgh	Prince of Wales's Island
Alpinia calcarata	?Mr Kerr	China

A number of these people, such as Carey, Campbell, and the young William Roxburgh, all sent more than one species, and some from more than one situation. He was also in touch with Banks and Dr Combe who had advised him on the availability of some of these species in apothecaries' shops in London; thus of *Costus speciosus*, he stated that 'Sir Joseph Banks informs me, that the root does not at all resemble the *Costus Arabicus* of the shops.' There was also Dr J. E. Smith, for 'I also take this opportunity of thanking Dr Smith for having (in consequence of his discovering, that my *Colebrookia bulbifera*, was *Globba marantina* of Linnaeus,) transferred that name to another new genus of East Indian plants.'

This article produced a response from William Roscoe, who sought the advice of Smith about it: 'As I find my Letter to you on Dr Roxburgh's paper is to be printed, I hope that you will read it with the jealous eye of a friend, & see that nothing goes before the public but what is founded on good authority.'[64] This got an almost immediate response from Smith who said that he could not 'see any alteration necessary in your paper …. I fear poor Roxburgh is dying at Chelsea.'[65] Roxburgh and Smith must have got on well, for the following month, Roxburgh wrote that his health 'has been so bad as to deprive of the pleasure I expected to derive from the society of my scientific friends. & the more so when I found you were one of them …. I therefore fear it will not be in my power to visit you at Norwich for some

61 Shorea robusta, *named by Roxburgh after Sir John Shore, later Lord Teignmouth, friend of Roxburgh in Calcutta. Copy of a drawing made, c.1796, by one of Roxburgh's Indian artists in Calcutta (Roxburgh drawing 1070).*

time, but as I now begin to hope for a few years longer life, trust your friendly invitation will not be thrown away on us.'[66] The letter from Roscoe to Smith referred to above, was published the following year and written up as a collection of fasciculi in the 1820s.[67] In the former, Roscoe gave due homage to Roxburgh, 'we find many new and splendid plants, now first introduced to our notice, accompanied by such descriptions and illustrations as induce us to hope that, by a further perseverance, this portion of the vegetable kingdom, which was left in the greatest disorder by both Linnaeus and Jussieu, will at length be thoroughly understood.' He then continued, 'I shall now proceed briefly to point out such parts of Dr Roxburgh's valuable Paper as seem to me to require observation.' Amongst these observations is the implication that Roxburgh was, certainly at times, very much working on his own: 'as there is no other figure or author referred to by Dr Roxburgh, we may presume the plant [*Curcuma zedoaria*] to be ascertained beyond doubt. But on proceeding to his next species, *C. zerumbet*, we find the same plate of the *Hort Mal.* xi. tab. 7, referred to by Dr R as a figure of this plant also; a circumstance which leaves us still in doubt as to which of the two plants is there represented.' What this does also highlight is the need for herbarium specimens, rather than having to rely on illustrations, however good these may be.

However, when the chances of herbarium material surviving the journey from India, and indeed surviving the ravages of insects and mould in the tropics, the importance of drawings for type specimens was important. The shortcomings that these could produce has been mentioned before, but the care that Roxburgh took in the detail of some of the microscopic parts of plants does suggest that he was well aware of their importance taxonomically. This care over naming by Roxburgh is shown in his description of *Curcuma zerumbet* in his *Flora Indica*, when he wrote that 'in 1785, I gave some of the sliced and dried bulbous, and planate tuberous roots of this plant to Sir Joseph Banks, which he gave to Dr Comb, who found that it was the real *Zedoaria* of our *Materia Medica*, and by the same means ascertained that the root of my *Curcuma Zedoaria*, is *Zedoaria rotunda* of the shops.'

Little has been said so far about the other monumental publication of Roxburgh's, his *Flora Indica*. When he left India, he handed the manuscript to William Carey who delayed doing anything with it until he realised that Roxburgh's idea of getting it published in England had ground to a standstill. Wallich promised to include his own findings in any such publication, and the first volume was published in 1820, but the next volume did not appear until 1824, partially because Wallich had been on an 18-month expedition to Nepal followed by illness and recuperation in the intervening years. Although a further volume was promised for 1825, it never materialised, thus leaving two-thirds of Roxburgh's manuscript still unpublished. Seven years later, Carey, at the request of Roxburgh's sons Bruce and James, published a complete, three-volume edition of Roxburgh's manuscript which appeared in 1832, excluding the additions Wallich had inserted in his edition. Again, a further volume was promised, of cryptogams, edited by William Griffith, but this did not appear until 1844.

An analysis of this work, suggests that Roxburgh named and described something approaching 2,000 new species and nearly 50 genera. A reflection of his importance can also be gleaned from the fact that approximately 400 species have been named after him, from a quick survey of the *Index Kewensis*.

A work of this magnitude took its toll on Roxburgh, who wrote in 1813 that

> during the last twelve months of my residence in Bengal, I employed every minute I could spare from describing & drawing new plants on making a new corrected & enlarged copy of all my former descriptions of plants, ever since <u>I began</u> after the dreadful storm & inundation at & about Coringa in 1787, arranged according to the sexual system. It forms a voluminous <u>Indian Flora</u> of 28 quires which I think might be published, either with or without figures. I will send it on now, under charge of Mr Mayne, the 2nd officer on the Castle Huntly.[68]

Roxburgh's *Flora Indica*, in the second, complete, edition was unchallenged as a national flora of sufficient standing that it was republished in 1874 as a single volume including the article of Roxburgh's on cryptogams,[69] by C. B. Clarke, and was deemed 'a model for all those botanists who subsequently attempted less ambitious regional floras.'[70] One of the great strengths of the *Flora Indica*, other than its range of descriptions of plants, was the emphasis that Roxburgh gave to the practical value of botany. Thus over and over again, the economic uses of the plants described are given, either for humans, as for instance for crops such as the sugars, of *Saccharum officinale* he wrote that 'it is much cultivated in the Rajamundri Circar, where they only make a coarse sort of brown sugar, which is sold on the spot for about three half-pence per pound'; for medicinal uses and alternatives for European herbs, for instance, he wrote of *Justicia nasuta*, 'the roots rubbed with lime juice, and pepper, are used, and often with good effect, to cure *ring worms*, or Herpes miliaris, which in India is a most troublesome disease, and very common'; as animal feeds, as he did for a number of grasses as forage crops as well as other species, such as *Cyperus rotundus*, of which he wrote that 'cattle eat it. Hogs are remarkably fond of the roots. Dried and powdered they are used as a perfume at the weddings of the natives'; but equally he was prepared to wax lyrical about the beauty of some of the flowers that he described, for example, of *Nyctanthes arbor-tristis* he wrote that 'the flowers of this tree are exquisitely fragrant, partaking of the smell of fresh honey, and on that account the plant is much esteemed'.

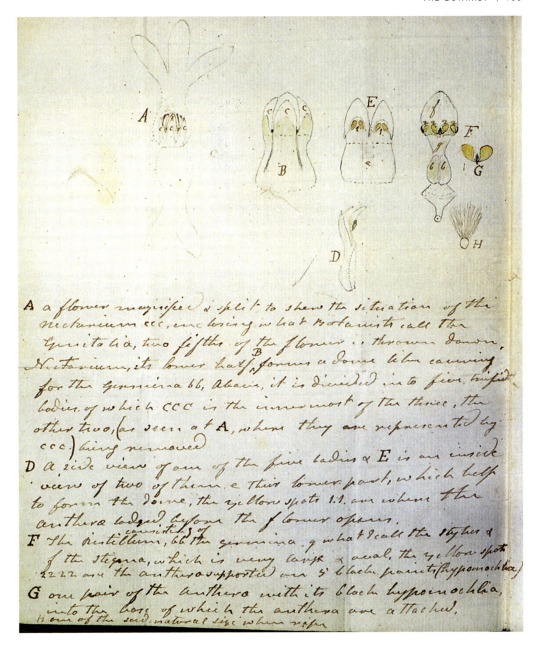

62 *Drawings and description of an unnamed creeper done by Roxburgh, sent to Baron Reickel at Fort St George, the accompanying letter dated 25 August 1789.*

Writing to Robert Brown at the New Year, again in a very frail hand, he was still working, with lists this time of drawings of plants and their dissections with descriptions, for inclusion in *The Plants of the Coast of Coromandel*, of some 53 species, but felt that he was too weak to complete his Indian Flora. If strength of hand is anything to go by, he was better by the middle of January when he wrote to Smith indicating, 'I much wished to have made a proposal to you to superintend the printing (under your own man Mr Taylor) my Indian Flora.' The idea was 'to translate the Generic & specific characters into Latin & leave the rest in English as it is. I am quite unable to undertake any active part in this work, nor do I see any prospect of effecting with your aid.'[71]

A number of points need stressing about this work. There were large tracts of India that were poorly represented, for instance northern India because it was hardly explored in Roxburgh's time. Secondly the locations attributed to species are sometimes erroneous because Roxburgh gave the location from which his collector sent the plant rather than where he had found it.[72] The largest problem stemmed from the delay in publication and the fact that by then a number of names had been

published for other plants, a point highlighted by Mabberley: 'the intricate tangle of names made ever more complicated by a small flood of books on Indian botany between [1813 and 1832]'.[73]

HERBARIUM

In spite of what has been said about Roxburgh's taxonomic work by a number of authors, he did keep herbarium specimens, even though some of these may have deteriorated as a result of insects and damp. To trace the full sequence and whereabouts of his extant specimens requires much further research, but a short description here taken with Appendix 2, will give a starting point, but among some of the more important collections are those at the Botany Department of the Natural History Museum in London, the Botanic Garden in Brussels, the Delessert Herbarium in Geneva which appears to have one of the largest Roxburgh collections, and the Royal Botanic Gardens, Kew. The problem was made particularly difficult because, when Nathaniel Wallich returned to England in 1828, he brought 'his' herbarium from Calcutta which included much of Roxburgh's. During his years in London, Wallich devoted his time, with the help of numerous collectors, going through both his own collection and that of the East India Company, distributing duplicates to interested collectors. In the process, most specimens appear to have been remounted, thereby losing the original details, as well as examples of Roxburgh's handwriting and the site and date of the collection: this is confirmed by Thomas Thomson, writing in 1857, who stated that 'the commencement of the present herbarium of the Calcutta Botanic Garden dates from Dr Wallich's return to India in 1832'.[74] However, Merrill stated that

63 Globba orixensis. *Hand-coloured engraving, based on a drawing by one of Roxburgh's Indian artists, c.1794 (Roxburgh drawing 504). Probably drawn from a specimen growing in the Calcutta Botanic Garden, to which it was introduced from Rungpur by Francis Buchanan (though found earlier in the mountains of the Northern Circars by Roxburgh himself). Published in* Asiatick Researches (*vol. 11, plate 8, 1810*).

64 Nyctanthes arbor-tristis. *Copy of one of Roxburgh's drawings made c.1814 (Roxburgh drawing 179, originally drawn c.1791) for an unknown botanist. This scented, night-flowering tree is sacred to Hindus; the flowers are the source of an orange dye.*

'the Superintendent has under his control the herbarium which has existed since the foundation of this Garden by Dr William Roxburgh, the Father of Indian Botany.'[75]

Although Roxburgh sent mostly drawings to Banks, he seems to have sent specimens to a number of other collectors, some of whose herbaria can be traced fairly easily. Thus, Robert Wight, writing in 1831, referred to 'a small Herbarium of the late Dr Roxburgh: no duplicates' amongst other herbaria still at India House.[76] The collection that Roxburgh sent to J. E. Smith is now held partly by the Linnean Society and partly in the Botany Department of Liverpool Museum who are in the process of conserving the former. The Linnean Society holding consists of approximately 78 specimens, some sent direct to Smith but 50 forwarded by Viscount Valentia, a few by Rottler and a few by Kindersley, Smith's nephew and friend of Roxburgh's. The Smith specimens in Liverpool's own collection came from both India, numbering some 66 specimens collected between 1789 and 1809 but they appear to have been remounted, and possibly from South Africa but more work needs to be done on these to check their provenance. Amongst the Royle collection at Liverpool there are also a number of specimens accredited to Roxburgh, 68 definitely and 103 possibles.[77] From two letters from Aylmer Lambert to Smith at the Linnean Society, the first dated 28 October 1805 and the second dated 5 January 1806, highlight some of the contemporary problems with these collections of Roxburgh: in the first Lambert stated that 'His [Roxburgh's] herbarium Jackson writes me word is from his account very small', but in January Lambert corrected this view, writing that 'He has brought me his whole collection of Plants to my Shop!!! Which is very large.' If it was the latter, then it would appear that a considerable portion of the Linnean Collection must have gone missing, for it is no longer 'very large'.

The largest collection of Roxburgh's herbarium specimens is in the Delessert Herbarium in Geneva which holds the material that came to Benjamin Delessert via Aylmer Lambert after the sale of his herbarium: 'l'herbier du docteur Roxburgh, maintenant en la possession de M. Delessert, se compose des grandes collections qu'il a fait dans l'Inde continentale, à la côte de Coromandel et à Banda, Amboine et autre îles de l'archipel Indien, et des plantes qu'il a recueille au Cape de Bonne-Espérance.'[78] There is also an important collection at the Brussels Botanic Garden Herbarium, bought by Martius at the Linnean Society sale of 1863, which holds about 40 of Roxburgh's type specimens and nearly 80 of his lectotypes as well as many holotypes. As this last collection has examples of William Roxburgh junior's handwriting as well as that of his father and possibly of his elder brother John, more work on them needs to be done.[79]

THE FLORA OF ST HELENA

This introduces his final piece of important botanical work, the floral list of St Helena. Although he arrived on 7 June 1813 in a state of health worse than when he left Bengal,[80] during the nine months he stayed on the island, he produced an annotated list of the island's flora. This was published as an appendix to Beatson's *Tracts* and remained the only accessible printed account of the flora until 1875.[81] The importance of this list is shown by the fact that it was reprinted in various forms a number of times over the next thirty years, and is described in Cronk's recent volume on the endemic flora of St Helena:

> The importance of Roxburgh's list (at least in the original) lies in the new species it described and the notes given on certain of the endemics. Its shortcomings reside in its incompleteness, the dubious nomenclature and (especially of the Asteraceae) its taxonomic confusion. The list is only partially backed by specimens, which found their way into the Banksian herbarium, and so to the Natural History Museum.[82]

A quick check though Cronk's book gives nearly twenty species first described by Roxburgh and the fact that he was deemed of sufficient importance to the island

65 View of the interior of St Helena, very much as Roxburgh would have seen it on his visits. Engraving by J. Greig after a drawing made by James Forbes in 1784, from Forbes's Oriental Memoirs *(page 270, vol. 4, 1834).*

flora, that one endemic species was named after him: *Wahlenbergia roxburghii*, by A. de Candolle. It is of interest that Darwin 'was to rely almost exclusively in the course of his visit to the island during the *Beagle* voyage' on Roxburgh's list.[83] For a sick man, he did remarkably well 'to identify 363 species of flowering plants and twenty-five ferns, of which only thirty-three were native species.'[84]

Among the Brown MSS at the Natural History Museum is a 'Florae St^{ae} Helenae' which claimed to have been prepared in 1813 and 1814. However, it refers extensively to Roxburgh's work which took place in those years, and so it is possible that Brown relied heavily if not entirely on his friend's notes. Dr William Wright, President of the Royal Society of Edinburgh, wrote to Brown shortly after Roxburgh's death, stating: 'Our Friend Dr Roxburgh was worn to a shadow before He died, I never saw any one so exhausted to last so long: and yet he was to the last writing a Flora of St Helena!'[85]

One other point is worth considering about Roxburgh's influence on St Helena and the insight this gives into his awareness of contemporary thinking. The island had long been an important port of call for ships sailing to and from the east, and as a result much of the native timber had been used. When Roxburgh requested permission to leave Bengal for St Helena, he expanded on what he had done for the island over the years:

> in consequence of repeated applications from the various Governors of St Helena for many years past and of orders from the Honorable the Court of Directors, plants, seeds, and roots of the most useful kind have been frequently sent from hence to that Island, for its improvement, but hitherto with much less success than was hoped for. Forest trees they want most of all, and I understand they have in general failed in rearing such trees from seed. Permit me therefore to request you will suggest to His Lordship in Council, that it might be advisable to send by each of the companies ships which form the next Fleet, one or two Chests filled with growing plants of all those trees as appear to promise most success. Such as Sissoo Buddam Lundry, Teak, Sumatra Cassis, &ca &ca which the state of the nurseries in this Garden can at present furnish. If this idea meet with the approbation of Government have the goodness to oblige me with the order for getting them ready as soon as you can. It is probable that I shall be at that Island at the time the plants will arrive, and in that case will with pleasure give the best advice and directions in my power for their planting and future management.[86]

66 *Bulbul in a Custard Apple tree. Hand-coloured engraving by William Hooker, after a drawing made by James Forbes in 1769, from Forbes's* Oriental Memoirs *(page 54, vol. 1, 1813).*

67 *Cashew tree* (Anacardium occidentale*). Hand-coloured engraving by William Hooker, after a drawing made by James Forbes in 1772, from Forbes's* Oriental Memoirs *(plate XXXVII, page 342, vol. 2, 1813).*

Forestry trees were of obvious importance for ships before the days of the 'ironclads', but there was another aspect of reafforestation and the natural flora of St Helena. William Burchell had been appointed schoolteacher and Acting Botanist on the island in 1805, positions he held until 1810, so Roxburgh might have met him on his passage through to England in 1805. Burchell, like Kyd at Calcutta, was aware of the idea of rarity; also, by the time of his arrival, many of the older plantations had become neglected and the earlier signs of reafforestation had disappeared. Tied in with the ideas of rarity was the concept of species extinction which was linked to the concept of death: with Roxburgh's awareness of the latter for humans after what he saw during the famines of the Northern Circars during the early 1790s, he would have been particularly sensitive to the issue at St Helena, and this may, indeed, have been behind his idea of making his descriptive list of island endemics.

Grove states that Burchell's contemporary influence 'seems to have been confined to making Roxburgh more aware of the possibilities of endemism and the threat of species extinctions in a more global sense. This was an important development, and one which may have influenced him in his stewardship of the Calcutta Botanic Garden between 1792 and 1813. This was a period during which the role of that garden as a nursery of plantation trees to replace those lost in deforestation was increasingly being stressed.'[87] Apart from the superficial error of the date at which Roxburgh arrived at Calcutta, of more importance is that Roxburgh started work on teak as soon as he arrived at Calcutta in 1793 when Burchell was only 11 years old, making it unlikely that the influence on Roxburgh came from Burchell. Of more importance is the relationship between Beatson and Roxburgh, for Beatson, Grove claims, 'put into effect a programme of "improvement", which included irrigation, tree planting and forest protection. A comparison with Pierre Poivre is not an idle one, as Beatson was certainly impressed by the efforts the French had made to develop Mauritius.'[88] When Roxburgh and Beatson met twenty years earlier, the latter was an army major with strong interest in surveying whilst it was Roxburgh who had pushed for the idea of canalising the Godavary to relieve the future possibility of famine and the planting of trees and other crops as a result (see Chapter 13 for a fuller discussion of this). Roxburgh's knowledge of Mauritius is reflected in his intention of visiting that island on his proposed return to India, and the fact that he had been corresponding with Céré, so it is probable that any ideas of 'active conservationist intervention' that Roxburgh may have had came direct from his dealings with the French rather than from either Burchell or Beatson. There is further confirmation for this theory from an undated letter to Wallich at Serampore, therefore between 1809 and 1813, in which Roxburgh intimates that he will give Wallich 'some seeds to carry with you to the Mauritius & a letter to Monsr Levi, the gentleman in charge'.[89]

REVIEW

Building on the teachings of Hope and the guidance of König, Roxburgh was responsible for the opening up of the knowledge of the Indian flora to Europeans, through his herbarium specimens as well as the drawings and descriptions, and the large number of living introductions that he sent especially through Banks to Kew. If he was the 'founding father of Indian botany' the question arises as to who were his children. Although the importance of his herbarium is still crucial to the study of the flora of India, reflected in the large number of his type specimens, the European legacy that he left were the people who developed his ideas, and here one must start with people like Francis Buchanan, Nathaniel Wallich, John Forbes Royle and Joseph Dalton Hooker. Of equal importance are his *Flora Indica* which was not replaced until Hooker produced his multi-volume *Flora of British India*, and his list of St Helena plants which Darwin used as his reference work when he visited the island over thirty years later.

Travel, both for plant collectors and for the living plants, was a major problem. Roxburgh built up a large set of correspondents which reduced his need to travel large distances, either by sea or by land in palanquins. For the movement of plants, he realised the disadvantages of the systems used for shorter distances, from the equator northwards, and thus developed a box of his own which tried to reduce the likelihood of his plants dying on their voyage to Britain. Equally, especially towards the end of his stay in India, he sent seeds and dormant tubers which had a greater chance of survival.

CHAPTER 7

Wider Scientific Interests

INTRODUCTION

From his time at Edinburgh University, under Hope, and then in London following the influence of Pringle, Roxburgh gained a much wider interest in science than just botany, as was standard for any well educated man of the time. What was almost more important for Roxburgh was the legacy of Hope's inculcation of a strong scientific methodology and the cautious and sceptical view of other people's work, and thus the need for a rigorous basis for his own published works. These also reflect his range of interests, from the articles on weather and meteorology, to the varied pieces that he wrote on various aspects of zoology, from lac insects used for dyes to new species of dolphin and ox as well as his care of shawl goats.

His position at the centre of the scientific community in Calcutta meant that he was intimately involved with the Asiatic Society and its publications. As a result of these contacts, he would have been very aware of the development of contemporary thought and *mores* on Indian culture and the changing attitudes of the Europeans towards this rich and diverse society.

THE BROADER SCIENTIFIC INTERESTS

Shortly after he arrived in Calcutta, he wrote to the President and Members of the American Philosophical Society in Philadelphia, with a collection of 'several of our most valuable East India, & China Plants' with a request that they would 'send us to India, such of your useful Plants and seeds as may be deemed most likely to add to our comfort' with particular reference to 'peaches, nectarines, apricots, Gooseberrys, Currants, Strawberries, Apples, &c.'[1] These plants were deposited with Mr William Hamilton in America, who was a fertile correspondent with Roxburgh over the next 17 years, providing 185 plants for the Calcutta Botanic Garden.[2] Hamilton nominated Roxburgh as a member of the American Philosophical Society in 1801 and this led to an interesting correspondence about the supply of the skeleton of an elephant for the Society. The first indication is a letter in which Roxburgh states that he 'will immediately write to my friends up the country to endeavor to procure for the Society the skeleton of an Elephant of a large size, & that you may rest assured that every thing in my power shall be done to procure, & send as complete a one as can be got.' Accompanying this letter is another, from Sergooga Nugur, giving the dimensions of one recently killed.[3] Meanwhile in America, there were discussions about how much of the skeleton was required, with preference for a complete one if possible, otherwise the head and feet, with just the head being the least.[4] The cost of shipment was beginning to worry John Vaughan in Philadelphia and the delay in its arrival annoying Roxburgh at Calcutta, 'all I have yet received of it is constant promises'.[5] In fact the skeleton was not ready before Roxburgh departed for England in 1805, but it did arrive, for it was displayed in the Boston Museum until recently.[6]

Roxburgh's knowledge of zoology was sufficiently wide for Dr J. E. Smith to ask him about the terns of India, for shortly after arriving at Calcutta, Roxburgh wrote to Smith that 'We have few Terns in India, but more here than on the Coast, as soon as time will admit, I will cause prepared specimens to send to you.'[7] However, there is no further mention of these birds and no record of the Linnean Society having received any, so presumably Roxburgh never got round to this. The first zoological reference, apart from the help with Russell over his snakes, is in a letter from the Rev. John, asking about a piece of bone, in which he states, 'As I am not so acquainted with the anatomie of Animals as You I might wish your Opinion of it.'[8] The following January, he wrote to Patrick Russell about rats, and was sufficiently conversant with the problems of these rodents to be sent a specimen preserved in spirit, and gave some of the diseases resulting from rat bites.[9]

His interest is further shown by three more examples. Roxburgh sent to Banks 'a charming Musk deer in very high order ... and ... another rare animal, the Lemur tardigradus, which is even here a very great rarity indeed.'[10] The value of the former can be gauged by the fact that the surgeon of the ship taking it to England was promised £50 'on presenting notification from you of your having received it alive'. Of more importance, however, was his discovery of a new species of dolphin, about which he first wrote to Dr Smith about a month later and sent the jaw to Banks next April who agreed that it was indeed a new species.[11] This led

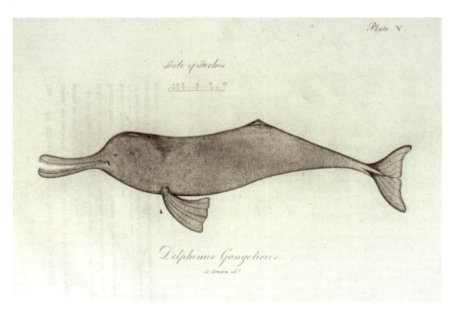

68 *The Ganges dolphin* Dolphinus gangeticus, *first described by William Roxburgh and published in* Asiatick Researches *(vol. 7, 1801).*

to the publication of the description in the *Asiatick Researches* in 1801: another example of this periodical publishing the description of an important new species.[12] Finally, Henry Colebrooke asked him to describe an ox, the Gayal (the Indian bison, *Bos gaurus*): 'herds of this species of cattle have been long possessed by many gentlemen, in the eastern district of Bengal, and also in other parts of this province; but no detailed account of the animal, and its habits, has been yet published in India. To remedy this deficiency, Dr Roxburgh undertook, at my solicitation, to describe the Gayal, from those seen by him in a herd belonging to the Governor General.'[13]

There are two other areas of science that he was involved in. The first came in a letter of 24 April 1798, in which he sent two shells to Banks because he felt that they were new to science. A further supply of shells was sent in 1802, mentioned in his letter to Banks of 26 November 1802. As a totally different area of science, writing to Charles Taylor on 18 April 1808, he was sending specimens of amber that he had received from the Malabar coast, so that their potential could be gauged, as 'the purest pieces are susceptible of a fine polish, and are here cut into beads, &ca which are much worn by the natives, as well as by the European Ladies. I once saw a very beautiful string of these beads, sent to England under the name of Amber beads. The most beautiful amber coloured pieces are therefore the most valuable and are sould [*sic*] for about a shilling the pound by retail in the Bazar.'

ROXBURGH'S SCIENTIFIC METHODOLOGY

One characteristic of Roxburgh's science is the rigour of his experiments. From the earliest publication, he was stringent in detailing how his experiments were carried out, thus making them as reproducible as possible. Three examples are worth mentioning at this point. The first is an unpublished article on the effects of 'black-body radiation', where he was explicit about the material used, the controls practised and the possible errors that might have affected the result.[14] A standard black cloth was always used as a control; two thermometers were used, 'that corresponded exactly, & made by Gregory & Wright'; having got one set of results, the two thermometers were then swapped and the experiment repeated; and a number of different colours were used, to get the results shown in Table 11. With the arrival of Lord Mornington as Governor General, there was a move away from wearing local dress to British clothes, so anything that might show the evil consequences of this would not have been politic, which probably explains why this piece of his work was never published. In all cases it does show the deleterious effect of wearing dark clothes, preferred by the Europeans and it is not surprising that so many of them died, one imagines a considerable proportion from heat stroke.

	Black 1	*Colour 2*	*Black 2*	*Colour 1*
White	132	118	142	121
Purple	141	134	144	134
Dark blue	132	116		

Table 11 The results of Roxburgh's experiments on Black-body radiation, using different cloths, based on his unpublished manuscript of 1793. The numbers are °F.

The second is the article 'A Description of the plant Butea' in which he described both the plant and its uses, but his 'Description of the Nerium Tinctorium would have been subjoined; but publication of it is delayed, until the Society have been favoured with the result of his farther experiments.'[15] The important points here are that he did not publish further material until the experiments were complete, but also felt that there might be a correlation between the colour of the lac insects and the colour of the resin which the plant exudes: 'it would require a set of experiments accurately made on specimens of lac gathered from the various trees it is found on, at the same time and as nearly as possible from the same place, to determine this point.' This last comment gives a good indication of his awareness of the variables which might affect the colour, and therefore the design of the experiment would need to take account of these parameters.

THE ASIATIC SOCIETY

Much has been written about the attitude of the British towards India and the Indians during the late 18th century, which will not be recapitulated here. What is germane to this study is the scientific milieu into which Roxburgh entered when he reached Calcutta, for he was only on the edge when he was at the Coromandel Coast although he does appear from his correspondence to have been eager to take part in the more global scientific world; his close relationship with König would have brought him into contact with this as König was one of the founder members of the Asiatic Society. Some Company servants had started taking an interest in Indian science and knowledge before Warren Hastings came to Calcutta as Governor in 1773, but he was the first man to make the acquisition of this knowledge, and thereby an understanding of the peoples of India, a major basis of his policies.

Hastings believed that even if the East India Company accepted full responsibility for the government of Bengal it should still govern in ways familiar to the masses of the population and should apply Hindu and Muslim law with only limited innovation.

If the Company was to govern in accordance with these Indian traditions, however, it had first to find out what these traditions were. Added to this, religion was still a major preoccupation of European intellectual life, and therefore a knowledge of it was seen as the key to understanding all things Indian. As Harlow argued, this attitude to the new Empire had to be seen in the context of the Treaty of Paris of 1763 and the loss of America: 'the prevailing mood was one of tenacity: to hold on (even a little more securely) to what remained, ... The new empire could be fashioned on a rational analysis of imperial needs, commercial and political alike.'[16] A small group of men had started to work on acquainting themselves with the Indian languages, laws and religions, three of the main workers in this field being Alexander Dow, Nathaniel Halhed and Charles Wilkins. The problem, as Halhed explained in the Preface to his *A Code of Gentoo Law*, was that the 'Pundits would not teach him Sanskrit, despite Hastings's official requests, so that he had to translate a Persian version'.[17] However, by 1778 Wilkins when he started studying Sanskrit was able to find a pundit of more liberal mind, and by the time of Henry Colebrooke's studies in the early 1790s, Brahmins were allowing access to their sciences and Vedas. The attitude by the 1790s is reflected in the diary entry of Lord Macartney when he was in China in January 1794, when he stated that 'we wear a dress as different from theirs as can be fashioned. We are quite ignorant of their language (which, I suppose, cannot be a very difficult one).'[18]

The key person who was the catalyst for Hastings's work to be systematised was, however, Sir William Jones. He arrived in India in 1782 as a High Court Judge and the first meeting of the Asiatic Society, that he helped set up with Hastings, took place in February 1784. Being a man of great integrity, he felt that if the Europeans could devote their leisure hours to the gathering of information and an understanding of the natives rather than the interminable drinking, gambling and taking concubines even by churchmen, it would be to the advantage of both peoples.[19] He was a man of enormously wide erudition, a linguist with more than twenty languages, botanist, classicist as well as an eminent lawyer.

The interests of the Society were restricted to Asia but were to include 'whatever is performed by man or produced by nature'.[20] Many of the underlying philosophies of the Society can be gleaned from Jones's 'Anniversary Discourses' which appeared in the volumes of the *Asiatick Researches* and the Introduction and Objects which were published in the first volume, in 1788:

> A mere man of letters, retired from the world and allotting his whole time to philosophical or literary pursuits, is a character unknown among Europeans resident in India, where every individual is a man of business in the civil or military state, constantly occupied either in the affairs of government, in the administration of justice, in some department of revenue or commerce, or in one of the liberal professions; very few hours, therefore, in the day or night can be reserved for any study, that has no immediate connection with business, even by those who are most habituated to mental applications, and it is impossible to preserve health in Bengal without regular exercise and seasonal relaxation of mind.[21]

> Objects: Man and Nature; whatever is performed by the one, or produced by the other. Human knowledge has been elegantly analysed according to the three great faculties of the mind, memory, reason, and imagination, which we constantly find employed in comparing, retaining and distinguishing, combining and diversifying, the ideas, which we receive through our senses, or acquire on reflection; hence the three main branches of learning are history, science, and art.[22]

Jones's attitude to Indian knowledge is summed up in his 'Second Discourse', in which he stated

> that reason and taste are the grand prerogatives of European minds, while the Asiaticks have soared to loftier heights in the sphere of imagination ... but the natural productions of these territories, especially in the vegetable and mineral systems, are

momentous objects of research to an imperial, but, which is of a character of equal dignity, a commercial people …. So highly has medical skill been prized by the ancient Indians, that one of the fourteen Retna's, or precious things, which their Gods are believed to have produced by the churning the ocean with the mountain Mandara, was a learned physician.[23]

To gain political importance, and make it more likely that these aims would be achieved, Jones arranged for the Governor General always to be the Patron, and he was himself President until his early death in February 1794, when both positions were held by the Governor General, Sir John Shore, later Lord Teignmouth. Although some modern writers claim that 'the new rulers sought political legitimacy by patronising Indian learning',[24] put on a wider context, the fact that the early volumes of the *Asiatick Researches* were sufficiently important to be published in London in a pirated edition, does suggest that there was a genuine scientific rigour to the work done by Jones and the contributors to his periodical, for instance Wilkins contributed five articles and Marsden one on Indian languages.[25] Similarly, under Jones's Presidency, articles were published in the *Asiatick Researches* rather than being sent home and published by the Royal Society as had occurred before.

Volume	Calcutta edition	London edition
I	1788	
II	1790	
III	1792	1799
IV	1795	1799
V	1797	1799
VI	1799	1801
VII	1801	1803
VIII	1805	1808
IX	1807	1809
X	1808	1811
XI	1810	1812
XII	1816	

Table 12 Comparison of the dates of the Calcutta and London editions of the *Asiatick Researches*.

This enormous breadth of interest was reflected in the range of articles that appeared in these early volumes, under the guidance of Jones. This was still initially at a level of recording findings, whether they were archaeological, literary or botanical. One of the emerging areas of concern was to fit the Indian cosmology within the accepted time frame of Christianity. Although there was a great respect for the religions of Hinduism and Islam, for the nature of Buddhism was not fully understood until later, the attitude was still very Euro-centric, which meant that any chronologies had to fit within the time frame that the Europeans brought with them, and this in turn was based very much on the Old Testament. Thus, John Bentley writing in 1796, gave the age of the world as 5,803 years old.[26] Roxburgh, as one of the leading scientists in Calcutta and on the Publishing Committee of the Society, would have been very aware of these discussions and the people involved.

Behind the need to understand the Indian culture was, nevertheless, a sense of superiority by the 1790s. This strong European and Christian position was shown particularly well in one of Jones's articles, in which he stated that

it is my design in this essay, to point out such a resemblance between the popular worship of the old Greeks and Italians and that of the Hindus: … . From all this, if it be satisfactorily proved, we may infer a general union or affinity between the most distinguished inhabitants of the primitive world, at the time when they deviated, as they did too early deviate, from the rational adoration of the only true God.[27]

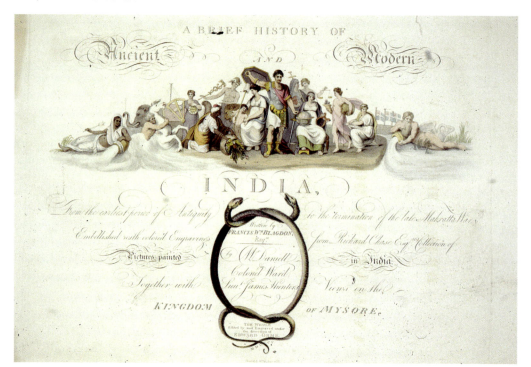

69 *Drawing of the sculpted group by John Bacon that once graced the pediment of East India House, Leadenhall Street, London, redolent with the symbolism of the East India Company, Rivers Ganges and Thames, the wealth of the East and the trade with the West, etc. Hand-coloured engraving, used as decoration on the title page of Blagdon's* A Brief History of Ancient and Modern India *(1805).*

Another reflection of this attitude is seen in the sculpture that adorned the pediment of the portico of East India House, the headquarters of the East India Company in London, carved by John Bacon in 1797-99. As Blagdon explained it twenty years later, 'The Sentiment of this Composition is – That a nation can be truly prosperous, only when it has a King, who makes Religion and Justice the basis of his Government, and a Constitution which, while it secures the Liberties of the Subjects, maintains a due Subordination in the several ranks of Society; and where the Integrity of the People secures to each Individual those advantages which Industry creates and cultivates.'[28] The roles of India and Britain are clearly symbolised, with the former bringing the commerce of Asia that has flowed from the bountiful Ganges, with the help of Mercury, to Britannia. While the Indian side has a European trying to tame the white wild horse, symbolising the untamed nature of the country, the King, Britannia, Liberty, Order, Religion and Justice are all calmly grouped round the centre and the right side, giving credence to the concept of growing European dominance and Euro-centric importance. However, one must remember that by the time that Blagdon was writing, the Evangelicals had gained a far greater power and their missionary work was undermining the earlier egalitarian views of Jones and Hastings.

SIR WILLIAM JONES AND ROXBURGH

Jones and Roxburgh met in a number of contexts. Jones must have known of Roxburgh at least superficially, for it was he who seconded Roxburgh as a member of the Asiatic Society in December 1793, virtually as soon as he arrived in Calcutta.[29] Possibly this knowledge came through the fact that König was also a founding member of the Society with Jones, and the fact that Roxburgh had contributed papers to the second and third volumes of *Asiatick Researches*. With Jones's keen interest in botany, it is not surprising that in the second volume he published an extensive article on the need for a book on Indian plants.[30] It is interesting to note that the format he recommended was very much that which Roxburgh was to follow in his *Flora Indica*, though it is likely that it was a practice he had already been using when writing his plant descriptions on the Carnatic:

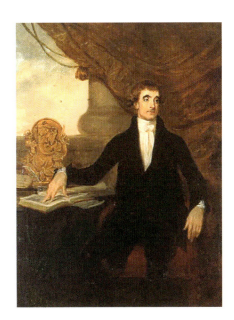

70 Sir William Jones (1746–94), by Sir Joshua Reynolds.

when the Sanskrit names of the Indian plants have been correctly written … the fresh plants themselves, procured in their respective seasons, must be concisely but accurately, classed and described; after which their several uses in medicine, diet, or manufactures, may be collected, with the assistance of Hindu physicians, from the medicinal books in Sanskrit, and their accounts either disproved or established by repeated experiments, as fast as they can be made with exactness.

Jones also suggested that the Linnaean system should be used, 'numerals may be used for the eleven first classes, the former of two numbers being always appropriated to the stamens, and the latter to the pistils.'[31]

Roxburgh took this one stage further, for he included the Indian names for the plants in *Flora Indica*, as well as those from surrounding areas. Thus not only were the Sanskrit, Bengali, Hindi, Tamil and Telegu names given but also those from the Persian, Malay and Hebrew. In fact the only areas for which the local names appear to have been omitted were for the plants from the Moluccas, Ceylon and some from Nepal and Silhet – which (except for Nepal which were probably collected by Francis Buchanan) could be for those plants collected by Christopher Smith. This could give a further link between the two men, for Jones was, as mentioned above, a great linguist, so that Roxburgh's interest in native names could have been instigated by Jones's philological studies. Roxburgh's interest in Indian languages is further shown by a letter from Andrew Ross, dated 24 March 1793, in which he stated that

Our friend Dr Anderson says that he saw a Gentoo & French Dictionary of much value in the hands of Mr Pybus – when he was at Masulipm & that it was sent at his demise to the British Museum to be deposited as a Curiosity. The doctor is very desirous that you should enquire after it & recover it if possible & also enquire after any others, & Gramars & vocabularys that have been compiled by these Ecclesiasticks, as the Protestant Missionarys have compiled & published in the Malabar Language. This is an important & Curious Subject if you can pursue to advantage, you would much gratify your learned Friend Sir Wm Jones & Coll: Kyd.

Current ideas of what was required of a modern flora in the late 18th century are to be found in a review of the *Flora Caroliniana* which appeared in the *Monthly Review* of 1789 a copy of which is in the Roxburgh archives, suggesting that he kept abreast of current thinking on this as well as so much else. The reviewer concluded that

The cultivation of Botany has been promoted by various means, but, perhaps, by none more successful than by the Publishing of Floras of different countries & climates. In these Publications, it is usual, not only to give a minute description of every indigenous plant, but to enlarge on their various uses, to specify the situations in which they naturally best thrive, and add such general remarks as may be serviceable to the Gardner and the farmer …. We are under the disagreeable necessity of informing our readers, that we suppose none of these particulars have been attended to by the author of the Flora Caroliniana; we have formed this supposition, because they are not noticed in this publication, which consists only of generic and specific characters of the native plants of Carolina; even the duration of the plant is omitted, except in a few cases where it forms a specific distinction.[32]

Secondly, there was already through Kyd, a connection between the Calcutta Botanic Garden and Jones. This had evolved from the mid-1780s when Sir George Young, who had set up the Botanic Garden at St Vincent, was asking for plants to introduce into cultivation.[33] The connection between Calcutta and St Vincent lasted until after 1800, for Sir George sent plants to Roxburgh in 1800 and 1801 (see Appendix 5). Jones wanted the Calcutta Garden to be more than Kyd's original idea, for

besides wishing to have all the Indian plants represented in Kyd's Calcutta Gardens, Jones hoped to introduce Arabian, Persian, and Chinese plants …. When Banks requested more help than he had been giving Yonge, ... Jones made his excuses: ... if you wish to transfer our Indian plants to the Western Isles, the Company must direct Kyd and Roxburgh to send them.[34]

At a more general level, both men were products of the Enlightenment: Roxburgh through his time in Edinburgh, meeting some of the great men, such as Professors John Hope, Joseph Black and Dr John Boswell; and Jones through his enormous breadth of knowledge, for which his contemporaries saw him 'no less as a man of law and literature, politics, and religion.'[35]

There was yet another area of common interest, for Jones published an article in the *Asiatick Researches* on the Spikenard (*Nardostachys grandiflora*) which lead Roxburgh to publish a further article on it, 'I need scarce attempt to give any further history of this famous odoriferous plant than what is merely botanical, and that with a view to help to illustrate the learned dissertations thereon, by the late Sir William Jones.'[36] The main point of this article of Roxburgh's was to clarify 'whether the roots, or stalks, were the parts esteemed for use, the testimony of the ancients themselves on this head being ambiguous.' An earlier article in the same issue by Roxburgh described a genus that he named after Jones: the *Jonesia*, which was 'Consecrated to the remembrance of our late President, the most justly celebrated Sir William Jones, whose great knowledge of this science, independent of his other incomparable qualifications, justly entitles his memory to this mark of regard.'[37]

Both Jones and Roxburgh had the well-being of the native Indians at heart, Roxburgh in his intentions to relieve the effects of famines, as is discussed in more detail in Chapter 13: doing what a botanist could to help those not able or willing to help themselves. Jones's attitude was reflected in 'his wish that the twenty million British subjects in Bengal should receive the full protection of British law, and that the natives of these important provinces be indulged in their own prejudices, civil and religious, and suffered to enjoy their own customs unmolested'[38] implementing these ideas through the Supreme Court where he sat.

71 Nardostachys jatamansi (N. grandiflora, Valeriana jatamansi). *Sir William Jones investigated this plant, the spikenard of the Ancients. Drawing by one of Roxburgh's Calcutta artists, c.1796, showing the root, which is still dug up from the wild in the Himalayas for its valuable medicinal properties (Roxburgh drawing 1017).*

THE WIDER SCIENTIFIC CIRCLE

As Superintendent of the Botanic Garden and a Committee Member of the Asiatic Society, Roxburgh was at the heart of the active search for information and explanation of the Indian scientific and cultural treasure chest. As mentioned above, he joined the Asiatic Society almost as soon as he arrived in Calcutta, and within three years was on the Committee of Papers, with the President, Sir John Shore, the two Vice-Presidents, Dr John Fleming and John Harrington, plus Dr James Dinwiddie, Francis Horsley, William Blaquière and William Hunter, and the Secretary, Codrington Carrington.[39] He was to remain on the Committee probably until he left for Britain on furlough in 1805 and he was no longer on it in 1808, but he remained a member of the Society at least until his final departure in 1813.

On the assumption that the plan of holding meetings once a month was adhered to and that all members of the Committee attended, Roxburgh would have had comparatively frequent dealings with the highest in the land. As a member of the Committee of Papers, he would have seen and presumably advised on the inclusion of papers for publication as well as refereeing those on botanical matters. Even if attendance was not so frequent, he would have been sufficiently well-known and regarded, to have had a respectable standing in Calcutta intellectual society; also reflected in the size of his mansion, as discussed earlier. The importance of these connections can be seen from the fact that Roxburgh, in his will, made Thomas Colebrooke (on the Supreme Council of Bengal and a member of the Committee of Papers since at least 1799), Samuel Dennis (Director of the East India Company) and Henry Colebrooke (a member of the Committee of Papers since at least 1802) responsible for the care of his botanical drawings and manuscripts 'from being lost to the public'.[40]

A study of his correspondents is also revealing, both for the geographical spread of his contacts and for the importance of many of them: a study of Roxburgh's *Hortus Bengalensis* gives the names of more than 140 people who sent him plants. Of particular interest in the context of the history of botany and its philosophical development, is the number of French and continental botanists connected with Mauritius: Céré, de Cossigny, Jannet as well as Jacquin; and there was also the correspondence with Sir George Young on St Vincent and later at the Cape. Nicholas Céré was the Curator of the Botanic Garden on Mauritius and had been the appointee of Pierre Poivre, one of the French 'physiocrats' and great proponents of the botanic garden movement – his influence on Banks's interest has been stressed by Grove.[41]

There are other important groups of people as well: a considerable number of medical gentlemen corresponded with Roxburgh, as might be expected with the emphasis given to botany during the study of medicine, for example, Drs Anderson, Berry, Fleming and Hunter. Secondly there were a large number of men who, by their ranks, were either in the army or navy (either King's or Company's), such as Captains Blake, Denton and Dickinson, as well as Thomas Hardwicke who started as captain and progressed to colonel, and at the senior level there were Generals Martin and M'Dowall. The gentry and nobility were also represented, by people such as Lady Ann Barnard, Sir C. Blunt, the Hon. C. A. and N. C. Bruce, Lady Amelia Hume and her husband, Lord Minto and the King of Nepal as well as the Hon. Charles Greville and Viscount Valentine.

One of the justifications by which Kyd persuaded the Court to take on the Calcutta Garden was the humanitarian need to safeguard the natives from famine: this should be done through the introduction of drought-resistant crops, an idea already well established in Roxburgh's practices, particularly after the disastrous famine of 1791-93 in the Circars. Charpentier de Cossigny had been on the island of Mauritius at least from the early 1760s, where he had been a proponent of afforestation and the conservation of endemic forests for economic use: echoed by Roxburgh in his ambitious plans for teak plantations as will be discussed briefly in Chapter 8. Joseph Jacquin was the son of Nicolaus whose herbarium in Vienna Banks had bought in 1777, both father and son being major botanists on the Continent.[42] Roxburgh was overtly aware of others, such as Pierre Sonnerat, the nephew of Poivre, but of whom Roxburgh said, 'Sonnerat was a miserable Botanist. I know him well, collecting specimens was all the good he could do in that line. Indeed I am well informed that his publications were stolen from the Missionaries at Pondichery', a comment showing a certain independence of thought.[43]

The importance of the Botanic Garden at St Vincent lay in two areas. The first was the involvement of Alexander Anderson, 'a radical Scottish physician, botanist and first curator'[44] who was behind the need for the Garden because of fears of soil erosion and the effects of climate. It was Sir George Young who arranged for Anderson's letter about the link between cloud and forestry to be published by the Royal Society. It is possible that some of these introductions may have been made through Banks, but the fact that it was Roxburgh himself who wrote to the American Philosophical Society, with his own interests in meteorology from his earliest days in India and his Scottish Enlightenment background, he must be given credit for taking the initiative. A further example of this appears in one of his later articles, 'On the land Winds of the Coromandel, and their Causes.'[45] This article takes a look first at an explanation of what the land winds are before considering the existing theories and their weaknesses, and then going on to explain his own theory, based on the latent heat of vaporisation and condensation, ideas proposed and developed by Black in Edinburgh as recently as the 1760s. This approach to science, studying the literature, exposition of results and observations from which hypotheses could be developed, very much followed the Boerhaavian tradition learnt at Edinburgh. As an aside to this article, Roxburgh also gives an indication on the extent of his travels once he was based in Bengal: in 1799 he was back in the Northern Circars, on 4 June 1800 he was at Madavaram near Bangalore and again in March and April 1804. One further fact comes out of this, that Roxburgh kept detailed notes, referring in this case to his 'Meteorological Journal'.

Another example of Roxburgh's connections with this wider circle was through Colin Mackenzie, who had enlisted in the Madras Infantry in 1783 before transferring to the Engineers to become a military surveyor. Although unrecorded, it is possible, with the comparatively small number of Europeans in the Madras Presidency that the two men could have met at this stage. In his role as surveyor, Mackenzie visited a number of previously unrecorded Hindu and Jain sites and began a collection of Indian curios, which developed into his becoming one of the most important archaeologists of the period. Certainly these two men were corresponding when Mackenzie, as Surveyor-General to the Madras Presidency, accompanied Stamford Raffles on his invasion of Java in 1811, for there are a number of letters in the National Library of Scotland between Roxburgh and Mackenzie during that year.[46]

72 *A young fisher boy of the Coromandel Coast with some of his catch. Photograph by the author's mother c.1955.*

Francis Buchanan was another of these Scottish polymaths, but this time the links between him and Roxburgh are better known and closer. Buchanan's first claim to fame in India was his journey through Mysore and Hyderabad which had first been requested by the Marquis of Wellesley of Mackenzie in 1799. Instead it was Buchanan who set out in April 1800, to study the culture, geography and natural resources of the area, and his report was published in 1807, and there is a collection of 84 drawings done for him on this tour in the British Library. His next trip was to accompany Captain Knox on his embassy to Nepal in 1802 to 1803, which although not successful diplomatically, proved of great value to Buchanan as he was able to make large collections of plants that he sent to Roxburgh, as well as studying the local religions: again, there is a large collection of 96 drawings of Nepalese plants in the British Library, published in 1819 as the *Account of the Kingdom of Nepal* and followed in 1825 with his *Prodromus florae Nepalensis*. Finally, there was Buchanan's survey of Bengal, which lasted from 1807 until 1814, little of which was published in his lifetime. However, these seven years also produced literally hundreds of drawings of plants, birds, animals and fishes. The closeness between Roxburgh and Buchanan was considered when discussing the succession to Roxburgh at Calcutta in Chapter 3.

METEOROLOGY

One of the great interests of the Royal Society during the 18th century was collecting meteorological data, so that the climates of those parts of the world that were being explored and opened up to trade up could be better understood, as well as helping the naval and merchant shipping in forecasting potentially dangerous times to sail in certain areas. Another reason for this increased interest in meteorology during the

18th century was the fact that the instruments, such as barometers and thermometers, were getting more accurate. As President of the Royal Society, Sir John Pringle was very involved with this and thus Roxburgh would also have been made aware of its importance, both intrinsically and if he wanted to be accepted amongst the metropolitan scientific world. An understanding of climate as a part of the habitat of plants had also been inculcated by Hope in Edinburgh.

In the Madras area, this understanding also had economic benefits, for cyclones, usually in May or November, occurred on average every ten to twelve years, and famines often lasting two or even three years, every twenty or so years, and Roxburgh was to experience both in his seventeen years on the Coromandel.[47] Although Roxburgh only published his meteorological diaries when he first arrived in India, it is evident from his letters and later articles that he continued to record data.[48] There was, however, an interesting gap, for in a letter to Sir Joseph Banks in December 1784, Roxburgh stated: 'The beginning of 81 we were obliged to evacuate Nagore in a great hurry, by which I lost my Thermometer & many things of value. Since I have not kept any account of the Weather, which I am now sorry for, as it appears by the note at the bottom of your Letter, that such observations are acceptable.'[49] This also has interesting connotations regarding the different expectations that Roxburgh deemed that Banks had compared with Pringle, and Roxburgh's intention of planning his scientific work to please these men. Although he did not start keeping his meteorological diary immediately afterwards, this had been kept for some time by January 1793, and he was referring to the rainfall measurements in 1791.[50]

His care with the recording of data is apparent from these earliest articles, for he studiously stated what equipment he used and where they were kept, reflecting the quality of his guidance of experimental methodology almost certainly learnt at Edinburgh under the tutelage of John Hope: the barometer used outside was made by Ramsden and the thermometer by Nairne and Blunt (interestingly, different from the ones that he used in his experiments on black-body radiation), and he had a rain gauge made for him which was kept 'on the roof of my house, which is about twenty-five feet high; at a considerable distance from any other building, &c, except the hospital, which is distant about one hundred yards, and of the same height; no trees above twelve feet high within many hundred yards.' This awareness of the 'shadow' of high buildings and plants in affecting the accuracy of meteorological readings is again a sign of the care to details that Roxburgh took into consideration when performing and describing his experiments.

73 Fishermen's huts thatched with leaves of the palmyra palm (Borassus flabellifer) on the Coromandel coast. Photograph by the author's mother c.1955.

The periodicity of these disasters was stressed in this description, when one correspondent wrote: 'I had indeed heard of a tradition among the Natives, that about a Century ago, the Sea ran as high as the tallest Palmira Trees, which I have ever disregarded as fabulous, till the present unusual appearance called it more forcibly to my mind.'[51] This letter also mentioned that, 'At Coringa, out of four thousand Inhabitants, it is said, not more than twenty were saved' although only three Europeans were drowned. This wholesale destruction covered much of this area, for 'Many of the Villages in the low Country, between Coringa and Jaggernaickporam were totally destroyed ... [and] the Inundation penetrated inland, about ten Coss from the Sea in a direct line, but did little more damage to the Westward of us, than destroying the vegetation.' Parsons also added that about 100,000 cattle were drowned. One of the side-effects was that among those drowned were the local fishermen, so even that source of food was removed, but 'the generous supplies [of rice] that have been sent us from The Presidency, will, I trust, secure us from serious want.' Roxburgh's approach to these climatic disasters is the subject of a more detailed study in Chapter 13.

Roxburgh, ever the practical scientist, even when describing the event shortly afterwards, sought an explanation:

> I conceive the cause of the Sea rising to such a great height, was, from the force of the Wind while at North East, which drove the water up into the Bay and over the adjacent low Lands: Point Godavery prevented its progress South west along the Coast, which I think will appear evident to you, so well acquainted with the Bay of Coringa.[52]

Briefly referred to above is the article that Roxburgh published in the *Transactions of the London Medical Journal*. This is particularly fascinating for the insight that it gives of his knowledge of the concepts of the latent heat of evaporation and condensation, and the fact that these were the same reverse processes. He then applied them in a practical way to explain the winds that come off the land at sea.

THE SHAWL GOATS

As head of the Botanical Establishment in Calcutta, Roxburgh was given some strange assignments, one such being the care of some shawl goats on their way to Europe. Before Roxburgh arrived at Calcutta, Edward Hay, the Secretary to the Government, had requested Edward Otto Ives, Resident at the Vizier's Court, to send some shawl goats. Ives sent nineteen, ten males and nine females, to Calcutta in January 1794 as his collection had

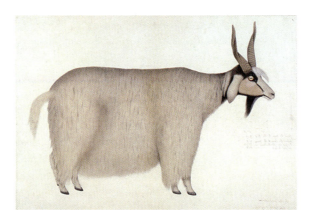

74 Shawl goat. Watercolour by Zayn al-Din, 1779. This animal was brought by George Bogle from Bhutan and kept in the menagerie of Sir Elijah Impey, Chief Justice of the Supreme Court in Calcutta, for whom this painting was made. Roxburgh was involved in trying to send some of these animals to England from Calcutta.

been too late for despatch from Lucknow the previous season to reach Calcutta in time for them to be included in the ships for England.[53] Their future in Calcutta seems to have been left very much in the air, for by August Roxburgh was reporting that 'more than half of the Shawl Goats, sent to the Botanic Garden in December last, are dead, and that I wish to know how to dispose of those that are still alive. If kept here, I fear the whole will soon die, as the wet weather seems to be the cause of the mortality.' The reply that came back was that 'it is left to Mr Roxburgh to dispose of them as he may think proper.'[54] By the end of September, Roxburgh wrote that 'the remaining Shawl Goats were brought to my House in town & every possible care taken of them, but almost in vain, for there remains only a single male alive.'[55]

By January he appeared to have more and was making arrangements for their despatch on two ships going to Madras, and was advising on the care necessary to be taken of them:

> permit me to recommend them being sent on shore during the time the ship remain at that place, so far as I know, they will not require any further care while on shore, than common pasturage, and a daily supply of half a seer of raw Gram to each. Fresh Hay ought to be sent on board with them for Madras, Gram sufficient for the whole voyage I will send from hence together with hay to serve them to Madras.[56]

This is confirmed by the letter a fortnight later when

> the 15 Shawl Goats were sent on the Asia and General Goddard, Viz[t] two Males and five females on the Asia, and two males and six females on the General Goddard,

together with grams sufficient for the voyage and Hay to serve them to Madras, where I requested a Stock of Fresh Hay might be procured, and that the Goats might be kept on Shore while the ships remain there.[57]

The fact that these were sent for Sir John Sinclair was confirmed by Roxburgh in 1796, when he wrote 'of animals I ought to except Shawl Goats, and fifteen of them were sent last year on the ships Asia and General Goddard'.[58] The trauma experienced by the goats being landed at Madras needs stressing, for they would have been transferred from the ship to massulah boats to get through the surf before being man-handled once more onto the shore.

MINERALOGY

Roxburgh's knowledge of zoology has already been touched upon, as has his awareness of geology. However, yet another string to his bow was mineralogy, for, when he was surveying the Northern Circars, he got Benjamin Heyne to produce a paper on iron ore and its associated industry.[59] The concomitant study of metallurgy is reflected in the ideas for copper coinage, which appear in letters he wrote in 1793:

> your paper containing the Correspondence between the Court of Directors & the Madras Board relative to the Copper Coin used in the Northern Circars, from Dec '86 to May '93, ... that you might use the best exertions in your Power, ... for promoting the good intentions of the Company & their Govt to put that important Business upon a better footing, for the relief of the lower classes of the people in the Circars, than it has hitherto been.[60]

A further aspect of earth sciences was a note attached to 'Benj. Heyne's MSS Memoir on the Copper Mines in the Callastry and Venketgherry District in the latitude of Ongole', signed by Roxburgh.[61]

It is almost certain that some of this scientific interest was based on potential for trade, because in a letter towards the end of his time in India, he wrote about amber: 'the purest pieces are susceptable of a fine polish, and are here cut into beads, &ca which are much worn by the natives, as well as by the European ladies. I once saw a very beautiful string of these beads, sent to England under the name of Amber beads. The most beautiful amber coloured pieces are therefore the most valuable and are should for about a shilling the pound by retail in the Bazar.'[62]

THE PETRIFIED SHELLS

However, one area which is especially interesting, and has not been studied before, derives from an undated description of two petrified shells, but their position in the archives suggests that they were attached to a letter from Roxburgh to Banks, dated 24 April 1798. This is of importance, for it throws new light on Roxburgh's zoological knowledge and secondly it gives a fascinating insight into his knowledge and ideas of geology. The Rev. John at Tranquebar sent him a specific character (ie a diagnostic description), which Roxburgh disagreed with, because it 'gives the One reason to think his shell was not so complete, as those in my possession, which I received from Mr Lambert of Calcutta, who got them from Sumatra some months ago.'[63] After describing in detail the parts of his shells, he continued:

> It is by these remains of the primitive substance, that we have undoubted proof of these specimens being originally testaceous, but how the progress of petrification came to have been so much slower in these parts I cannot pretend to say, for every other part thereof, has arrived at the most perfect stage. The whole, except the parts just mentioned, being one sparry mass, consequently the time those shells have remained in the ground must be very long, as the period which nature employs in petrifying bodies, of even an ordinary size, is found to be very great. Many of the wooden Pillars of the Bridge built

by Trajan over the Danube near Belgrade, are still remaining, which could scarce happen without they become Petrified. At the request of the late Emperor, the duke of Lorraine, one of them was taken up, & after a period of 1500 years, the progress of Petrification was found to be have only advanced three fourths of an inch, what length of time must we then allow the large, now well known Tamarind trees, at Trevi-carry near Pondicherry, to have been rendered so completely Petrified, is a question of great importance, for altho no man can for a moment imagine the progress thereof equal in all parts of the world, yet to ascertain a few facts of the kind, would be gaining much, & lead to still further discoveries.

There are various points that emerge from this piece. First is the breadth of Roxburgh's scientific interest and reading, especially when added to it is the fact that he was aware of the 'dentalic found by Abbé de Sauvage in Lanquodoc'. Although there are apparently no purely geological volumes in his library, both from the regional descriptions of such places as Paraguay, Mexico and Forster's *Observations*, Roxburgh would not have been geologically naïve. To this must be added the fact that James Hutton had read his theory of the Earth to the Royal Society of Edinburgh,[64] which was first published in 1785, shortly before Roxburgh himself was elected to that body.

Another thread to put into this argument is that as an active and important member of the Asiatic Society, Roxburgh was aware of the current arguments and discussions which were circulating in Calcutta scientific circles. With the common botanical interest with Sir William Jones, particularly with their articles on spikenard, Roxburgh and Jones would certainly have corresponded and indeed may have met during the short time they overlapped in Calcutta. One of the contentious areas discussed in a number of articles in the *Asiatick Researches* in its early years, was the concept of time, although the great emphasis was on the equivalent ages given to the earth by Christianity and Hinduism. The Christian conservatives tried to set the Hindu concepts, and historical and legendary timescale, within that of the origin of the earth in 4004 BC, a feat that was done only by the denigration of the accuracy of the Hindu texts. Roxburgh, with his Scottish Presbyterian background coloured with the medical questioning of Boerhaave, does not appear happy with Jones's strict Christian teaching:

> If the human race then be, as we may confidently assume, of one natural species, they must all have proceeded from one pair; and if perfect justice be, as it is most indubitably, an essential attribute of God, that pair must have been gifted with sufficient wisdom and strength to be virtuous, and, as far as their nature admitted, happy, but instructed with freedom of will to be vicious and consequently degraded.[65]

It would seem sensible, therefore, to suggest that Roxburgh was aware of the debates on the age of the earth and with his wide botanical knowledge and non-conformist connections may well have questioned the concepts of species origination, especially when linked with his knowledge of plant rarities and extinctions, especially when bracketed with his knowledge of island endemics.

The interest in all these areas of science reflect the knowledge of a well-educated man of some intelligence. It must be remembered that, at this time, it would have been expected of such a person that he would have this breadth of interest, confirming Roxburgh's position among the top scientists in India at the time.

75 Amomum aculetatum. *Hand-coloured engraving, based on a drawing by one of Roxburgh's Calcutta artists, c.1810 (Roxburgh drawing 1761). Probably drawn from a specimen growing in the Calcutta Botanic Garden, to which it was introduced from the Malay Archipelago. Published in* Asiatick Researches *(vol. 13, 1812).*

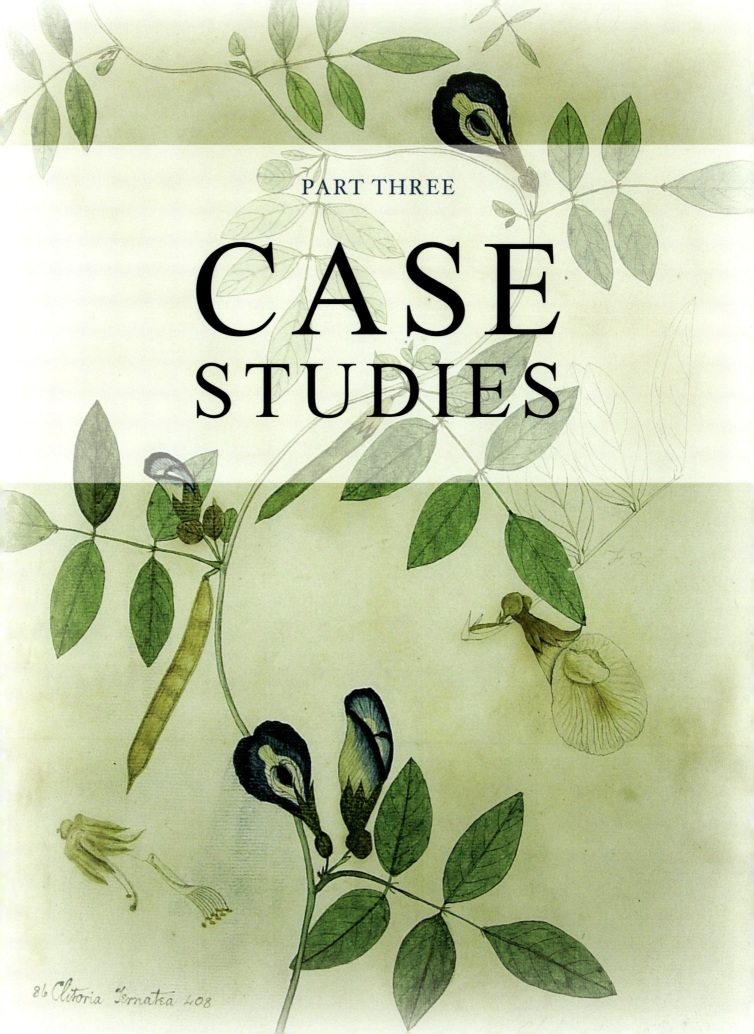

PART THREE

CASE
STUDIES

86 Clitoria Ternatea 408

CHAPTER 8

Background

INTRODUCTION

As has been mentioned already, William Roxburgh spent significant parts of his life in the company of educated, professional men, many of them also being scientists with a medical background, a training at Edinburgh often being a common link. The importance of his time under Hope has been studied above as have the influences of such men as Sir John Pringle, Sir Joseph Banks and Sir William Jones. It is thus not surprising that Roxburgh should devote so much of his time not only to the description of the Indian flora but also to ensuring that the facts behind the uses of the economical plants were based on a strict experimental foundation. The work he did on tea and the attempts to introduce the bread fruit tree are discussed below. However, many other plants were also studied, which can be broadly fitted into a number of categories.

As a trading company, the East India Company was ultimately responsible to its shareholders to produce dividends; thus the driving force in any exploration, whether botanical, zoological or mineralogical was commercial gain. Added to this, especially as far as Roxburgh was concerned, from the early 1790s Britain was at war with France and therefore needed numerous articles of trade that had previously been acquired from Continental Europe or via indirect trade links through Europe. Therefore it was essential that alternative sources for these articles should be found, and a particularly good example of this was hemp for cordage used on ships. From almost the very beginning of his time in India, Roxburgh was involved in searches for economically useful plants, with his work at Samulcottah being a perfect example.

It is generally accepted that it was Sir Joseph Banks who was the promoter of colonial botanic gardens.[1] There are grounds for believing, however, that the idea of a Botanic Garden in the Madras Presidency was suggested by Roxburgh, for in a letter of August 1790 he said: 'I have mentioned a Public Garden to Mr Dalrymple, & I think you might get Sir Joseph Banks also to recommend it to the Directors.'[2] This in turn suggests that he may have been influenced by his knowledge of the work done in Mauritius. As was mentioned above when discussing his varied correspondents, Roxburgh was in touch with Jean Nicolas de Céré who was director of Le Jardin du Roy on that island from 1772-1812, and who continued the tradition of Pierre le Poivre.[3] This very much ties in with some of the ideas of Richard Grove who emphasised the French influence on Banks as well as the importance of the Scottish Protestant doctors.[4] However, this letter of Roxburgh's, together with the environmentalist training at Edinburgh, does raise the interesting possibility of whether he got the idea direct from Mauritius rather than from Banks. What must also be remembered is that Roxburgh was brought up in Edinburgh where he came under the influence of Dr John Hope, with the long tradition of a physick garden which Hope had been instrumental in developing more into a botanic garden, as well as Hope's worldwide range of botanical contacts.

Company to send out books 'on the Arts & sciences, the cost would be trifling, & they would enable their servants to make improvements, Discoveries, &c to the honour & advantage of the Company. For the same reason a Botanical Garden would be of infinite use.'[12] He then went on to describe some of the plants in his own garden: 'I have done a good deal at my own private expence. Some thousand sappan wood plants I have raised, some hundred Anotta or Arnotta plants, Teak plants in great numbers, several [eleven added in] fine Cinnamon Plants, two fine Nutmeg plants, &c but as these are in my own private Garden, they will most likely be lost should anything remove me from hence.' It is possible that his thoughts on leaving India following the death of his first wife in 1788 may have been influenced by worries over his future career in India with the experimental work, especially as he continually regretted the lack of books after the hurricane in which he lost virtually everything.

BREAD FRUIT AND THE AGRICULTURAL BOARD

The study of two particular areas show the respect in which Roxburgh was held. Banks first came across the bread fruit (*Artocarpus altilis*) on his circumnavigation on the *Endeavour*, when he was made aware of its value as a source of food: much effort went into the introduction of this plant to the West Indies as an easy and cheap food especially for the plantation workers. In 1787, Banks was instrumental in arranging for Arthur Phillip, Governor at Botany Bay, to gather these plants from Tahiti under the ill-fated command

78 Sorghum bicolor (Andropogon sorghum). *Drawing by one of Roxburgh's Indian artists, c.1794 (Roxburgh drawing 898). One of the important grain crops of drier parts of India.*

of Captain Bligh who returned ignominiously in 1790. Roxburgh was ahead of Bligh on this, for in 1789, he had written to the Court of Directors offering to send bread fruit trees.[13] This had been arranged by the end of the year, when Andrew Ross wrote to Roxburgh that 'the Bread Fruit Tree is in excellent order, & shall be sent to St Helena by the Genl Goddard,' possibly from a supply from Tranquebar, for the Rev. John wrote 'all my Breadfruit Plants are gone by the late inundation but I will send you Roots & lay out another Plantation which will answer our hopes when the hot season is over.' This was obviously successful for John wrote in November 1792 that he had 'sent you a large Chest with Breadfruit Plants & Roots', and more were ready in April 1793.[14] Roxburgh sent six of these plants to Banks and 'this ensuing season I shall be able to send home twice that number, such annual supplies will I hope prevent the sending to the South Sea Islands for them.'[15] It does appear, however, that these first consignments suffered as so many other plants, for Roxburgh wrote: 'I am concerned for the fate of the Bread Fruit Plants, as well as the Box with others for the Kings Garden at Kew. Mr Aiton writes me that they all perished.'[16] With Roxburgh's move to Calcutta, no more is recorded about his attempts to help out with this venture, but by this time Bligh had been successful in bringing some plants back himself.[17]

The second area was the request from Sir John Sinclair, founder and first president of the Board of Agriculture, set up in the autumn of 1793, whose secretary was Arthur Young; its objectives included the promotion of improved agriculture in England.[18] Sinclair must have approached the Court of Directors about a year later, for in October 1795 Roxburgh had received a request via the Court for plants and seeds.[19] Roxburgh devoted some thought to this request, but commented that 'natives of the warmer parts of Asia, few if any can be expected to thrive, nay scarcely to exist in the open air in England. Flax and a few umbelliferous plants are the only exceptions I know of'.[20]

Thus he sent a small bag of flax seed, with the additional support that 'the present great scarcity, and consequent high price of Flax-seed over Europe may render the importation of it from Bengal an additional object of Utility.' Of the Umbelliferae, he felt that many had come to India from Europe, but one, a new species of *Ligusticum*, he sent with the flax, adding that it requires 'soil and mode of cultivation similar to what is proper for Caraway'. He included with this despatch some jute, remarking that 'it is not likely that ... [it] will grow in England out of the conservatory. However as it may gratify Sir John Sinclair to have some seed for trial, I have also sent with the rest a small quantity of it.'

A year later, another collection of seeds was sent, with the comment that the attempt to send Nepal rice had failed because the rice had not ripened, 'which shows it to be of a very different nature from the other sorts in general cultivation over these Provinces, and gives me the more reason to hope it would grow well in England.'[21] With this letter went 'a Package containing seeds of various kinds of vetches, and Tares, the produce of this and the Upper Provinces, where the seed is employed by the Natives in their Diet, and the straw to feed their Cattle with.' This time there is a list of the names of the plants that Roxburgh sent: *Lathyrus sativus, Vicia lathyroides, Ervum bispermum, E. filiforme, Medicago esculenta* and *Andropogon martinii*;[22] a further list includes some of these and gives the barest description of how they might be cultivated (the first six are legumes, followed by two grasses and a final chenopod):

1. *Pisum arvense*, seems only a variety of the common field Pea. Is sown about the close of the rains in October, on a rather elevated light soil.
2. *Lathyrus*, probably *sativus*, however I cannot be perfectly certain for want of good figures; it is called Keysarie by the Hindoos, and also sown about the close of the Rains, thrives well on a low strong soil.
3. *Vicia lathyroides* Ankari of the Hindoos, requires a similar treatment to the last mentioned, but the seeds are seldom eat so that it is reared for Cattle only.
4. *Ervum*, probably a new species which I have called *filiforme*, the Natives call it Juajume Ankari; requires a rich soil, and rather high; is sown about the close of the Rains. Captain Fraser, the Superintendant of the Honourable Companies Stud, praises this sort much for feeding horses with.
5. *Ervum bispermum* Mussoor of the Hindoos, is much cultivated by the Natives during the Cold Season, requires a rich loose, elevated Soil, seed much used in diet by them, the straw held in a little estimation for Cattle.
6. *Medicago esculenta* – Peering of the Hindoos. This new species of Medick is cultivated by the Natives about Calcutta; as a pot-herb, I cannot reduce it to any of those hitherto described, therefore consider it a non descript, is also reared during the Cold Season on a rich Loose elevated Soil, about a pound of seed, they say is sufficient to sow an Acre, it produces several cuttings during the Season.
7. *Agrostis linearis*, or Dub-Grass, a small creeping remarkably nutritious, sweet grass, grows readily in almost all soils and situations, yielding the best food for Cattle at all Seasons. A description and drawing (No 819) have already been sent to the Honourable the Court of Directors.
8. *Andropogon*, a new Species, which from its introducer into Bengal, General Martin, I call *Martinii*.
9. *Beta Bengalensis*, Palang of the Hindoos, by whom it is much cultivated during the Cold Season.

This package accompanied another which contained the seeds of 208 species of plants for the Court of Directors.

Sinclair sent his account of the progress of the Board to Sir John Shore which arrived early in 1798. The purpose of this was to ask for 'any suggestions that can tend to the advantage of these Kingdoms, or of the Countries over which you preside, By which you confer a particular favor on the Board of Agriculture ... . The other copies ... I beg you would have the goodness to circulate among those Gentlemen of your acquaintance who are attached to such inquiries.'[23] There was a further request of seeds from Sinclair in 1798 which arrived in Calcutta at the end of 1799. Roxburgh was unable to send any seed then because they 'cannot be had of this years crop till February, and last years seed will certainly not keep good to England. I will not therefore think of sending any of them till those of the present season are ripe.'[24] The last collection of seeds sent for the Board of Agriculture was despatched in 1803, comprising 24 species of plants, many duplicates of those already sent.[25]

TEAK AND MAHOGANY

The first mention of teak (*Tectona grandis*) came in a letter from Roxburgh to Colin Shakespear in 1796 in which he stated that 'in 1787 I was directed by the Government in Madras to send Lord Cornwallis as many of the Rajahmundry Teak Plants, & Seeds, as could be procured, which I did'.[26] From then on, Roxburgh pressed for the introduction of plantations, which he was able to undertake himself once he arrived at Calcutta. Thus he was able to write in 1799 that

> at the close of 1793 I began & have continued to the present time, to distribute both seeds and plants of the Teak tree not only officially by order of Government to the different Collectors but to every person who would accept of and promise to take care of them, and from various reports as well as my own observations, am happy to say that many thousand Young Trees are now in a forward state.[27]

Roxburgh's innate caution is reflected in the approach he took to the testing of the best conditions for the growth and use of teak. Writing at the end of December 1793, he commented that

> as I am unacquainted with the interior parts of Bengal, I cannot say what provinces the Teak Tree will grow best in, but conceive the higher lands between hills and mountains the most favorable, at least it was in such situations where the fine Rajahmundry Teak grows, however as there is plenty of seed, I could wish to send a portion to each Collector or to whoever else Government may think proper and if I am favored with the address of those Gentlemen I will send the seeds in small packages by Dawkes with such instructions as I may think necessary.[28]

This was agreed and the Collectors warned that they would be receiving seed; one, Thomas Parr at Tipperah, acknowledged this warning by the end of January, and that he would 'attend to any observations I may receive from the Superintendant [*sic*] of the Botanical Establishment'.[29] By the end of 1797, Roxburgh was able to state that 'the Teak continues to thrive well here and that several thousands are already planted out in the Botanic Garden, and great numbers of both plants, and seeds distributed over many parts of Bengal, and Bahar [Bihar].'[30] By 1803, Roxburgh was writing that he had sent small supplies to a number of people, as set out in Table 13, below, but the need for more ground was becoming more important as the distribution of these plants was not going ahead as fast as they were germinating: in September 1805 the Board of Revenue was authorised to purchase more ground at Burdwan.[31]

Individual	Number of teak plants
Mr Ernst, Collector at Midnapore	1620
Mr Middleton, Judge at Jessore	100
Mr Elliot, Judge at Tipperah	100
Mr Massie, Collector at Patna	473
Mr Barton, Collector at Benares	500
Mr Crommelin, Commercial Resident at Gazypore	326
Mr Arbuthnot, Judge at Benares	100
	3219

Table 13 Recipients of teak plants with the number sent by Roxburgh.[32]

79 Soymeda febrifuga
(Swietenia febrifuga).
*Hand-coloured engraving
after a drawing by one of
Roxburgh's Indian artists,
c.1791 (Roxburgh drawing 27),
published in* Plants of the
Coast of Coromandel *(vol.
1, plate 17, 1795). Roxburgh
used an extract of this plant
against malaria.*

An example of the care in ascertaining the financial value of a commodity is found in the case of teak, for when a teak tree which had been grown from the plants sent in 1787 was blown down in 1796, with a girth of over 40 inches, he sent a plank to 'Mr Gillett, the ship builder, requesting he would be so good as examine it & favor me with his opinion of its quality'.[33] Attached to this letter was Gillett's reply, in which he commented that, having compared it the teak from Malabar and Pegu, he found it 'equal if not superior to either ... I think it well worth the attention of Government to encourage the cultivation of so valuable wood'. Gilbert's position was important, for Huggins writing in 1824, spoke of 'the teak ships built in them [the Calcutta dockyards] are remarkable for strength, and last a much longer period than English ships'.[34]

The care that Roxburgh took in putting both the economic arguments to the company was always supported by his views as to the importance of such measures to the natives and how they could be persuaded to introduce them: in this case

it is much to be wished that the native Landholders could be brought to think that both their health and Interest will be promoted by planting all such places with useful Trees. Probably a translation into their own language of these instructions for raising Teak trees, & of Mr Gilbert the ship builder's report of the quality of a piece of that timber which grew in the Botanic Garden might be some inducement. An honorary or pecuniary reward of small value for the greatest number of trees reared within a given time might also have a considerable influence among the Natives.[35]

As a result of Roxburgh's experiments and the success of the trial at Rampore Bauleah by the Assistant Surgeon there, Henry Barnett, the Company approved the rental of over 850 biggahs for the development of a teak plantation under the supervision of Barnett in 1809.[36] Two years later, when Roxburgh was asked to comment on this, by then under the supervision of Mr George Ballard, he reiterated the economic importance of the teak, compared its growth favourably with English oak, and returned to an old idea of his, of planting these trees in hedge-rows and other unused places.[37] Ballard did not agree with Roxburgh on the advisability of the latter idea, nor that there would be no cost of labour, but did give an indication of the success of the plantation, stating that in the 715 biggahs planted, there were about 36,000 trees, whose average girth was slightly over 3 feet 1 inch taken from a sample of 20 trees.[38] The economics of teak were explained by Roxburgh in which he deduced that each biggah should produce a clear profit of 1104 rupees.[39]

Roxburgh also grew the mahogany tree (*Swietenia mahagoni*), well known for its use in European furniture at the time, and there is a record of the growth of one of the trees in the Calcutta Botanic Garden, when the girth at three feet above the ground grew from 25½ inches after seven years to 54 inches after 15 years.[40]

ROXBURGH AND TEA

As we have seen, Roxburgh's involvement with the introduction of new plants at Calcutta had started before he arrived, with such plants as the sago palm from Ceylon, sugar from China and the cultivation of pepper. In October 1793, shortly after he had arrived in Calcutta, Ross was advising him:

When you are in Bengal you will enquire whether the Tea [*Camellia sinensis*] plant [grows there] (but do it as quietly & privately as you can) as it may not be cultivated in some

80 Camellia sinensis, *tea.*
Watercolour by an anonymous
Portuguese artist, c.1848.

of the Upper Provinces, similar to the climate & soil of those parts of China where it grows – so as to produce the large Crops that will be required in England to rival the Trade of China.[41]

The secretiveness implied in this letter does reinforce the possibility that Ross and Roxburgh had been working together in some way, and Ross was trying to pre-empt anybody else getting the benefit from the introduction of such a valuable commodity as tea.

In fact Banks had already approached the Court about the suitability of introducing the plant and Kyd had supported this idea, as he had had some plants growing in his own garden as early as 1780, although he admitted that it was a most unsuitable soil and climate. Some of these weak plants were still growing when Roxburgh took over the Garden in 1793.[42] Banks appears to have been aware as early as 1788 that tea grows best in temperate rather than sub-tropical regions. However, a more powerful embassy had already gone to China, to the Emperor Qianglong, under Lord Macartney and his diary of 17 November 1793 states:

> We have quitted the mountains and got into a charming fruitful country. Here the tea grows in great abundance on the dry rising ground. The mulberry flourishes most on the loamy flats. I have given directions to have some young tea plants taken up, if possible, as also the varnish tree [*Rhus verniciflua*] and tallow tree [*Sapium sebiferum*], with an intention of sending them to Bengal, in hopes of Colonel Kyd's being able to nurse them and bring them to maturity, so that one day or other they may be reckoned among the commercial resources of our own territories.[43]

Dr James Dinwiddie brought these plants to Calcutta but 'all except some plants of the Tallow Tree, and two said to be Varnish Tree, were to all appearance, dead, Doctor Roxburgh is however not without Hopes that some of the Tea Plants may still shoot up from the Root.'[44] The Court were obviously highly interested in the success of this venture, and asked to be kept informed about Roxburgh's success in nurturing the sickly consignment.[45] Macartney was aware of the need to record the climate that these plants grew, for on 21 November, he wrote: 'The place where we procured our tea plants is nearly twenty-eight degrees north latitude. The summers here are very hot and the winters extremely cold, but not attended with frost or snow.' Macartney was careful of the state in which this important consignment was despatched, for he said that

> the Viceroy ... allowed us to collect seeds ... and to take up several tea plants in a growing state with large balls of earth adhering to them, which tea plants I flatter myself I shall be able to transmit to Bengal, where I have no doubt that by the spirit and patriotism of its Government an effective cultivation of this valuable shrub will be undertaken and pursued with success.

Roxburgh tried to get a further supply of tea plants in March 1796, when he requested 'to have a few of the various sorts of Sugar Cane Cultivated in China and Cochin China or at least as can be procured, also a Number of Plants & kinds of the Tea'.[46] A year later he was complaining that 'neither Tea Plants nor seeds thereof have yet been received. I therefore beg leave request that the Supra Cargoes at Canton may be put in mind of them'.[47] These attempts proved unsuccessful, but when he was at the Cape, 'The Tea Plant I have found in one Garden growing luxuriantly. Seeds & plants I have secured for Bengal, where it is not. The Government often wrote to China for them. I mean to have it tried on the N.E. frontier of Bengal. Here there is no doubt of it thriving well, & may become an object if we keep the Cape.'[48] This suggestion of intending trials in north east Bengal, not far from where wild plants were found growing thirty years later, is almost haunting in its forethought. In view of the emphasis that has been given to Banks's interest in introducing tea into India, the fact that he did not follow Roxburgh's idea further would seem to argue against the great man's genuine involvement in such an important crop.

These ideas and suggestions seem to have been lost, although a few plants continued to grow at Calcutta, for when Royle wrote his report, he started with the comment, 'Though unacquainted with the fact, I was in the year 1839 informed by Mr Greene, that Sir Joseph Banks had many years previously recommended the cultivation of tea in the Himalayan Mountains.'[49] This short study of Roxburgh's involvement with the attempt to introduce tea commercially into India highlights one aspect of his methodology which has been mentioned before. He learnt from John Hope, while he was at Edinburgh, the necessity of studying the literature, and this can be seen from a glance through the titles that he had in his library: in this case, there is a publication by Lettsom on the natural history of the tea trade (see Appendix 6). What is of particular note, however, is that Roxburgh told Banks about the possibilities of where to try to grow tea and yet no action was taken to develop what became such an important export of India: this does question the way in which Banks was liable to subsume other botanists' findings to his own views and interests.

CHAPTER 9

Dyes

INTRODUCTION

In an age before chemical dyes, most colours for cloth were derived from plant material, with a few from mineral sources. The bright colours of much of the Indian cloth invited research into the plants that produced this exotic range and the potential profit that would follow from success. It is therefore no surprise that the East India Company was prepared to devote money for its employees to experiment with various ideas, and there are records of one of the earliest Company traders dealing in indigo as early as 1608, less than ten years after the Company had been founded.

The research that Roxburgh did on dyes highlights a number of his working practices. The first was his awareness of the need to produce crops that were of economic importance. In a country that produced much coloured cloth, the need to find cheap and readily accessible supplies was self-evident. The problem, however, was that the plant sources were not known and so experiments on the potential of various species was necessary, backed up with searches through the contemporary literature for ideas. This is shown in his work on both a dye from madder and his work on indigo.

The development of an alternative for the Spanish monopoly of the scarlet lac from cochineal shows a different approach. Here Roxburgh was involved in what might now be referred to as industrial espionage, with the theft of the cochineal insect and the attempt to find a suitable host plant.

When his work on indigo is considered, the concern which emerges is the stress that Roxburgh put on producing a system that was easy to work within Indian culture: as labour was cheap, and provided employment outside agriculture, this was preferable to the mechanised systems being developed in Europe. The taxonomic botanist, however, was never far away, for Roxburgh was always interested in describing and giving a name within the Linnean system to the plants that were of economic importance, a fact that comes out from all these studies.

MADDER

With the demand for printed cloth, there was always going to be a major requirement for dyes, especially if cheaper alternatives could be found, either from new plants or from new techniques using existing sources. This was particularly the case when the source was controlled by some other country: madder, for instance, 'was, until the introduction of the coal-tar colours of the nineteenth century, the principal source of all red dyes',[1] but for which 'the English had come to depend for this material upon the Dutch, who did not fail to take advantage of the monopoly'.[2] It is not surprising, therefore, to find that Roxburgh was involved in looking for an alternative, spending many years working on the introduction of the lac insect from which the red dye cochineal is derived.

81 Silk moths, drawn by Roxburgh and sent to Smith at the Linnean Society in London. Roxburgh and his second wife did various experiments to study the possibility of breeding silk moths as a commercial operation.

One of the main problems, however, with vegetable dyes was getting them to take to the cloth, for which some knowledge of chemistry was, and still is, crucial: some mordaunt is necessary. To inform himself on this subject, Roxburgh had a number of publications in his library: there are references in his correspondence to the following:

Anon, *Natural Shorts Dyeing* details of which have not been traced
J. A. C. Chaptal, *Elements of Chemistry*, translated from the French 2nd edition, 3 vols. (London, 1795).
P. J. Macquer, *Art de la teinture en soie* (Paris, 1763).
W. Nicholson, *The First Principles of Chemistry* (London, 1790).
Tancroix, *Chemistry*, details of which have not been traced.
A. F. de Fourcroy, *Elements of natural history and of chemistry*, translated by William Nicholson, 4 vols. (London, 1788).

Added to these, he at various times referred to the *Transactions* of the Royal Society, the Royal Society of Edinburgh, the Linnean Society, the Society for the Encouragement of Arts, Manufactures and Commerce, as well as the publications of the Asiatic Society, the Batavian Society and the American Philosophical Society. The fact that his own publications appeared in Nicholson's *Journal* and Tilloch's *Philosophical Magazine* would suggest that he may well have had access to these as well. A quick glance through Nicholson's *Journal* reveals articles by such eminent men as Berthollet, Coulomb, Count Rumford, Davy, Priestley, Lussac, Dalton and La Place, indicating that Roxburgh was at least potentially aware of the latest developments in the physical sciences – which was very much the publishing policy of both Nicholson and Tilloch who reproduced articles by such men for wider circulation. Add to this wealth of material some manuscripts which are in a collection of Roxburgh's:[3]

An untitled copy on chemistry and dyeing pigments	3ff
English Review, Jan 1790, Article IV on Dyeing	4ff
Extracts of papers relating to the East India trade, compiled April 1788, on the Indigo trade	2ff
Critical Review, Mar 1788, review of *Elements of Natural History and of Chemistry* by M. de Fourcroy	4ff
Monthly Review, Oct 1788, directions for dyeing with red	4ff
Report from the Court of Directors, dated 30 May 1792, relating to the favourable state of the market for Bengal Indigo	2ff
Recommendation of the Bengal Board for the package of Indigo sent to Europe, 19 October 1792	1f
Monthly Review, Apr 1792, review of Chaptal's *Elements of chemistry*, translated by W. Nicholson, including the theory of colour	6ff
Monthly Review, Jan 1792, on chemical experiments	2ff

The fact that this large collection, now bound into a volume of some 65 items, was all carefully folded and indexed by Roxburgh, suggests that they were an important part of his reference material. The first in the above list is 'From Kyds MSS' and refers to a method of extracting 'Red, yellow, and white Pigments for painting from Calx of lead by means of common Salt. At this present time a Mr Turner has a patent for the exclusive privilege of preparing these colours in a Similar manner.' Whilst it has not yet been possible to determine who this 'Mr Turner' was, it is interesting that, once again, Roxburgh was reading up about both local practices as well as keeping in touch with the latest European advances. Similarly, the excerpt from the *Monthly Review* of October 1788 gave directions for the 'Printing or Dyeing Cotton Cloth into Red, as practised at Madras.'

With this breadth of resource, it is not surprising to find that one of the first extant letters of Roxburgh's is about plants which could be used for a dye, the yellow from the caducay gall. These galls he first performed experiments on and

82 Coromandel coat, referred to as a Banyan, painted cotton, dating from 1750-1800. Roxburgh did much research on the dyes for just this sort of garment.

it will prove the strongest vegetable astringent known. I always make my infusion with it. The Fruit and Galls I use promiscuously, but I believe the galls are best. The Galls are of great use in Clients paintings. It is the best standing yellow they have, and a weak Decoction is used to prepare the cloth for printing. Without it the colors would run like Ink on Blotting Paper.[4]

What is tantalising is a letter from the Rev. John in which he stated, 'I congratulate you & the Public to the high success you make in Botanicks & to the most usefull Discovery of a new Dye'.[5] The problem is that there is no indication of what the dye is or the plant from which it was developed.

In the review of *The Art of Dyeing Wool, Silk & Cotton* in the *English Review* of January 1790, there is the enlightening comment that 'though an illiterate dyer may accidentally stumble on an useful improvement, the art of dying will never reach the perfection of which it is capable until those who possess it shall make themselves acquainted with the chemical theory on which its operations are founded'.[6] This article continues by mentioning that Macquer emphasised the different skills required for different materials, devoting 39 chapters to wool, and stating that 'silk is a glutinous matter formed in the body of the worm, & which hardens in the air while the animal is spinning'.

COCHINEAL, LAC AND THE SEARCH FOR A RED DYE

By the end of the 18th century, some 340 tonnes a year of red cochineal dye was being imported to Europe from New Spain. Of this some 20 per cent was for Britain, some used for fashion but also used by the military for its uniforms, so that the dye would mask the blood spilt in battle! Like madder, the scarlet cochineal dye was subject to a monopoly, in this case by Spain, through its South American colonies until 1777 when a Frenchman, Thierry de Menonville, transferred two varieties of the cochineal insect, each with its specific host plant, from New Mexico to Port au Prince in San Domingo. Banks acquired a copy of de Menonville's description which appeared in 1787 and 'made an admirable technical digest in English and from it and other sources compounded a clear account of the two forms of cochineal known to the English trade – the grana fina, granilla or fine cochineal, and the grana sylvestra or sylvester cochineal .... By January 1788 he [Banks] had obtained samples of each kind from John Maitland, with the current prices of 5 shillings per pound for the granilla and 3 shillings for the sylvestra; lower than the figure of 18-19 shillings per pound he had elsewhere determined as general.'[7]

Once again, there appears to be some doubt as to the exact sequence of Banks's interest in a scientific subject. Carter continued his study of Banks's involvement in the spread of cochineal with 'since February 1787, he [Banks] had been in correspondence with Dr James Anderson, ... about a form of insect apparently native to India. The substance of Anderson's ideas was printed in the *Madras Courier* and later published in quarto.'[8] However, Patrick Russell in March 1787, wrote to Sir Archibald Campbell, President in Council at Fort St George,

I have the honor to transmit the accompanying Packett containing a Piece of Flannel Sattin, and Shawl for your Inspection, which have been Dye'd with the Cochineal Insects, I gave you an account of last November. As I have struck this Dye on white materials, without assistance of any other colouring Drug whatever in this, which is but the Second experiment I have made with them, and seeing they farr surpass my expectation, I have thought it incumbent on me, to lay these specimens before you.[9]

This means that Russell, a close friend of Roxburgh's, was working on the cochineal in late 1786, well before Banks was in correspondence with Anderson, in February 1787. As Anderson had acquired these specimens of *Opuntia* species (the host plant for the cochineal insect) from China via Manila and had already sent specimens to Col. Kyd at his garden in Calcutta, there seems strong evidence that the work in India preceded Banks's involvement from Soho Square. This appears to be confirmed by the fact that Anderson had been working on his own alternative to cochineal, even though 'a chemical analysis of these specimens indicated that they were commercially worthless but Anderson, who had been commended for his efforts, was given a garden at Marmelon in 1788 to grow *Opuntia* and known as the Company's Nopalry. His nephew, Dr Andrew Berry was appointed Superintendent of this garden'.[10] The fact that at least some of the impetus came from India is supported by the fact that it was not until March 1790 that Banks persuaded the Company to offer a £1,000 reward for the clandestine transfer of the true cochineal, grana fina, to London where they could be nurtured before despatch to India.[11]

This early importance is reflected in the presence of an extract in the manuscript collection of Roxburgh's, dated 1788, which stated that improvements in cochineal 'which if effected would transfer to the Company a further sum of 350,000£ per annum, said to be paid to foreigners'.[12] This is another example of Roxburgh using the economic arguments for pressing for the introduction of plants. The early interest for introducing cochineal into India came from Dr Anderson, but by the end of 1791, he had transferred his enthusiasm: 'that you have transferred, for the present your attention from the culture of Cochineal to the cultivation of Silk.'[13] In fact, it is probable that all three of these men, Russell, Anderson and Roxburgh, would all have worked together in some way, as they were all part of the medical department at Madras.

To return to the development of cochineal itself, there is, once more, a discrepancy between the archives and Carter's version. Banks, as mentioned in Chapter 3, arranged for Christopher Smith to sail to Calcutta as the new nursery man, with Peter Good who would return with plants for the King and Kew. Shortly after Good left in the *Royal Admiral*, Captain Neilson had arrived at Madras in early May with 'some of that sort of Cochineal Insect, called in America Sylvester, which he procured at Rio Janeiro'.[14] To this was appended a footnote, giving detailed knowledge of de Menonville's description, which included of the Sylvester form 'that rearing the last form is attended with infinitely less trouble than the former [grana], as it endures storms, &ca vicissitudes of climate much better and yields a die equally good only requiring a larger quantity of the drug to produce the same effect'. In other words, Roxburgh was well aware of de Menonville's work, probably from the original but possibly through Banks's later English version.

The biology of the cochineal insects is complicated by the fact that the insect that produces the true dye, *Dactylopius coccus*, has to live on the cactus *Opuntia coccinellifera*, while the grana sylvestris, which can be made from either *Dactylopius indicus* or *D. tomentosus*, can be grown on a number of species of *Opuntia*: *O. monacantha, O. stricta, O. elatior* or *O. ficus-indica* and also *O. dillenii*. While importations of the true host plant failed, that of the other species have 'left a legacy of opuntias rampant across the sub-continent.'[15] I remember as a child being shown the effect of the growth that produces the dye, when this was crushed between one's fingers to produce a bright scarlet liquid.

83 Opuntia cochenillifer (Opuntia cocinellifera), *the true host plant for the cochineal insect in Mexico. Hand-coloured lithograph from* Abbildung und Beschreibung Blühender Cacteen *by L. Pfeiffer and F. Otto (plate 24, vol. 1, Vienna, 1842).*

The host plants on which Neilson's insects arrived 'were so completely decayed, that the Insects must have perished had the passage been one week longer, and from the same cause, it was impossible to even guess what species of Opuntia, or rather Cactus it was'. Roxburgh then tried two host species which had proved satisfactory to the true insects, but 'every one of the young brood brought forth upon them perished in the course of a day or two'. Two further host species were tried, one from China and the other from the West Indies, with an equal lack of success. 'There was now little prospect of being able to preserve the valuable present we had just received, without some other more favorable species could be found, and fortunately at this critical moment, I discovered for the first time another species of Opuntia, which I find is indigenous here, and am surprised at not having observed it before.' According to Carter, this work was performed by Smith but unfortunately did not give any sources which can be checked. The fact that Smith had only just arrived at Calcutta and that he was sent to the Malaccas before the end of the year, makes his involvement with these experiments doubtful. This would appear to be supported by the fact that there was an unease in the relationship between Smith and Roxburgh, which has already been considered.

The suggestion which is put forward here is that Smith was trying to build up his own involvement in the development of what could prove of great commercial value to the Company, thereby hoping to get some reward, hence his probably false claim. Against this was the recommendation following Roxburgh's letter to Edward Hay, that the latter was ordered to write a letter to Captain Neilson in which it was stated: 'The Governor General in Council that neither his nor your expectations of the advantage to be derived from your exertions will be disappointed, and in this expectation it will afford him peculiar satisfaction to inform the Hon'ble the Court of Directors that a valuable article of commerce has been introduced into this Country, by your means, and to recommend a remuneration in proportion to its importance.'[16] This interpretation would seem to be further supported by the comment in the Public Letter to the Court: the cochineal insect 'is the kind called in America, Sylvester and was procured at Rio Janeiro, by Captain Neilson, of his Majesty's Service, from whom Doctor Roxburgh received it'.[17] Desmond's version would also support this view of Smith, as he stated that 'Christopher Smith took the credit. "I flatter myself that I can manage the cochineal insect now equal to any man in South America".'[18] The irony of this letter is that the insects were of the grana sylvestris rather than grana fina, showing that he did not know the difference between the two, a weakness admitted by Roxburgh.

As Superintendent of the Garden, Roxburgh immediately took 'every possible pains to increase our stock of plants and insects, that I can think of having already formed a plantation of several hundred plants'. The aim was to be able to supply 'from these Nurseries the Insects ... to such as may wish to speculate in this new branch of commerce, and I hope, like the Indigo planters, they will find it their interest to extend their labours so far as to be able to supply Europe with cochineal at a cheaper rate, than it can be had from Spanish America.' As a taxonomist, Roxburgh had also produced detailed drawings of *Opuntia*, which immediately point out their specific differences, and on the same piece of paper 'figured the Cochineal Insects themselves, in their various stages, as mentioned in the references annexed to the drawings'.

As soon as Roxburgh had discovered the appropriate host plant, he sent some to Dr Anderson at Madras, who had replied in time for Roxburgh to include his comments in this long letter: 'it is found plentifully about Amboor, Pondicherry, and several other places West and South of Madras.'

Before sending the insects out, Roxburgh was 'desired to continue the management of the Insect' and to arrange for experiments to be made on the dye, with the help of Dr Dinwiddie, if necessary. Roxburgh's reaction was typical of the man: 'I am too ignorant of the practical part of the art of dyeing to presume to make any experiment that I could offer to Government. I will therefore be happy to have Dr Dinwiddie's assistance.' Then Roxburgh added the interesting comment, 'The best chemists have found it difficult to produce colours equal to the most illiterate artist.'[19] Roxburgh's awareness of factors that might affect the success of these experiments again comes out in this letter: 'I conceive it will not be giving the quality of the Cochineal a fair trial if we employ for our experiment those Insects rear'd under cover during the rains, at least the colouring matter of the vegetables under such circumstances, would be exceeding defective.' Roxburgh had, however, already started distributing some of the insects, 'so that we shall the sooner be able to know the best Stations, &ca for rearing them, an essential point to ascertain'.

Roxburgh, true to his word, got Dinwiddie to perform the necessary experiments, which were completed towards the end of October. The latter's report set the parameters of the experimental design: 'decoctions were made of the Bengal and Mustique or East Mexican cochineal perfectly similar in every respect.'[20] The results of this set of experiments were that

> two pounds of the former [Bengal] are fully equivalent to one of the latter [Mustique] .... These experiments seem to prove that the Bengal Cochineal is of the Sylvestris or second kind, and also pretty high in this class, which varies much in its quality, being some times as low as thirty and seldom rising higher than sixty per cent of the value of Mustique.

By February of the following year, Roxburgh had been able to produce sufficient insects and plants to supply them to 'about forty different People ... to stock their Plantations with' as well as supply Dinwiddie with those necessary for his experiments, and 'a small quantity eight ounces' as a sample for the Court of Directors.[21] However, there had presumably not been enough to include in a shipment of plants in December 1795, but two 'Cactus nag-jenny' plants were included in the chest of plants for Sir Joseph Banks by the *Amelia*, in February.[22] There is no record, sadly, of these plants arriving at Kew. By the end of February he had supplied another ten 'Gentlemen with sufficient stock for beginning their Plantations besides considerable quantities for Dr Fleming & Dr Dinwiddie for their various experiments'.[23]

Russell and Roxburgh were well acquainted with one another, and Roxburgh succeeded Russell as Company Naturalist in the Carnatic. Roxburgh would therefore almost certainly have been aware of Russell's work on cochineal, and probably helped in the search for alternative sources of red dye. Certainly by 1790, when Russell had returned home, Roxburgh was writing, 'Some time ago I found pieces of Lacc with the Insects. I transmit you a copy of that part of my Diary which describes them. You are at liberty to make what use of it you please'.[24] This was followed by a second batch, for, at the end of December, Roxburgh wrote to Banks: 'My descriptions and drawings of the Lacc Insect were sent to Dr Russell in September last, I hope he has presented them to you, for your perusal.'[25] What he had sent to Russell was a detailed description of the insect, with some ideas as to its life cycle, based on observations which had taken from the previous December through one complete cycle, but with allusions to similarity with the parthenogenic stage of aphids.[26]

Further research on the Lac was hampered by the famine for, writing in 1792, Roxburgh stated: 'I have not been able to add anything to my observations on the Lacc Insects, nor do I see any prospect of being able to finish that account, this is now the third year that a most dreadful Famine has raged throughout this province, which makes it both difficult & dangerous to go in amongst the mountains where they are found.'[27] The earlier results of these researches were published first in the *Asiatick Researches* followed by a slightly expanded version in the *Philosophical Transactions of the Royal Society*, which was read by Patrick Russell[28] (see Fig. 84). Some idea of the precision with which Roxburgh approached his observations can be gleaned from some excerpts of the longer article:

84 Laccifer lacca (Chermes lacca), *the lac insect. Engraving by Basire, after a drawing by Roxburgh, published to illustrate his paper on 'Chermes lacca' in* Philosophical Transactions of the Royal Society (*vol. 81, 1791*).

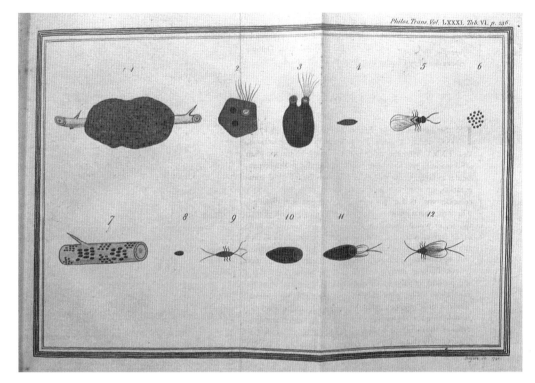

Some pieces of very fresh-looking lac, adhering to small branches of *Mimosa cinerea*, were brought me from the mountains, on the 20th of November 1789. I kept them carefully in wide-mouthed crystal bottles, slightly covered; and this day, the 4th of December, fourteen days from the time they came from the hills, thousands of exceeding minute red animals were observed crawling about the lac and the branches it adhered to, and still more were issuing from small holes on the surface of the cells. By the assistance of glasses, small imperforated excrescences were also observed ... . I opened some of the middle-sized [pupae], and found they contained a thick, deep blood-coloured liquid.

When it came to their potential

œconomy of these little animals, I must, for the present, be silent; having little more than conjecture to offer on that head. The eggs, and dark-coloured glutinous liquor they are found in, communicate to water a most beautiful red colour, while fresh. After they have dried, the colour they give to water is less bright; it would therefore be well worth while for those, who are situated near places where the lac is plentifully found, to try to extract and preserve the colouring principles by such means as would prevent them from being injured by keeping. I doubt not but in time a method may be discovered to render this colouring matter as valuable as cochineal.

This is a good insight into Roxburgh's awareness of the potential of what he was describing: an alternative to the valuable red cochineal dye.

In this article, he then went on to state: 'Mr Hellot's process for extracting the colouring matter from dry lac deserves to be tried with the fresh lac in the month of October, or beginning of November, before the insects have acquired life; for I found the deepest and best colour was procured from the eggs while mixed with their nidus.' The implications of this quotation are that Roxburgh must have done a number of experiments in which he tried different stages of the insect, taken at varying times within their life cycle, as well as comparing it with the dried material, probably also taken from the varied times within the life cycle. These experiments were performed after he had read up Hellot's procedures, to try to improve upon them and apply them to a slightly different situation. This is confirmed by the fact that Hellot used a

decoction of comfrey root ... in India comfrey roots are not to be had; but any other mucilagous root, gum, or bark, would probably answer equally well. On some parts on

85 Butea monosperma
(B. frondosa). *Hand-coloured engraving by one of Roxburgh's Indian artists, c.1791 (Roxburgh drawing 67), reproduced in Roxburgh's* Plants of the Coast of Coromandel (*vol. 1, plate 21, 1796*).

the Coromandel coast, if not over it all, a decoction of the seeds of a very common plant (Cassia tora of Linnaeus), which is exceedingly mucilaginous, is used by the dyers of cotton cloth blue, to help to prepare the blue vat.

In the next volume of the *Asiatick Researches*, Roxburgh published an article on *Butea frondosa* (now *B. monosperma*), which

> from natural fissures, and wounds made in the bark of this tree, during the hot season, there issues a most beautiful red juice, which soon hardens into a ruby-coloured brittle astringent gum: but it soon loses its beautiful colour, if exposed to air: to preserve the colour, it must be gathered as soon as it becomes hard, and kept closely corked in a bottle.[29]

Towards the end of the article, he stated:

> the Lac insects are frequently found on the small branches and the petioles of the leaves of this tree: whether the natural juices of its bark contribute to improve the colour of their red colouring matter, I cannot say: it would require a set of experiments accurately

made on specimens of lac gathered from the various trees it is found on, at the same time and as nearly as possible from the same place, to determine this point.

This is another neat exposition of Roxburgh's awareness of the need to control experimental variables, in this case by reducing the effects of environment by taking samples at the same time and from the same locality.

The end of the story of Roxburgh's involvement with the lac insect came in 1801, when Andrew Stephens was awarded the Silver Medal by the Society for the Encouragement of Arts, for his work on the insect, and that of Dr Bancroft. The latter had taken note of Roxburgh's work, but had also received a sample of 'colouring-matter', which was referred to as 'East-India cochineal', about three years before the publication of Roxburgh's paper, from 'a gentleman who had lately returned to this country from India'. Stephens then went on to say that he had made a number of experiments with this, uncannily similar to those recommended by Roxburgh.[30] This is an example of a Roxburghian trait that appears from time to time, particularly in the earlier archives: a certain naïvety, of which he seemed not unaware, for he often wrote of the fact that all he wanted to do was what was best for science and the poor.

Roxburgh does appear at this stage to be more interested in the descriptive taxonomy of the lac insect rather than its commercial development, and certainly not for his own gain. This seems to be confirmed in his long obituary, which stated of Roxburgh's recommendation for its use, 'this idea was afterwards followed up, and the liquid in question, at the present day, forms a valuable branch of export from Calcutta to London, under the name of lacc-lake. It is chiefly employed as the substitute of cochineal for the dyeing of scarlet.'[31]

The success of this venture can be gauged by the fact that the Board of Revenue decided to encourage 'the natives for the introduction of cochineal' by purchasing it from them at the rate of one Pagoda per pound in August 1796, and by February 1798 the Board had bought 21,744$^9/_{16}$ lb. In view of this large quantity, the Board was suggesting that they should reduce the price 'to two Rupees per lb for the best, and one Rupee per lb for the second sort of Cochineal'.[32] The reply to this was somewhat disappointing, but took over two years to be written:

> The Specimens of Cochineal forwarded by Dr. Roxburgh and Mr. Taswell from Bengal, as also two Parcels sent from your coast by Dr Anderson and Dr Berry, the whole raised from the Insect procured at Rio de Janeiro, by Captain Neilson, have been shown to some of the best Judges of the Commodity in the Kingdom, who all agree, that it is the Sylvester or wild species, and that there is little prospect of its being cultivated to any advantage for the supply of the Europe Market, unless it could be afforded at about one third of the Price of the Grana, or say, from five to six shillings per Freight, and all charges included. We enclose a Number in the Packet, an account shewing the issue of our concerns in the article of Cochineal, from the first of its import to the present period; as also the relative proportion the Sale prices have borne to the genuine sort at the various periods. The parcel received by the Thetis in the present year, appears to be of a very inferior quality, and it is doubted whether the Insect is the same as that supplied in the former Seasons; if it is the same, some defect has occurred in the process of preparation. Your Consignments have hitherto yielded a sufficient degree of Profit, and probably may continue so to do, while the War lasts, and the quantities are in some measure limitted; but we have no reason to suppose it will ever make any great inroad upon the Spanish assortment, when things shall again revert to their ordinary channels.[33]

The increasing production and deteriorating quality had a devastating effect on the price, which sold in London at 19s. 4d. per lb in 1797 against a price of 8s. 8¼d. in 1799. But the variability of the price had been shown even in 1792, when 'the Pursuit of the Culture of Cochineal, which is of infinite importance, the Cost of it being from 10s to 20s Sterling per pound'.[34] A full view of the changing prices is given in the table below. However, the host plant and insect were being grown by May 1796, for David

Scott (Chairman of the East India Company) wrote to one of his correspondents in that month, saying, 'I shall be anxious to have a clear report on the cochineal'.[35]

Year	Ship	Quantity	Sale value	Price per lb (d)
1797	*Airly Castle*	177	171.. 3.. 7	19s. 4d
	Princess of Wales	4,214	3,528.. 18.. 2	16s. 9d
1798	*King George*	5,946	3,211.. 4.. 7	10s. 9½d
	Marquis of Lansdown	11,355	5,391.. 11.. −	9s. 6d
	Lord Macartney	9,865	4,875.. 7.. 10	9s. 10½d
	Lord Hawkesbury	15,168	7,634.. 8.. 10	10s. 0¾d
1799	*Dover Castle*	8,471	3,683.. 6.. 11	8s. 8¼d

Table 14 The amounts of cochineal imported into Britain from India, its sale price and price per pound.[36]

As a result of this falling price, the Court decided that it was no longer worthwhile economically, and that the quantity purchased should be fixed at 8,000 lbs for 1809, reducing to half that the following year and then ceasing its purchase altogether. The obverse of this decision was that there should be an incentive to introduce the *grana fina*, or real Mexican insect, set at Pagodas 5,000 or £2,000, providing it came on an English vessel. This manuscript also recorded that Captain Neilson was granted a reward of £500, no mean sum at that time.

Roxburgh's part in the success of the cochineal trade can, therefore, be traced quite clearly, from the first gift by Captain Neilson when Roxburgh had to determine which plant it could grow on, to the development of a plantation of the necessary hosts. From this, he followed his initial remit of distributing potentially economically useful plants to growers who could develop them as profitable crops. At the same time, he provided detailed drawings and descriptions of both the host plant and the insect, with full explanation of the various stages of its life cycle. In fact, the only place where the genuine cochineal insect, the *grana fina*, was ever introduced was the Canary Islands.

INDIGO

One of the early crops that Europeans grew on a commercial scale in India was the plant *Indigofera tinctoria*, from which the indigo dye was extracted. English planters in the West Indies, Carolina and Georgia, however, had developed successful crops of the Guatemalan indigo, so indigo imports from India consequently slumped until about the middle of the 18th century when the West Indies moved over to the more lucrative growth of sugar and coffee. The Company tried to revitalise the Indian market by enticing European planters from the Americas, and built the first indigo factory in Bengal in the late 1770s. However, this was a very erratic market, for the price varied enormously from year to year. For example, David Scott, in two letters about three weeks apart, wrote first that 'you will find that the indigo had a sudden rise which would prevent so many suffering' and then 'indigo had a great fall in the London market but then recovered again'.[37] Scott continued his letter to Harris, 'Dr Roxburgh is a most valuable acquisition, I therefore hope he is perfectly well again', the train of thought possibly relating in part at least to Roxburgh's importance in this field.

The seminal work on indigo was published in 1779, J. F. C. de Cossigny's *Essai sur la culture et la fabrication de l'Indigo*, an English translation of which appeared in 1789, *Memoir ... on the cultivation and manufacture of Indigo*. Roxburgh contributed to this translation, a 'Process for making Indigo on the coast of Ingeram',[38] which appears to be his first published botanical work. It describes briefly the way in which the dye was extracted in the Northern Circars shortly after his arrival there, and shows a familiarity with the Linnean nomenclature and descriptions of plants, as well as the more popular production of this: 'the rest of the process is the same as described by Pere Labat.'

86 Clerodendrum japonicum (Volkameria kaempferi). *Drawing by one of Roxburgh's Calcutta artists, c.1795 (Roxburgh drawing 965). An attractive flowering shrub native to Malaysia, but widely cultivated and first described from Japan.*

The first reference in the Roxburgh archives to the dye occurs in a letter from Ross:

It is long since I have had it in my mind to write to you about Indigo, but neglected it. I now send you 2 Papers on the subject – & a muster of a good quality made in Bengal. You will make every – & the early the enquiry about it in your parts – & likewise at a distance – especialy to the Westwards as far as you have opportunities – & acquaint me with the result whether it is – or soon may be, cultivated – & made of a tollerable good quality – where – in what quantity, & at what price, & if you meet with any send me Musters – the sooner the better. If a proper attention is given to it – it may be rendered of consequence – & therefore – I wish it to rest with you & I assuring you that I shall not be inattentive in promoting it.[39]

This would seem to suggest that, initially, Ross tried to get Roxburgh to help in trading in the dye, a theory supported by another letter at the end of 1788: 'In the mean time – if you receive this in time to enable you to procure any quantity of the best sort of Indigo – that it is not too dear – suppose at about 5 or at 6 Madras Pagodas – for our Md of 25th I should be obliged to you to purchase it – & to bring it with you.'[40] This help for Roxburgh was also evident through Ross sending, amongst a List of papers for Dr Roxburgh, an 'Instructions for the Manufacture of Indigo – together with an Advertisement of the Bengal Govt for encouraging the cultivation of it – 5th Jany 89'.[41]

Roxburgh was ever the descriptive taxonomist, for a year later the Rev. John wrote, 'I wait very much your Descriptions, Drawings & Specimens of 8 distinct Species of Rottbola & of your Indigo Plant which all shall be transmitted faithfully to Professor Rottboll.'[42] The description was sent to the Court who passed a copy to Dalrymple who published it in his *Oriental Repertory*, and this gives another glimpse of the way Roxburgh worked. The new source for indigo was a tree, *Nerium tinctorium* (now *Wrightia tinctoria*, see Figure 89, p. 150),[43] of which he said: 'it is not taken notice of, by the great Sir C. Linnaeus, nor by his Son, in his last Botanical Publication, The Supplementum Plantarum, published in 1781. It comes nearest to Nerium Antidysentericum, the Tree which yields the Conessi Bark, of our Materia Medica; Cadoga Pala of the Hortus Malabaricus, Pala Cadija of the Hindoos.'[44] This highlights the research that Roxburgh did through the literature as well as checking the local names and uses of plants before publishing, and quoting his sources. The actual experimental process is also worth reproducing in full, for it shows the thoroughness with which he approached a problem:

The method I took to extract the Colour, was, by collecting, promiscuously, the large and small leaves, while fresh; I put them on the fire, in common unglazed earthen pots, with soft well water, and, when scalding hot, strained off the liquor, which had acquired a deep green colour, with something of the violet coloured scum, that is observed on the common Indigo, not towards the end of fermentation; with little agitation, this liquor began to granulate; to promote the granulation as well as the precipitation, I tried various liquors, *viz.* cold Infusion of Jamblong bark (*Jambolifera pedunculata* of Linnaeus)

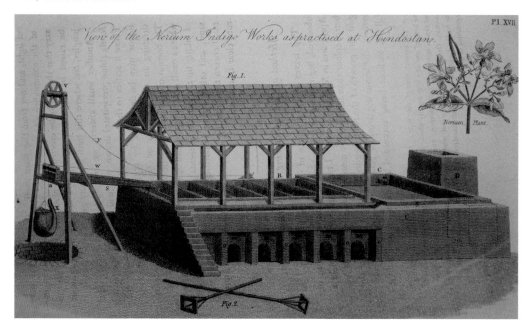

87 *Indigo works for producing dye from* Wrightia tinctoria. *Engraving by S. Porter after a drawing by J. Farey, reproduced in Roxburgh's 'Account of a new Species of Nerium, the leaves of which yield Indigo',* Transactions of the Society for the Encouragement of Arts, Manufactures, and Commerce (*vol. 28, facing page 262, plate XVII, 1810*).

which is what the Hindoos use universally on this part of the Coast, to precipitate their Indigo; I also tried Lime-water, a lixivium of wood-ashes, a mixture of Lime-water and lixivium of wood-ashes, and also a ley, made of equal parts of caustic vegetable alkali and quick-lime; these five I have repeatedly tried, and as often found that lime-water and lixivium of wood-ashes, mixed together, answered best, the faecula was washed, filtrated and dried, in the usual manner.

Roxburgh was aware that he was not a skilled dyer, stating, 'I have no doubt but a person skilled in the manufacture of Indigo, would by the common process of fermentation, &c extract a much finer colour from the leaves of this Tree.' He was also aware that there were already a number of blue dyes available, but put forward a strong economic argument for its consideration:

It may be said that we are already in possession of a sufficient number of blues, consequently, unnecessary to attend to this new Indigo; to obviate this objection, let me observe, that the common Indigo plant is only to be brought to perfection, by nice, expensive, and laborious culture, is liable to many accidents, from changes of weather, and other causes, that no human foresight can prevent: these are well known facts to every one that cultivates Indigo to any extent: only last year a considerable manufacturer, in Bengal, wrote me as follows on the 10th October: 'The wet season has hurt my Indigo very much, it has not yielded one fourth proportion of colour, so that instead of making near 300 Maund (24,000lb) as I expected, I do not get above 80 Maund (6,400lb).' This Tree is not subject to these inconveniences, besides it requires not the smallest care, is found in the greatest abundance, growing wild, in the most barren tracts that can possibly be imagined, and, from what I have seen and learned from the natives, it requires only to be cut down, once every year, to make it produce a large supply of young shoots, with very luxuriant leaves the following season; besides, the colour that this Indigo may give to cloth, &c may be different from any hitherto known, consequently, may prove of considerable benefit to a commercial nation like Great Britain.

My experiments do not yet enable me to determine, with precision, the proportion of pure colouring-matter these leaves yield, but have reason to think, it will be in about one in two hundred, that is, two hundred pounds of fresh leaves, produce one pound of Indigo, which is rather more than the leaves of the common Indigo Plant yield.

Although the samples that Roxburgh sent to Ross were not of the best quality, Ross was not to be put off:

Yet as I should think it not impossible, that with the Industry, Sagacity, attention & Scientific Knowledge, which you are in possession of, & can still more improve, in this

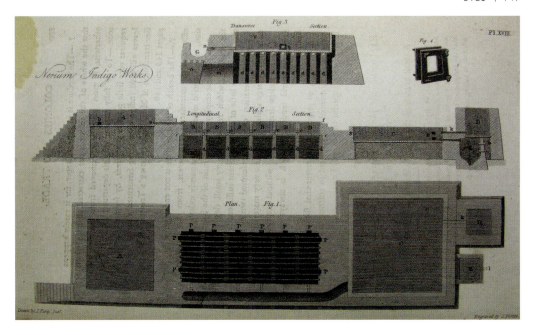

or any thing else, that you may give your mind to, it may not be impossible to over come every difficulty, & obtain success, if not immediately under your own eye, where the disadvantages may be too many, yet some other situation, & particularly (as I have heard my sensible & very intelligent friend Mr De Souza – say) in some parts of the Chicacole District, & I suppose near the Hills, where the Dews, promote the growth, & quality of the Plant, (& which you may enquire about directly, when you go to Vizagapatam, and you may also send intelligent people further on, or inland, to bring you the best information) I am not willing to give up the practicability of the article being improved, and the best (when the quality is increased by a greater demand) procured at such an abatement in the price, as may make it answer, especially as the consumption in Europe is so great, and the demand so perpetual, that it is an object of such importance; and therefore I send you a Muster, (of very good quality, tho' it may be improved) lately made by Mr De Souza's people near Tanjore; and I also send you two very valuable papers, one from the Court of Directors, lately published in the Calcutta Gazette, of which Mr De Souza, Dr Anderson and Dr Berry speak highly; and another, Instructions for the Manufacture of it, sent to me by Mr Fergusson.[45]

Ross was still very anxious to continue using this as a trading article, persuading Roxburgh to look for new plants and new situations where they might be grown. Roxburgh, in turn, had been looking and performing experiments, for in his description of *Nerium tinctorium* he mentioned that it had been found in 'the lower region of the Mountains, directly North of Coringa'. Ross continued the above letter, 'as to your experiment in this way from a new Plant, which I have seen with Dr Anderson it does not promise so well'. Roxburgh gave what might be a partial explanation for this disappointing result, in a letter to Patrick Russell:

The leaves of a nerium, which I believe is called in Your Herbarium, nerium antidysentarium, yields a very fine Indigo: this I have discovered by accident, within these few months. I have made some of the Indigo, but as the leaves of the Shrub have not been in season for extracting their best colouring matter since I made the discovery, I am therefore obliged to postpone transmitting to the directors, a description, drawings of the shrub &ca till I can make a quantity of the Indigo at the time the leaves are in season. Mr Harris in Bengal, who is one of the Greatest Indigo makers, says it is very fine Indigo, if so when the leaves are old and half dried what will it be when they are in full force.[46]

Once more, we get an example of the cautious approach with which Roxburgh attacked new problems: looking at the possibilities of the plant in all seasons, and not divulging the results until he had completed as full a range of experiments as possible, both in the field and in the laboratory.

By August, he was able to write to Russell again, stating, 'my new Indigo Tree promises exceeding well, a description & drawings & a sample of the Indigo goes by the Houghton to the Directors.'[47] Ross must have been informed of this progress, for in January, he was asking for 'some ounces of the different sorts of your new Indigo, & what it will cost when it is on board of the ship for England'.[48] The interest became evident a few months later in a letter that also showed the generosity of Roxburgh, for he sent 'Seeds of the Teak Tree, & of your new discovered Indigo Tree which you sent me' for St Helena, where he felt it would grow well and would be useful if for no other purpose than to replenish the stocks of firewood.[49] Ross continued, 'nor did I neglect to show your letter on the subject of the Ground which you wish to rent for the cultivation of Indigo, to Sir Chas Oakely, (supporting your request with the best arguments that occurred to me) and I had the satisfaction to meet with the best encouragement that I could wish.' By this time, Ross was asking for '10 or 12 Pounds of the new Indigo, ... that the trials may be made without delay, as they are very material.'[50]

Ross continued to make enquiries among the local dyers:

> Whether a much larger Plant than the Shrub which usually produces that Drug – and more like a tree in size & quality was known to them – and they not only answered in the affirmative, but in a few days brought some Branches from the Country which I shewed to Dr Anderson, Dr Berry & Mr De Souza, who all agreed that it corresponded with your description – And Mr De Souza has since learnt, that many of these trees grow in different parts of the Country particularly near to Ambore – & in the Marlbeubar & Cuddalore Countrys where a great deal of Indigo is made, & a part of it from these Trees – the Quality of which is equal to that produced by the Shrub, which last as it is produced with less expence & trouble, is, & will always be, preferred; tho' Dr Anderson made this sagacious remark, that the Tree might be so suitable to home soils to encourage its cultivation, in the idea, that the extra labour & trouble which it will of course require untill it increases in size, (perhaps in the 3rd year) may be abundantly repaid, in Seasons when a failure in rains Keeps back the growth of the Shrub – & then, the supply from the Tree will become valuable. The Box of Musters of this sort, which you sent me I put in the hands of Mr De Souza & he has repeatedly employed the Indigo makers & the Dyers to try it, the result of which is that it is found equal in quality to the other sort, & on repeated trials, of Dyeing, both Cotton & Woollen Cloth, the Colour is found to be as clean strong & durable, as what was done in the other sort.[51]

This extensive quotation is important, for it shows the way that Roxburgh worked very closely with Ross, who in turn went to Anderson, Berry and De Souza for advice. The two former were eminent botanists while the latter, being a major indigo dyer, was able to help with testing the quality of the new indigo. Having said that, Ross continued,

> it will – I should suppose be proper – that you undeceive those to whom you reported the sort extracted from the Tree, as a new discovery; which does not seem to be the less proper, that the circumstance was not known to you – that candour, is what you owe to yourself & to the public, nor can it lessen your merit to avow it.

This ethical advice at a time when such openness and honesty were not always common, is also interesting, and probably accounts, at least in part, for the high regard in which Roxburgh came to be held as a result of following it. Ross went on, however, to say that he did not consider it 'an article that will answer so well in your parts as in many other situations in this Country' which is the likely reason for Roxburgh looking for alternative crops for his tenanted land at Corcondah. Two years later, however, Ross wrote, 'as to the still more important Articles of Sugar & Indigo – which the District of Corcondah is eminently calculated to produce – they are ... doubtless the two great stapels on which this Govt (& particularly also the Sugar ...) & the Court of

	1782	1783	1784	1785	1786	1787	1788	1789	1790	1791
Denmark		112								
Russia										
Sweden										
East Country						400		336		
Germany	2330	10000	100	3039	3624	15870	1793	7962	15000	2797
Holland		117235	16130	1027	7296	5340	4300	4276	57258	8472
Flanders Austrina	78070	83955	10270	11851	3800	3000	2400	240339	19420	1752
France		91980	157627	40691	11452	17231	18764	60748	51222	104953
Portugal	27308	21411	33148	79258	123796	167662	81808	96647	50392	95428
Italy				8000	900			1153	65	3125
Straights										
Turkey										
Ireland	6373	883	2256	1556	1567	2500	300	5131	2108	4006
Islands of Guernsey, Jersey &	1120	1975	4000				40			
States of America	161216	518980	701938	678911	765241	941927	1060164	528194	626042	588805
West Indies	64309	204645	54583	301761	86845	39872	94550	35597	126220	38406
British Continental colonies	128640	90000	21150	16580	55012	22839	5640		5610	462
Africa								300		
Asia	25535	93047	237230	154291	253345	363046	622691	371469	531619	
Spain	200	50340	257947	398100	666979	300643	204461	319066	355859	287389
Total	495101	1284563	1496379	1695065	1979857	1880330	2096911	1671218	1840815	1135595
		1214563						1971218		

Indigo Exported from Great Britain

	1783	1784	1785	1786	1787	1788	1789	1790
	263979	293731	605304	542454	559933	508209	744601	861908

St Domingo production

	1783	1784	1785	1786	1787	1788	1789
	1868720	1555142	1546575	1103907	1166177	930016	958626

Table 15 The Imports of Indigo into Britain from 1782 to 1791, together with the re-export of Indigo from Britain, and the production of Indigo in St Domingo.

Directors will put a dependence to be reared by you.'[52] Earlier that summer, Roxburgh had written that these two crops 'promise the most essential benefit, not only to great Britain, but to her territorial acquisitions in India',[53] having started the letter with one of the few philosophical comments in any of his letters: 'By an extensive well regulated commerce between India & Great Britain the nation derive the greatest advantages, for as old commodities become less wanted new ones (at least new from India) start up which probably do more to keep up these advantages, by replacing such as have fallen into decay.' There was a reason behind this statement, for the letter continues with a plea for the tenancy of Corcondah, on which to grow these crops.

Although interested in the idea of developing an indigo farm, Roxburgh was still searching for new plants, and in January 1793 sent two new species of indigo to England, followed by a letter to Dr James E. Smith in which he stated,

I enclose you a Specimen of an Indigofera the leaves of which yield the most beautiful Indigo I ever saw, & in a very large proportion. I have sent Home to the directors a Drawing & Description under the name of Indigofera caerulea. I wish you would let me know whether you think it has been before described or not. I do not imagine that König ever saw it, as it does not grow to the Southwards, where he chiefly was.[54]

This search for better forms of indigo was not limited to India, for, in July, Ross wrote: 'I am well pleased at what you write the improvement of Indigo from the Guatemala Seed. I can have no difficulty probably next year to get abundance of it from Manilla where I have particular friends.'[55] This shows that the Spanish colonies were just as keen to spread their successful plants within their territories, and to keep out competition.

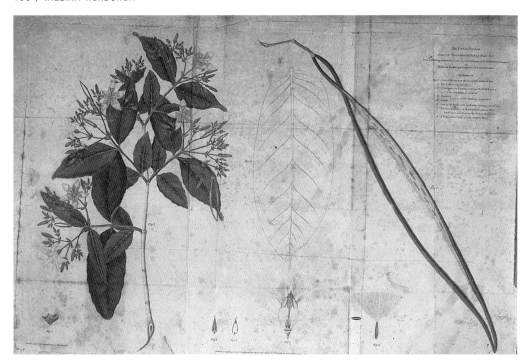

89 Wrightia tinctoria (Nerium tinctorium). Engraving by J. Walker after a drawing by one of Roxburgh's Indian artists, originally drawn c.1790 (Roxburgh drawing 18), reproduced in Dalrymple's Oriental Repertory (page 39, vol. 1, 1793).

The optimism that Roxburgh had for the work that he did with indigo is reflected in a letter to Banks that December: 'The Indigofera, which I wrote to you once or twice about some time ago, proving to become valuable. Some of the Bengal manufacturers think it will soon be the only sort cultivated. Hitherto it is preserved for seeds only.'[56] Research on the plants continued, for he wrote again some sixteen months later, 'in looking over the originals, when my writer was copying the 17 [*Nerium tinctorium*], I blushed at the numerous errors, & have the utmost reliance on your kind attention during the publication.'[57]

Roxburgh, as has been frequently stressed above, was aware of the need for commodities to be profitable if they were to be taken on by the Company, and the depth of this interest is shown in a number of his reports that are in his manuscript collection. Not only was the price and care during transport important, but so was the need for the chests for packing the indigo to be of a consistent size:

> If the different proprietors of the Indigo were to have their Chests carefully made to run of one weight, it would then only be necessary to turn out a few of each Persons, to ascertain a tare which might be allowed as an average for the whole of each persons. That not being the case, each Chest is now obliged to be turned out to have the tare of it taken, by which there must, notwithstanding the greatest care, be some waste, & some of the Indigo must get broke, which is an injury to the sale of it …. As loss of space is occasioned in loading the ships, by the Indigo Chests being corded, the Board of Trade request the indigo contractors, & others who may send indigo in the Companys ships, not to cord their Chests; but, should they apprehend them to be otherwise not sufficiently strong, to secure them with Iron Clamps.[58]

There was also a general awareness of and interest in the economics of the product. Among the manuscript collection is a copy of an article from the *Calcutta Gazette* of December 1792, detailing British imports of indigo. The reproduction of a letter from the Court of Directors starts with 'the articles as to quality is still increasing in reputation: it has already surpassed the American & French; & there is no doubt but by perseverance and attention on the part of the Planters, it will effectively rival the Spanish'.[59] This was followed by a table of indigo imports and exports into and out of Britain from a number of countries (see Table 15). To increase interest in improving the quality of indigo, the articles then went on to give examples of what happened to those planters who could produce the best: 'A parcel of five Chests per *Prince Henry*, belonging to Messrs Gilchrist & Charters, was declared to be superior to the Spanish, & sold at a higher rate.' The reduction of the St Domingo production, from whom the French imported most of theirs, was put down to the fact the planters on that island had gone from indigo into coffee: 'By an account published in France in 1770, it was said that St Domingo at that time yielded of Coffee lb 5,000,000 & of Indigo lb 2,000,000. In 1789 the produce of Coffee had increased to upwards of 176 millions; while that of Indigo had decreased to under one million.'

The success of the indigo industry in India was such that it did become one of the world's major suppliers, and the Company, having suffered a number of losses, could withdraw the subsidies and grants to indigo planters in 1802 without destroying the industry.

YELLOW DYES

The dyeing of Indian chintzes was an important use of plant dyes, and Roxburgh appears to have been aware of this from one of his earliest extant letters to Sir Joseph Banks, sending him some caducay galls. These came from the tree he knew as *Mimosa arabica* (now *Acacia nilotica* ssp. *indica*) and from which a yellow dye was made.[60] Although the subject is mentioned occasionally over the years, the only major piece of writing on it is in a letter to Dr Taylor, the Secretary to the Society for the Encouragement of Arts, Manufactures and Commerce.[61] In this Roxburgh gives an interesting insight into the growth strategy of plants: 'the legumes were gathered when near full grown; at which period I am inclined to think they possess most tan and astringent matter, without being so much encumbered with the seed as when full grown ripe; when they would add greatly to the bulk of the commodity, and weaken the powers for which I hope this substance will be found useful.' This appears to imply that the plant contains an optimum amount of the 'tan and astringent matter' before the seeds are ripe when the extra nutrients would have gone into the seeds.

The tree is widespread in India and grew commonly on the Coast, but the galls from which the dye was extracted occurred only on plants growing further inland, where Roxburgh had 'never ventured so far in amongst the mountains as where the galls are found'. He had gleaned a good understanding of the parasitic life cycle of the insect that causes the galls, stating that 'it seems that an insect punctures and deposits its eggs in the young tender leaves of the tree, which causes them to swell into the various forms the galls assume'. Later, he explained that 'upon the leaves of this tree I have found an insect, which I take to be the larva of a coccus, or chermes'. After describing them, he continued, 'the whole insect is replete with a bright yellow juice, which stains paper a very deep and rich yellow colour. Could these insects be collected in any quantity, I am inclined to think they might prove as valuable a yellow dye as the cochineal is a red.'

Having said that, he also stated of the galls, 'they are sold in every market, being one of the most useful dyeing drugs the natives know. Their best and most durable yellow is dyed with them, and fixed with alum. With ferruginous mud they are used to dye black. They are also the chintz painters best yellow.' At the same time as this letter was written, Roxburgh sent a sample of the galls as well as a number of other samples of useful materials. These were shipped on the *Hope*, but 'from subsequent information which the Society have received, the ship Hope, containing the legumes of the Mimosa Arabica, was taken by the French, and no other specimens of that article have yet been sent.'

90 Acacia nilotica *subsp.* indica (Mimosa arabica). *Hand-coloured engraving from a drawing by one of Roxburgh's Indian artists, originally drawn c.1794 (Roxburgh drawing 489), reproduced in Roxburgh's* Plants of the Coast of Coromandel *(vol. 2, plate 149, 1805).*

SUMMARY

With the development of aniline dyes in the 19th century, Roxburgh's work on natural dyes was superseded by the introduction of artificial dyes. However, his work on cochineal did help to identify and separate the various species of *Opuntia* involved as host plants for the different insects that produced the various lac colours. Equally, his work on indigo viewed this industry from the point of view of what was best for the Indian labour force at the same time as trying to make the collection and transport of the material more efficient for the indigo planters.

Roxburgh's work on cochineal also emphasises the awareness that Roxburgh had of his own shortcomings, using the expertise of men such as Dinwiddie to perform those experiments that he felt unqualified to carry out. The short-term effect of this work was successful, in that both cochineal and indigo became major contributors to the East India Company's sales. The effect of other dyes is more difficult to determine without looking at more possible sources.

CHAPTER 10

Hemp and Fibrous Plants

INTRODUCTION

The interest in hemp (*Cannabis sativa*) started when the Secretary to the Court of Directors (Mr Martin) wrote to Banks in April 1785 canvassing the idea that he might hold experiments on Chinese hemp on behalf of the East India Company at Kew.[1] As will be discussed below, although the Company was interested for experimental work to be carried out for potentially economic plants, they did not always use their existing botanical garden. It is interesting that Roxburgh must have been thinking about Indian fibres at the same time, for in his letter of 18 December 1784, he had sent some seeds of flax (*Linum usitatissimum*) although he did not know what the natives used it for, other than bow-strings: he is unlikely to have heard of Martin's request for he thanked Banks for his letter of November 1783. There were a few further references to jute (*Corchorus capsularis* and *C. olitorius*) and other fibres before the interest in hemp developed. This does introduce a problem that has not been fully addressed: the exact plants which were referred to. There is first the fact that hemp is variously referred to as Chinese, English, Russian and Indian, although all were thought probably the same plant, a fact that Roxburgh himself mentioned. When we come to jute (*Corchorus capsularis*), the problem is more difficult, for this and Bengalese sun (*Crotalaria juncea*) appear to be used by some contemporary authors synonymously.

Finally, there is Roxburgh's work on hemp (*Cannabis sativa*) and other fibrous plants. The need that drove this research was based on the cordage used by ships and that, with the Napoleonic War with France, the supply from Russia had dried up: Roxburgh's awareness is shown in a letter to Robert Wissett, Secretary to the Society for the Encouragement of Arts, 'The discovery of a substitute for Russian hemp is certainly an object of first magnitude'.[2] He performed comparative experiments over a number of years on 26 different species, of which those which have since been used for fibre include coir (*Cocos nucifera*), hemp (*Cannabis sativa*), sun hemp (*Crotalaria juncea*) and flax (*Linum usitatissimum*). The care that he took with the experimental design is shown by comments such as the results 'cannot be deemed any thing more than a first essay', because of differences in the lengths and degree of twisting, and that 'from 100 to 200 additional experiments' were performed for 'every thing depends on their [the samples of fibre] being exactly of the same size and degree of twist'.[3] Also the importance given to this work by the East India Company is shown by the fact they were prepared to pay for Thomas Douglas to run Roxburgh's experiments at Reshira.

The results were published through the Society for the Encouragement of Arts, which gives an insight into the changing allegiance that Roxburgh had with the placing of his publications, starting with the Royal Society, under the patronage of Pringle, moving to the Asiatic Society as well as using the Linnean Society through the influence of his friend Kindersley's uncle, Dr James Edward Smith, and moving to the Society for the Encouragement of Arts when his publications were more on the uses of plants rather than their scientific description. Smith certainly had a high regard for him, for he wrote in 1814. 'I fear poor Roxburgh is dying at Chelsea. He is one of the "salt of the earth", truly amiable, good, mild, & liberal, – god bless him & <u>all</u> such.'[4]

HEMP

A study of hemp poses a number of problems, ranging over the whole field of this study. First there is the dilemma for a Scots scientist between the scholarly solution to the botanical description of a substance which was avowedly used as a narcotic drug, against the practical and religious pressures of possibly making this knowledge of such a plant more widely known. Secondly there was the purely botanical problem of what plant was really meant by hemp, for both English hemp and that from the East were, in fact, a number of different species. There was, therefore, the purely scholarly and practical problem of identification. Thirdly, there were the problems attached to the cultivation of the crop on an economical scale, producing problems of horticulture and economics, and possible conflicts between the scientific study of the plant, the moral problems related to its narcotic addiction, and the mercantile requirement to find suitable alternatives to Russian hemp at a time when shipping was being destroyed in the war with France. The fact that Roxburgh experimented with so many different plants reflects the sheer size of the problem.

The need for Britain to find a new supply of hemp resulted from the war with France, as hemp was used in the cordage on ships, and the main supply came from Russia. Thus, Roxburgh wrote to Robert Wissett, Secretary to the Royal Society for the Encouragement of Arts, 'The discovery of a substitute for Russian hemp is certainly an object of the first magnitude'.[5] In a letter published at the same time, he also stressed the patriotic duty of looking for such a substitute. Roxburgh had started researching alternatives at a much earlier stage, for writing in 1799, he stated, '... the cultivation of hemp, a thing I had begun some time before. Even on the coast of Coromandel, ten or twelve years ago, I made a most successful trial.'[6] The Royal Society of Arts did what it could to encourage the production of several fibres, primarily for its use in ships and therefore national defence, but also to reduce the drain on foreign currency in time of war. The Napoleonic War created an urgent demand for hemp replacements or substitutes, and the Society encouraged work on it both in Canada and India.[7]

Roxburgh's concern for finding a suitable alternative which would be both practicable in terms of strength and durability as well as economically feasible, is shown first in a letter to Banks. He reported that although hemp was common throughout India, it was 'only used, or rather abused medicinally'.[8] He then went on to describe an alternative, the Bengalese sun, *Crotalaria juncea*, which was 'fully as strong as hemp'. However, it would 'not bear tar' so the London rope-makers would not consider it for naval purposes. The Indian fishermen used it 'universally' for their nets, lines and fishing tackle, 'but they never use it without being first tarred with some vegetable astringent'. At the same time, he was looking at other alternatives, for in a Public Letter from the Court of Directors to Fort William there is the statement about 'the article of jute'.[9] Equally, when he was at the Cape of Good Hope in 1798, where he had gone to recover his health, he wrote to Banks, 'I darsay [sic] Flax will also grow exceedingly well here' but he was worried as to whether the Cape would long remain in British hands.[10]

91 *The Indian sun hemp,* Crotalaria juncea, *one of the fibrous plants that Roxburgh investigated. Hand-coloured engraving from a drawing by one of Roxburgh's Indian artists, originally drawn c.1792 (Roxburgh drawing 361), reproduced in Roxburgh's* Plants of the Coast of Coromandel *(vol. 2, plate 193, 1805).*

Number of voyages by ships of increasing age

Year	1	2	3	4	5	6	7	8	9	10	11	12	13	Total
1789/1790	4	16	3	5	3								1	32
1790/1791	5	5	5	9	0	1								25
1791/1792	0	2	8	7	7	4								28
1792/1793	4	5	15	4	14	0	1						1	44
1793/1794	0	5	8	13	11	7	2							46
1794/1795	3	3	1	13	2	12								34
1795/1796	13	0	5	8	11	6	1	2						46
1796/1797	11	2	0	1	13	6	0	1						34
1797/1798	5	3	5	2	2	5	4							26
1798/1799	5	11	1	4	6	6	2	0	1					36
1799/1800	16	6	4	4	1	2	2	0	1					36
1800/1801	10	10	8	1	4	15	2							50
1801/1802	16	11	6	3	2	0	0	1	0	1				40
1802/1803	21	12	9	7	3	2								54
1803/1804	5	19	10	8	2	4	4	0	1	0	1			54
1804/1805	12	13	10	8	4	2	1	0	0	0	0	1		51

Table 16 The number of ships that went to India each season, by the number of voyages taken by each ship. Before 1803, ships were supposed to have a maximum of six voyages which was increased to eight in 1803 and indefinitely in 1810. The final column was for the same ship that was on its eighth voyage.

92 *Partial inflorescence of the coconut palm,* Cocos nucifera. *Roxburgh also investigated coir, the fibre produced by this plant, to test its suitability as an alternative to hemp. Hand-coloured engraving from a drawing by one of Roxburgh's Indian artists, originally drawn* c.1794 *(Roxburgh drawing 448), reproduced in Roxburgh's* Plants of the Coast of Coromandel *(vol. 1, plate 73, 1796).*

EXPERIMENTAL DESIGN

This search for alternatives led him to do detailed experiments on 18 different fibrous plants, the results of which he published in 1804 with additional results two years later (see Table 17).[11] These articles show the rigour of Roxburgh's experiments and also his awareness of the weakness of some of his results. All the different fibres were treated in as similar a way as possible, with any deviations being given in some detail; thus the time when the fibres were macerated was described as well as the reasons for doing it at that time: in the hot rainy season it only took one day compared with eight in the cold of December to February, for hemp. Equally, so that people in Britain would understand what he had done, he 'prepared their fibres ... as with hemp and flax in Europe'.[12] All the samples of cord were made to the same length, four feet, and were twisted 'as nearly of the same size and hardness as a Hindoo rope-maker could make them', but was aware that these men twisted the fibres 'always too hard to be of the greatest strength'. The reasoning behind this was based on research done previously, which was referred to in a footnote, again supporting his academic rigour in the pursuit of his results. To make sure that his results were statistically acceptable, he performed six sets of experiments, 'when there was a sufficient quantity of the fibres' giving the results as an average of 'several trials'. If this was not the case, then he left a blank in the results.

When considering the weakness of the experimental design, he stressed that the results 'cannot be deemed any thing more than a first essay', because of differences in the lengths and degree of twisting. The experiments would be repeated, 'for every thing depends on their being exactly of the same size and degree of twist'. These additional experiments were published two years later, by which time Roxburgh had performed 'from 100 to 200 additional experiments'.[13] As some of the ropes had been immersed in 'stagnant fresh water in a rather putrid state' for 116 days, they 'must be as severe a trial for vegetable fibres, as can be well found in any country', it gives some idea of the time and thought that he took over these experiments. Further controls had been added into these experiments, for instance, as 'the cords now employed were made of three single yarns'. This time the results of 16 different species were tabulated, although he did give results of the same 18 as well. The latter were similar to those of 1804, and detailed in Table 17 are those species which appeared in the two sets of results.

Roxburgh's comments about jute are interesting, for he considered it as a substitute for flax, reinforcing its usefulness, 'on account of the length, strength, and fineness of the fibre, and from the durability and strength of it, after 116 days maceration'. In his *Flora Indica*, Roxburgh states that it is widely grown in Bengal and China for its fibres, from which rice bags were made.[14] This is in contrast to the much fuller emphasis he gave to Indian hemp, for which he described it, its preparation and uses in some length, ending with the comment that it was 'not inferior to the best Russian hemp'.[15]

The East India Company took considerable interest in the growth of hemp, arranging for Banks to send quantities of seed to Roxburgh, although with disappointing results over a number of years. The two quarters of seed that Banks sent in 1798 did not grow, and Mr Sinclair, whom the Company sent out at the same time to run a separate experiment, died and one of Roxburgh's gardeners took over the management. The different wording used by Roxburgh to Banks and to Wissett on this must throw some light onto the relevant relationships. To Banks, Roxburgh wrote, 'but the arrival of a Mr Sinclair from the Court of Directors for the express purpose induced them, here, to give him a separate establishment in a distant part. Poor Mr S- died soon after.'[16] While to Wissett, he is far more outspoken, 'I was rather surprised, on my return to Bengal, to find the Directors had sent out a person (Mr Sinclair) to establish the cultivation of hemp, a thing I had begun some time before …. Eighty pounds weight is all, I believe, that is yet forthcoming, and costs from 10,000 to 20,000 rupees.'[17]

The timing of these two letters possibly throws some light on Roxburgh's reaction, for in December 1800 Banks read to the Council of the Court of Directors an 'Extract from Dr Roxburgh's Observations on the Sun Plant of Bengal, called in the Circars Ishanamoo'.[18] This was followed by a statement from the British Consul at St Petersburg, 'Mr Shairp [*sic*]', who described the Russian hemp trade, and ended by stating that, had there been no ban on the importation of hemp, the trade would have been much as before. He was followed by Arthur Young, whose stress was on the possibility of growing it in England, but at a price that was double that for Russian hemp.

In spite of possible antipathy from the British merchants and rope-makers, Roxburgh continued to press for the trial of Indian hemp, shown in a letter to Mr Lambert in March 1801: 'Your Russian Merchants & probably Ropemakers may throw obstacles in the way of any other hemp than their own being brought to market, but every patriot will exert himself to counteract them. Should soon [Indian hemp, sun] be really found as good as Russian hemp, and I am strongly inclined to think it is, all circumstances considered.'[19] His aim to get it accepted led him to a quandary: to send the first

93 Sesbania cannabina (Aeschynomene cannabina). *Drawing by one of Roxburgh's Calcutta artists, c.1795 (Roxburgh drawing 904). This and Figs 94 and 95 were sent by Roxburgh to East India House London to accompany the results of his work investigating the fibres of these plants.*

94 *The tossa jute plant,* Corchorus olitorius. *Drawing by one of Roxburgh's Calcutta artists, c.1794 (Roxburgh drawing 901). One of the species from which jute is obtained, this one tending to be grown in more upland regions than* C. capsularis.

95 *The jute plant, ghee-nalta-paat (Corchorus capsularis). Drawing by one of Roxburgh's Calcutta artists, c.1794 (Roxburgh drawing 902). This lowland species is the main source of jute.*

samples or wait until the new season brought a better quality. This was highlighted in a letter to Banks in September 1801: 'Every exertion has been made on my part, & am still continuing them, to improve our Indian hemp as it is commonly called …. much goes home, and I fear in general of a bad quality till the new crop comes in. This I beg of you to keep in your mind, that you may not be too hasty in condemning its general quality.'[20]

ROXBURGH'S ACHIEVEMENTS

The success of this pressure led to Banks arranging for six gardeners (Edward Porter, William Holgate, Simon Benstead, William Reynolds, Mark Everson and Joseph Seymour), all from Lincolnshire, to be indentured for the cultivation of hemp, for a period of three years. Banks was to pay for their passage and a weekly wage of 30 shillings.[21] This may not have been an entirely successful venture, for at the end of the following year Roxburgh wrote to Banks, 'Not a single grain of your hemp seed grew, & your men are perfectly idle.'[22] However, the gardeners gave a rather different view, suggesting that Roxburgh had not given them the assistance that they had expected. They continued to complain: 'we think our case a very hard one being sent from our Native Country in hopes by our industry to gain a competency in a few years, instead of which we are lingering away our valuable time upon a bare allowance of fifty Rupees per Month which is not sufficient to support us.' They continued complaining of ill health, which had killed William Reynolds, and then stated that 'since our first arrival in Calcutta, we to our very great loss have been totally out of employ,' before finishing

No:2238 Bauhinia scandens. Willd.

96 Bauhinia scandens.
*Drawing by one of Roxburgh's
Calcutta artists, c.1813
(Roxburgh drawing 2238).
This is another of the fibrous
plants tested by Roxburgh
as an alternative to hemp.
Reproduced from Wight,*
Icones Plantarum, *plate 264.*

off with the statement that Calcutta was not a suitable place for hemp to grow, but they would be better off experimenting at Benares or Patna.[23]

    What finally happened to these men from Lincolnshire, I have not yet been able to ascertain, but Roxburgh continued to experiment and correspond about it. Shortly before he finally left India, he received a letter from William Moorcroft, describing the work that he had been doing in the areas around the 'southern front of Boothunt [Bhutan]', where it appears that the plants must have been unusually large, 'I fear a strict account of the size of the Hemp-plant ... might cause my veracity to be called in question', and continued by describing how the natives prepare it for use.[24] Roxburgh had also been corresponding for some time with Joseph Cotton, a Director of the

97 Pyrrosia angustata
(Polypodium coriaceum).
*Drawing by one of Roxburgh's
Calcutta artists, c.1808
(Roxburgh drawing 1744).*

Company, enough 'to compose a large volume', who was awarded the Royal Society for the Encouragement of Arts Silver Medal for his work on hemp, for which his collection of letters published in the *Transactions* gives due reference to Roxburgh.[25] (I have not yet traced the rest of this correspondence.) This was for the introduction of calooee hemp (*Urtica tenacissima*) and Roxburgh's description of it published in 1815 is probably the earliest account of it to be found in western literature.[26]

By the time that sufficient research had been done to show that India could produce hemp and other alternatives for it, such as jute, the war with France was over and the immediate need for its cheap supply from British controlled sources had disappeared, realised at the time by Cotton, who stated that 'the return of peace has for the present affected, not only the importation to any extent, but the culture of it'. It was not until the middle of the nineteenth century that the importation of hemp from India almost matched that from Russia: in 1851, 590,923 cwt came from India against 672,342 cwt from Russia.[27] However, the work done on hemp and alternative fibrous plants does show the central part played by Roxburgh as well as his determination to get results which would stand up to rigorous inspection. Having set up a detailed experimental design which was improved over time, he was aware of weaknesses and was honest about exposing them. By looking at a range of hemp substitutes, he was able to produce a number of suggestions, some of which like

Species	Common name	1804†	1806	Flora Indica
Agave Americana		11		No uses
Aletris nervosus*	Murga, Bow-string hemp	12		II, 161-64
Asclepias tenacissima			2,3	
Bauhinia scandens		19,20		No uses
Cannabis sativa	Hemp, Bang	1,2	1	III, 772-73
Cocos nucifera	Coir	3		III, 614-15
Corchorus capsularis	Ghee-nalta-paat, Jute	9	11,12	II, 581-82
Corchorus olitorius	Jute	8	13	II, 581
Crotalaria juncea	Sun	7	4 to 10	III, 259-63
Hibiscus abelmoschus	Kalee-kustooree		23	
Hibiscus bifurcatus			25	
Hibiscus cannabinus	Meesta-paat		19,20	
Hibiscus esculentus	D'heroos		24	
Hibiscus mutabilis		17		No uses
Hibiscus nova specium		18	21	
Hibiscus pilosus			26	
Hibiscus sabdariffa			22	
Hibiscus strictus		16	17,18	No uses
Hibiscus tiliaceus	Bola	15		No uses
Linum usitatissimum	Flax	10		
Musa superba			27	
Robinia Cannabina	Dansha	5,6	14	
Saguerus Rumphii	Ejoo	4		III, 626-29
Sterculia villosa		21		No uses
Theobroma augusta	Woollet-comal	13	15,16	III, 156
Theobroma guazuma	Bastard cedar	14	27	

* This species was also referred to by Roxburgh as *Sansevieria Zeylanica*.
† The figures in these two columns refer to the number of the species in the Tables in the original publications.

Table 17 Table showing the species of fibrous plants on which Roxburgh performed comparative tests on their strengths, as published in 1804 and 1806. The column headed *Flora Indica* indicates those pages where the plant has been described if it has uses included; a gap indicates that it was not described. The names used are those used by Roxburgh and published in the *Transactions of the Society for the Encouragement of Arts, Manufactures, and Commerce.*

98 Abroma augusta (Theobroma augusta). *Drawing by one of Roxburgh's Indian artists, c.1793 (Roxburgh drawing 415). One of the fibrous plants that Roxburgh tested as an alternative to hemp.*

Indian hemp, jute and coir, would be developed by later generations. Equally, he was prepared to take the long view in terms of his own research, an approach helped by the time it took to get responses from London and his other correspondents. The full implications of Roxburgh's work on jute is another area still to be researched.

CONCLUSIONS

These experiments show Roxburgh to be a meticulous and patient scientist who took all the precautions that he could think of to ensure the validity of his work. Equally he appears diffident, in that he did not want to publish until certain of his results and at the same time cautious, not to upset those in positions of power, as shown through his work on heat absorption by different coloured cloth, which was never published.

In terms of the introduction of these crops, events overtook Roxburgh, in that with the end of the war in 1815, the market for Russian hemp opened up again and the immediate need for alternatives disappeared. However, looking further ahead, the work he did laid the foundations for the jute industry that was to be such an important part of India's exports and the economy of at least the city of Dundee for decades.

CHAPTER 11

Pepper and Spices

INTRODUCTION

Over the years that Roxburgh was in India, there is mention of a number of spices, from pepper in his early days, to pimento and cinnamon as well as many references to his work in connection with nutmeg and cloves. From the very early days of British interest in India, people had been trying to introduce various spices into India, and with Colonel Robert Kyd's initiation of a Botanic Garden in Calcutta this was tried again, with requests in March 1787 from the Bengal Government to the Moluccas for cloves and nutmegs.[1] This attempt does not appear to have been successful, for Roxburgh appears to have started trying to import these plants in earnest as soon as he arrived at Calcutta. The reason for this effort was twofold. First the Dutch East Company exerted an effective monopoly over this trade from the Moluccas. Secondly there was enormous profit to be gained from the import into Europe of these spices, seen from the fact that while the total invoice value of spices imported by the Dutch East India Company in the three years 1778-80 was 3.1 per cent of all the goods they imported, this produced 24.43 per cent of the sale proceeds in Amsterdam.[2] It was not surprising, therefore, that the British East India Company was prepared to spend both time and money to try to break into this very lucrative market.

Before looking in detail at Roxburgh's work on pepper, and nutmeg and cloves, it is worth spending a little time on two other spices that occupied his attention: cinnamon and pimento. Cinnamon (*Cinnanomum* species) grows naturally in southern India, and was recorded by Strabo, the Greek historian. By July 1791 Roxburgh was writing, 'I expect daily 30 additional Cinnamon Plants, which with those I now have, will form a very good beginning for the Culture of this species, they are the best sort from Ceylon.' These had arrived by the end of the following month, giving him in all 54, last years are thriving well and seem to glory in having obtained their liberty from the Belgic yoke'.[3] It is not until Roxburgh was at Calcutta that more information comes, and two years later, after he must have started a new set of experiments, he wrote that the trees were still too young and small to be cut. In view of the lack of knowledge in Calcutta of these experiments, he continued,

> but as much must depend on the season the Bark is taken off at as well as the method or preparing it. I beg leave to Suggest to the Hon'ble the Governor General in Council, the propriety of having a person from Ceylon, who may be practically acquainted with the cultivation and mode of preparation. The footing we now have on that Island, may enable the Commanding Officer there to send such a man, or at least to procure the fullest directions on that head. At the same time, a number of Plants of the best sorts, or of Cinnamon may be procured from thence, & distributed over various parts of the Companies Territories for as there are several varieties of this Tree not less than nine are mentioned by Burman in his Thesaurus Zeylanica, it is not, on that Account, clear that we are in possession of the best.[4]

He then went on to say:

> Permit me further to observe that this tree thrives infinitely better in the Northern
> Circars than it does here. I resided there for many years and reared quickly to a fruit
> bearing state a considerable number in the course of a few years, some of which have
> already yielded two crops of Seeds, which has produced many thriving plants. There they
> are not near so subject to perish while young as they are here, but whether these are the
> best sort or not I am equally at a loss to determine, having never been on Ceylon nor have
> I ever met with any description of this tree sufficiently accurate to enable me to judge.

This ties in well with the fact that it is a plant of the more equatorial regions than as
far north as Calcutta. The success of this planting can be gleaned from the fact that
a large consignment was sent to Mr McRae in Chittagong in February 1799, by Dr
Fleming when Roxburgh was at the Cape. McRae was also working on nutmeg, and two
years later Roxburgh was able to say that McRae 'has made progress in cultivating this
valuable tree [nutmeg], as well as the Cinnamon'.[5] This movement of cinnamon from
Calcutta was to other places further east, for young William Roxburgh asked, in July
1805, for a bill of freight to be paid for two boxes of cinnamon sent to Fort Marlbro'.

Pimento or all spice (*Pimenta officinalis*) is a native of South America, and Roxburgh
first appears to ask Sir Joseph Banks to ask the Court of Directors of the Company to
procure him some trees around 1791, when he was still at Samulcottah. This interest
was emphasised in one of his early letters from Calcutta, when he wrote to Banks,
'The Pimento I consider as the first object of our care'.[6] The following week he wrote
to Edward Hay that 'this valuable Arramatic appears to me to claim the first attention
on the Coast.' This was followed the next year with a request to Banks for 'as many
more of Myrtus Pimenta (Jamaica all-spice) as can be sent, some hundreds if possible'.[7]
These requests do not seem to have been successful, but the experiment was, for at
the end of February 1796 he wrote

> The Pominto, or Jamaica Pepper plants succeed in this climate uncommonly well, every
> plant we have yet received or been able to rear from layers (none of them being yet of
> age to yield seed) are in the most healthy state, so there is scarce a doubt of that valuable
> spice succeeding perfectly well, in this climate, its agreeable aroma resembling that of
> a mixture of cinnamon cloves & nutmeg is highly grateful to the Mahomedans who are
> great consumers of this species, I should therefore imagine that it will be much sought
> after whenever it can be supplied fresh from plants reard in India, I could therefore
> wish that it may be again recommended to the Hon'ble Court of Directors to send out
> annually a supply of the plants from England or the West Indies, I took the liberty of
> addressing the late Secretary Hay on this, & other similar subjects, about two years ago
> but as I do not find any of the Plants mentioned therein have been sent in consequence
> of that application, I beg leave to trouble you with a copy of the part of that latter
> which mentions the plants & seeds applied for.

By December 1799 the trees were producing fertile seed that Roxburgh was using
to propagate further plants, 'so that we may now consider [it] established in India'.
This is confirmed in a letter of February 1802,

> since my last report dated 16th January 1801 about 600 additional Plants have been
> reared from seeds, in all above 1200 from the single tree, which I brought with me
> from the Coromandel Coast in 1794 ... some hundreds of the plant have already been
> distributed over the Country the rest are planted in the Botanic Garden, or remain
> there for distribution.[8]

Although Banks continued to send him plants, 15 in very high order arriving in April
1802, this was now well-established, so that in June 1804 Roxburgh was able to write:
'Cinnamon, Pimento, Arabian Coffee, Teak, China Sugar, Saguerus, a species of the
Sagoo Palm which also produces the Ejoo for Cables, and Cordage, Mahogany, Fruit
Trees, &ca from most parts of the world, continue to be successfully reared in this
Garden, and distributed freely to all those who apply for them.'[9]

PEPPER

The depth of Roxburgh's botanical experiments can be seen from looking at two of the many plants with which he was involved: pepper and hemp. Need for the first derived from the monopoly which the Dutch East India Company exercised on the export of all spices from their East Indian factories, and the demand for hemp built up with the removal of the easy supply of Russian hemp for the navy with the wars with the French, already discussed. In both cases there are some problems of scientific nomenclature: Roxburgh described two peppers, *Piper nigrum* and *P. trioicum* in his *Flora Indica* (vol. 1, pp. 151-4 for *Piper trioicum* and p. 150 for *P. nigrum*) which are now deemed to be the same, under the former name, a practice I have used.

The first mention of pepper by Roxburgh was in a letter to Banks in 1784: 'I have long wished to be in the same degree a means of introducing the cultures of common black pepper in this part of the Companies Territories, called the Circars; & with that view I have made particular enquiry regarding the soil Etc where that Plant thrives best and am induced to believe that it would grow fully as well here as in any part of India.'[10] By April 1786, he had found black pepper growing wild in the hills to the north-west of Coringa, confirming his opinion that the climate was right, and wrote to Andrew Ross, his agent in Madras, 'I must now request of you to take a favourable opportunity of mentioning the whole Plan to our new Governor'.[11] Ten days later, he wrote to say that he was getting peppers in the bazaars and at his own expense would get about 1,000 plants: however if the government wanted it introduced on a commercial scale, he would need about 5,000 which would be able to be doubled by

99 Roxburgh's problem over nomenclature can be seen from these two paintings. On the left is one labelled Piper nigrum *and on the right one called* Piper trioicum, *yet they appear to be copies of the same plant.*

careful nurturing. He also acquired a copy of Marsden's *History of Sumatra* which contained a description of how the plants were grown in that part of the world. A month after the first of this group of letters, Ross was writing to the new Governor, Sir Archibald Campbell, that pepper that Roxburgh had supplied was deemed by the native vendors to be 'at least 10 per cent better in quality'. So, by mid-1786, Roxburgh had checked out the horticultural requirements of growing pepper, the quality of the product and its availability. This information, together with samples, reached the Court of Directors in June 1787, who, after checking with merchants in London, approved the measures taken but delayed any decision until they had more information: 'we trust to receive further particulars respecting the undertaking, as may lead us to form some judgement how far it may be a measure meriting our Attention, which will entirely depend upon the rate at which it can be provided.'[12]

In June 1788, Ross wrote to Roxburgh to say that Alexander Dalrymple was interested to hear about the pepper project, and Ross suggested that Roxburgh should write an article for Dalrymple and send him a sample.[13] By December the article was written and on its way to Dalrymple who was asked to use his influence with the Court of Directors to give suitable encouragement.[14] This seemed to work, for in April 1789 the Court wrote in the Madras Despatches, 'we now order & direct, that you afford Dr Roxburgh every encouragement & assistance in your power in the cultivation of such useful Articles of Commerce, particularly that of Pepper.'[15] By November, 1789, he had nearly 50,000 pepper plants in the Zemindaries of Pittapore and Peddapore (see Table 18),[16] but the Council at Masulipatam were still not being very helpful.[17] The heavy hand of bureaucracy was evident again in December, when Andrew Scott at Masulipatam wrote:

100 Abrus precatorius. *Drawing by one of Roxburgh's Calcutta artists, c.1797 (Roxburgh drawing 1157).*

> I am directed by the Chief & Council to acknowledge the receipt of your letter of the 2nd Ultimo & 13th Instant a copy of the former of which has been referred to the Superior Government, conformably to Your request, as you mentioned that some Bunches of Pepper were already well formed, and as its quality may of course soon be ascertained, Government have Resolved to Postpone the extending of the present Plantations until this can be done, and the price at which Pepper can be raised be known, for which purpose they have directed that the value of the Ground now occupied be minutely enquired into not only with a view of guarding against any unreasonable claim of the Zemindars but also to ascertain the Prospect of success in the culture of this article and the Chief and Council will accordingly give the necessary directions for that purpose.[18]

However, Scott did send 500 Madras pagodas to help cover the costs, which by this time had reached 900 M.P.[19]

Later that year, he found a major problem: 'I was astonished to find that one vine produced male Flowers & another female, which is constant, unfortunately a very large proportion of my vines are male, not less I imagine than 9 in 10, this will retard the produce much till I can rear Female plants & extirpate most of the males.'[20] This information was given in more detail in his botanical description which he had sent

Account of the number of Prop Trees and Pepper vines in the Companies Pepper Plantation, in the Zemindaries of Peddapore and Pittapore, also an account of the quantity of land they occupy with its value per annum, taken 1st November 1789

	Number of Prop Trees	Number of pepper Vines	Vissums of Land	Countas of Land	Value of the Land per annum	M.P.	F.	C.
The plantations under the villages of Irwada & Mallum in the Zemindary of Pittapore contain	11400	22800	10	23	at 5MP per vissum	53	23	
The Plantation at Samulcotah in the same Zemindary contain	4680	9360	4	23	at 5 Do per Do	23	32	
In the last mentioned plantation is some ground so bad as to be unfit for the culture of Pepper. I have therefore alloted it for grass to feed the cattle employed in cultivating the ground bringing manure of each the quantity may be			2	16.25	at 2 Do per Do	5	2	
Total in the Zemindary of Pittapore	16080	32160	18	-		82	18	
The Plantations under the villages of Mongotoor in the Zemindary of Peddapore contain	7200	14400	6	24	at 5MP per vissum	33	40	20
Total	23280	46560	24	24	Madras Pagodas	116	10	20

The ground I have had repeatedly valued by different cultivators. They disagree much, viz from 4 to 6 ps per vissum. The medium 5 will be as near the truth, I imagine, as can be ascertained.
There are two Pepper vines to each Prop Tree. The prop Trees are planted 6 cubits as under, but as they are planted in quincunx, the rows are little more than 5 cubits from each other. However I have, in the abovecalculation, reckoned 6 cubits each way.
The vissum of Land contains 31.25 square Countas of 32 square cubits each.

A quarter of the above plants are about two years old, one quarter about one year and half, the other half from 2 to 12 months old, I mean since planted out from their nurseries to the Prop Trees, beside the above, I have as many more plants in their nurseries, ready and nearly ready to transplant.

Table 18 Roxburgh's Pepper results, showing the extent of his plantations by November 1789.

to Patrick Russell in October and which appears almost verbatim in *Flora Indica*.[21] Further problems were to arise when Roxburgh discovered that there was a third version of the plant, one which he called a hermaphrodite plant (monoecious), and it was only this plant which produced the economically useful peppers. This meant that the plants brought to him by his peons were possibly knowingly of no value but there had been no way of discovering this until the plants fruited, by which time the collectors had gathered their pay and disappeared. In his letter to Banks detailing this state of affairs, he was quite honest, to the extent that he asked Banks to 'order a copy of this Letter to be sent to the Court of Directors, and if accompanied with a Note from Yourself they will be most perfectly convinced that my Description of this valuable vine is true, and will also be satisfied that the tardiness of return from the Young Plantations thereof under my care, does not originate in careless improper soil or climate but to my having as yet scarce any of the proper fertile vine.'[22] A new drawing had to be done, and this was sent to Banks in February 1792, with his third hundred drawings and descriptions.[23]

These set-backs led him to review his plantations, and by the end of the year he wrote to Banks:

My pepper plantations will become valuable, very much so if extended on the plan I have now adapted, which is to plant only the hermaphrodite vines, and instead of using that useless tree, Erythrina corallodendrum, for host trees, make use of the Sappan wood tree, which grows most easily & luxuriantly, and will by the time the rains have done leaving, be worth considerable sums.[24]

101 A legume in flower, whose planting is arranged as Roxburgh suggested for his pepper plantations.

102 Legume growing in Rajahmundry District in 2005, in the pattern that Roxburgh recommended for his pepper plantations.

At last he had found the plant which would produce the black pepper, but at this point he was called to take over as Superintendent of the Calcutta Botanic Garden on the death of its founder, Colonel Kyd. His successor at the plantation at Samulcottah was Dr Benjamin Heyne, whom Roxburgh recommended for the job.[25] Heyne's explanation for the failure of the vines to succeed, 'inappropriate methods of cultivation',[26] would itself seem not to understand the problems that Roxburgh had in first growing the plants successfully and then discovering that the plants which he had been fraudulently sold by his collectors were the wrong form.

As a final note on this study of the work that Roxburgh did on the black pepper, his article in the *Oriental Repertory* was corrected by a second shorter one, stating the new conditions he had laid out. By this time (1793), he had also been able to ascertain the best growing conditions for the vine and had changed the areas appropriately. It is worth mentioning in passing that he did not put all his eggs in one basket, growing coffee, cinnamon, bread-fruit, sappan wood, arnatto and mulberry trees at Samulcottah as well as pepper.[27]

During the nine years that he devoted to the development of pepper as a potential crop, Roxburgh first identified its economic potential. This was followed by studying the requirements of the plant and studying whether it could be grown in the Circars. Once this had been proved, he set about acquiring plants and developing the experimental farm on a small scale. When these plants flourished, he expanded the area under pepper, and got the approval and support of the Court of Directors. On discovering that the plants that he was growing were not the productive ones, he admitted his error and set about correcting the mistake by growing the hermaphrodite plants in sufficient quantity to make it a worthwhile project. It was unfortunate that he was asked to leave before this last stage had become profitable, and thus could not see the project through to its logical conclusion. The final test of the efficacy of his work must be reflected in the value of the crops produced. Although there is no readily available source for these figures, the table below (Table 19) does give some indication of the changing amounts produced.[28] In spite of the comments of Benjamin Heyne, once the problems with getting the right plants had been overcome,

Year	Quantity (lb)	Prime cost (£)	Sale amount Co. trade (£)	Private trade (£)	Dutch trade (£)
1786/87	3,299,865	53,028			
1787/88	1,968,789	35,433			
1788/89	2,286,936	26,875			
1789/90	1,372,245	24,748	76,438		
1790/91	2,264,956	46,599	143,097		
1791/92	2,263,141	60,025	153,132		
1792/93*	2,000,000	42,000	109,600		
1793/94			234,948		
1794/95			189,996		
1795/96			335,415	2,593	8,060
1796/97			254,194	11,163	86,775
1797/98			235,118		
1798/99			331,625		
1799/1800			260,299	26,106	1,581

* The figure for these years were all estimates, but the accuracy can be gauged from the fact that the estimate for the following year (sale amount £224,000) was so close to the actual amount.

Table 19 The amount of pepper produced and sold by the East India Company during the years 1786/87, and those immediately following, that Roxburgh was experimenting with its production. It must be borne in mind that the three years, from 1790 to 1792, the Northern Circars experienced one of their periodic devastating famines as a result of the failure of the monsoons.

the Company's production increased considerably, as did the private trade, suggesting that Roxburgh's work was successful, in good years being nearly five times greater than before the experiment started, in terms of the sale amount.

When I visited the area in early 2004, I was fascinated to see the red peppers growing in just the way that Roxburgh had recommended, with the ripe pods set out to dry in the heat of the sun.

103 Red peppers drying on the Coromandel Coast. A modern photograph of a scene that must have been familiar to Roxburgh.

SPICES – CLOVES AND NUTMEG

The first mention by Roxburgh of either of these spices, cloves and nutmeg, came shortly before he left Samulcottah for Calcutta, when Andrew Ross said that he sent two drawings of nutmeg and cloves to Sir Charles Oakley. The following month, Ross wrote to say that he hoped to receive nutmeg trees from Mr Light at Prince of Wales Island and it looks as though this may have been successful, for in September Ross wrote: 'I have rec'd within 8 days from Capt Light of Prince of Wales Island a favourable account of his [success] in the acquisition of the Nutmeg and Clove plants.'[29] Captain Light had negotiated the cession of Prince of Wales Island from the Sultan of Kedah to the Company in 1786 and became the island's first Governor.

Once in Calcutta, Roxburgh took a far more dynamic approach, writing to Banks: 'Pray let me beg you will now & then give the Directors a hint to encourage me, & I will encourage others over India, by which means we may do much, for there is still much to do. The true Nutmeg & clove plants are still want tho we possess most of the other useful plants of the East.'[30] It was not until Christopher Smith arrived in 1795 that there was any degree of success in procuring large numbers of seeds and plants, but even then they started arriving only in small quantities initially: at the beginning of September Roxburgh was still writing that he hoped Smith 'would soon put us in possession of both clove & nutmeg plants'. However, Banks supported Roxburgh, writing at the end of May, 'we wish for the Nutmeg & Clove here as much as you do. The latter is in the West Indies & succeeds very well under the care of one Man only who has discovered that it grows freely upon one kind of soil which he says is reddish clay on which the usual produce of the Island Sugar &c will not thrive & upon which hard wood Trees only will grow.'[31]

The first consignment appears to have arrived by the end of April 1797. The following week, Roxburgh wrote that he had received 'about fifty Plants of the true Banda Nutmeg from Mr Smith'.[32] However, Roxburgh was not confident about Smith's ability to look after the chests of plants that he sent to Pulo Penang to await his arrival, for Roxburgh continued his letter:

104 *Cloves,* Syzygium aromaticum (Caryophyllus aromaticus). *Engraved and drawn by James Sowerby for William Woodville's* Medical Botany (*vol. 2, plate 135, London 1792) – a book that Roxburgh had in his library.*

> he has already sent to Pulo Penang upwards of fifty Chests chiefly filled with those of Nutmeg and Clove to be kept there until his arrival from the Eastward with an additional collection; & am just informed by Captain Carnegy of the Nancy Grab, that a very great number of those Plants had actually arrived before he left that place about the beginning of the Month. I have therefore to request the Hon'ble the Vice President in Council will be pleased to direct the Superintendent at prince of Wales Island to forward all those Plants that may be in Chests & in a healthy state, as also seeds, or other Parcels for the Botanic Garden, by such good conveyances as may offer, directing the native Gentlemen who Mr Smith carried with him, or some other to come with the Plants to take care of them during the Passage. I am the more induced to make the above request, because it is scarce possible that Mr Smith can meet with any one Conveyance sufficiently large and commodious to bring what he may be expected to have under his own care on his arrival there from the Moluccas, and those already at Penang, I could therefore wish they may be forwarded as above-mentioned while the season may continue favorable.

This was agreed. The time that Smith was taking and the ideas that Roxburgh was developing for the spice plantations come to the fore in a long letter to Sir John

Shore, the Governor General, dated 15 May 1797.[33] This letter also highlights many of the problems of moving plants around, for the previous February Smith had sent

> by the ship Union Captain Macall fourteen Chests, containing upwards of one Thousand Nutmeg plants and Seeds, and am happy to say that a great many of them arrived safe, notwithstanding the long circuitous course Captain Macall was obliged to steer, during which he was so very attentive to the plants, that he gave up his own Cabin the better to enable him to preserve them during so tedious a passage.

Roxburgh then continued: 'By the Cartier Captain Nash, I also received two chests with the same sort of plants in excellent order, so that we may now reckon on having received about five hundred plants of the true Banda Nutmeg in good health, and about as many more in sickly state, most of which I still think will recover.'

In terms of the development of spice plantations, Roxburgh also included in this letter that

> Pullo Pinang also appears a very favorable place for the growth of those trees probably the best belonging to the Company, and therefore hope that you will be pleased authorise a small establishment under a European Gardiner, with a few small spots of ground situate on different parts of the Island, that we may the sooner learn from actual experience, what will suit them best; the same establishment will also serve for collecting, rearing, and forwarding to this and the other Presidencies the plants of the Eastern Peninsula of China, and the Malay Islands. To keep the expences of this proposed establishment as low as possible, particularly in a Country where labourers are scarce, consequently expensive, a few of the native convicts confined on that Islands, might be employed to clear and bring these spots into order, &ca laborious work – And as I have reason to think from the accompanying Memorandum Marked No. 3 that many Nutmeg and Clove plants are already there, and probably not much taken care of, beg leave to recommend as little delay as possible, for fear of their being neglected.

This, once more stresses the importance that Roxburgh gave to the careful experimental procedures before going into full production, as well as looking at methods of reducing costs, by using convict labour. In case of any problems about the care of the plants he also went into some detail about the conditions in which they should be grown, including soil and climatic preferences as well as mode of propagation. The quantity of plants that Smith was sending to Calcutta can be gleaned from Table 20.

The fate of some of these can be traced. In a letter dated 17 April 1802, Roxburgh included details of many of the shipments; those relating to the ones given in Table 20 are detailed in Table 21.

Although plants were slow in arriving at Calcutta, Roxburgh was nevertheless keen to get the nurseries for the clove and nutmeg at Prince of Wales Island under way. He wrote at the end of July:

> The sooner the Nurseries at prince of Wales Island are begun, the better, as there are already at that place many of both sorts [of cloves and nutmegs]. I have therefore to request being furnished, as soon as possible, with orders to be carried by Mr Allen Bowie, the European Gardiner ... to put him in possession of such spots of land as may be deemed most proper for the Growth of those Plants.[36]

Bowie was experienced, having 'been long in the Botanic Garden, and is the best man I can at present find to employ at Prince of Wales Island'. The care Roxburgh took over Bowie's move is further shown by the request that his salary should be increased from the 50 rupees a month he was getting to 'any sum between one hundred and one hundred and fifty', and then continued that he would need to be furnished 'with such Garden tools as may be thought necessary, as he may not [be] liable to procure them there'. However, as so few had arrived by November, Roxburgh wrote: 'I do not at present conceive it necessary to send from hence an European Gardiner to Cultivate them there.' In view of this, he suggested that to sustain the plants they should be sent

On the Suffolk

1	Chest containing	Clove plants	108	
1	Do	Chocolate Nut Do	27	
1	Do	Ganemoo Do	<u>76</u>	211

On the Gloucester

2	Chests containing	Clove Plants	305	
1	Do	Casambee Do	21	
1	Do	Mangosteen Do	10	
		Chocolate Nut Do	<u>7</u>	343

On the Sloop Swift

2	Chests containing	Clove plants	225	
1	Do	Nutmeg Do	<u>27</u>	252

On the Fly Brig

2	Chests containing	clove plants	306

On the Ervir

1	Chest containing	Canary Plants	18	
2	Do	Melaleuca Leucudendron	90	
1	Do	Inocarpus edulis	<u>224</u>	132

On the Eliza

2	Chests containing	Clove plants	<u>444</u>

Total for Madras — 1688
Of which 1,388 are Clove Plants

On the Ship Jane Captain Stewart

23	Chests containing	Nutmeg plants from Great Banda	853	
10	Do	Do from Banda Neijra	231	
11	Do	Do from poola Aij	<u>256</u>	
		Total Nutmeg plants		1340
9	Chests containing	Clove plants	631	
4	Do	Sago Do	146	
1	Do	Cossia lignea Do	100	
1	Do	Cajaput Oil Do	34	
1	Do	Long Nutmeg Do	8	
1	Do	Chocolate Nut Do	13	
2	Do	Kanary do	41	
1	Do	Barringtonia speciosa do	12	
		Hernandia Sonora do	6	
		Inocarpus edulis Do	6	
		Athrodactylis Spinosa Do	<u>12</u>	36
1	Do	Royal Dammer plants	5	
		Katakooke Do	4	
	Malay names	Atho Do	5	
		Tomy-Tomy Do	<u>4</u>	18

65	Chests on the Jane containing Plants		2367

On the Cartier

2	Chests containing	Nutmeg Plants	100

On the Union

12	Chests containing	Nutmeg Plants	33
		and of seeds put in the earth when the Ship sailed from Banda	1332
2	Chests Containing	Malacca Jambo Do	100

Table 20 Plants sent by Christopher Smith between 1795 and 1797. The first group went to Calcutta (on the *Suffolk*, *Gloucester*, *Swift*, *Brig*, *Ervir* and *Eliza*) and there is a note that 1212 plants, including 1172 clove, 33 Cajput and 7 nutmeg were sent on the *Yarmouth* for Calcutta but all were dead by Pulo Penang and the chests were thrown into the sea. Roxburgh commented that he had not heard from Captain Stewart of the *Jane* about the success of his cargo, and, for the third group, Roxburgh wrote that 'of the first mentioned two chests, Sixty plants arrived in good condition, and are thriving well, of the second twelve Chests, all the 33 Plants and between four and five hundred of those rear'd from seed during the Passage, arrived in good condition, and about as many in a sickly state'.[34]

Ship	Cargo	Plants	Result
Suffolk	Clove plants	108	No information
	Chocolate Nut	27	
	Ganemoo Do	76	
Gloucester	Clove Plants	305	[No entry]
	Casambee	21	
	Mangosteen	10	
Swift	Clove plants	225	[No entry]
	Nutmeg Do	27	
Fly	Clove plants	306	Nothing known
Ervir	Canary Plants	18	[No entry]
	Melaleuca Leucudendron	90	
	Inocarpus edulis	224	
Eliza	Clove plants	444	Nothing known
Jane	Nutmeg	1340	
	Clove	631	Believe to have landed on Prince of Wales Island
	Sago	146	
Cartier	Nutmeg Plants	100	
Union	Nutmeg Plants	33	
	Nutmeg seeds	1332	450 nutmeg in good condition

Table 21 Results of some plants sent by Christopher Smith.[35]

to whoever may be interested in growing them, particularly those who had plantations of peppers and other spices on the Island.[37]

By November, there were still no cloves at Calcutta, but the nutmegs were thriving, with 'above eight hundred in the Botanical Garden, and nearly two hundred have been distributed to such Gentlemen, over various parts of the country, as appear the most likely to take proper care of them.' The following month, Roxburgh was able to write '61 Nutmeg plants out of 297 that were shipped; also 7 Clove plants out of 200 shipped. Mr Smith also put on board the same ship 124 Sago plants, and 241 of various other useful Trees, native of the Moluccas, but unfortunately, not one of them reached this Port', giving survival rates of 20 per cent and 3½ per cent respectively which were below the usual even for Smith.[38]

In the last of the letters that Roxburgh wrote before leaving for the Cape, he emphasised once more the poor state in which the plants that Smith was sending arrived at their destinations and the care that they were receiving: 'nearly three fourths have perished during the passage, but I cannot impute this grievous misfortune to want of care in Captain Patrickson, for I went on Board while the ship was below this, before the plants were landed, & found them carefully placed between decks, well aired, watered, & in short carefully attended to.'[39] The success of the nutmeg plants (over 50 per cent) against the failure of the cloves (almost 90 per cent) bears out Roxburgh's concern.

It almost appears that Smith took the opportunity of Roxburgh's time at the Cape to persuade Dr Fleming who was the temporary Superintendent at Calcutta to allow the plantations to develop, without the care that Roxburgh would have given to this enterprise. Thus, Fleming, writing in July 1798, gave a list of the plants collected by Smith, including 41,936 nutmeg, 10,064 clove and 12,654 sago plants: what was not stated was how many arrived in a viable state. By this time, Fleming had 21,486 nutmeg plants at Calcutta awaiting disbursement.[40] In October, Fleming was being a bit more phlegmatic about the chances of the nutmeg in Calcutta even though it appeared to have propagated so well, writing, 'I confess I am not very sanguine in the hope that the Cultivation of the Nutmeg will succeed in Bengal, being afraid that a plant which hitherto has been found only to flourish well in a few degrees of the line [equator] will scarcely bear the cold of our winters mild as they are.' He then continued that they might do better in the southern parts of India, nearer the equator, but these

Nutmeg
Plants & sown seeds Shipped		1806	
Do received in a healthy state	661		
Do Do in a sickly state	270	939	
Do Dead & deficient		867	48%

Clove plants shipped		2180	
Do received in a healthy state	197		
Do Do in a sickly state	31	220	
Do Dead & deficient		1952	90%

Various other plants shipped		672	
Do received in a healthy state	145		
Do Do in a sickly state	48	193	
Do Dead & deficient		479	71%

Total shipped		4658	
Do received in a good condition	1003		
Do Do in a sickly condition	357	1360	
Do Dead & deficient		3298	

Table 22 Analysis of plants sent by Christopher Smith on the *Ganges* to Calcutta and their state of arrival on 18 January 1798.

experiments would not require the whole stock of 'upwards of nineteen Thousand' the rest being available for trials in 'a more favorable situation'.[41] By the beginning of the following year, when Fleming had handed over the Calcutta Garden to Francis Buchanan, the latter was writing to say that 6,000 nutmeg plants were ready to send to Chittagong. However, Buchanan was soon aware of the problems of transporting plants by sea, when a complete shipment of spice plants from Banda under the care of two slaves 'acquainted with the cultivation of the spice plants' arrived dead, because of 'their having got Salt Water'.[42]

On his return to Calcutta, Roxburgh reviewed the state of the various experiments with clove and nutmeg, stating

> The Clove plant, I am sorry to say does not thrive here. Out of above two hundred brought from the Moluccas only one or two sickly plants remain alive. At Pullo-pinang the few that had been received, I am informed were in a thriving state. At Fort Marlborough they had not been long received when Mr Charles Campbell's report was made out on the 1st January last ... I have lately received from Sir Joseph Banks a treatise on the cultivation of this plant in the West Indies [this was William Buee's *A narrative of the successful manner of cultivating clove trees in the island of Dominica* (London, 1797)] where it is now common, and thrives so well as to have enabled them to send some of their Cloves to London, where they have been judged equal to the best sort imported from the Moluccas so that failure of that plant in Bengal, is now a matter of less importance at least so far as it relates to the Europe market.[43]

This gives an interesting view of Roxburgh's method of working and an insight into the wider view he had of the worth of the work he was doing: the need to read round the subject, that has been mentioned before, and the economic standpoint of the Company for the products under review. The problems with transporting the clove were again highlighted a fortnight later in a letter to Sir Joseph Banks: 'Smith has not succeeded in transporting them.'

Returning to his letter at the beginning of December, his views on the potential of nutmeg at Calcutta were equally pessimistic:

> The nutmeg plant, as yielding two sorts of spice (namely nutmeg and mace) is of infinitely more consequence than the Clove, and am happy to say promises success particularly on the Islands of Sumatra and Pullo-pinang, as well as at Chittagong and about Tiperah. Here in the Botanic Garden they thrive but very slowly and many have died notwithstanding every attention, and with the additional advantage of two Malay

Gardeners from Amboyna to look after and manage than as at the spice Islands. I would therefore recommend that the greatest part of the plants in the Garden may be sent to those places where they are found to thrive best and as the communication between the Moluccas and Sumatra is precarious, I beg leave to suggest sending by the Lushington, as large share of what remain at present in the Garden, should this be approved I beg to be favored with an order to the Commander of that ship to receive the plants, together with two native Gardeners to take care of them on the passage.

105 Nutmeg, Myristica fragrans. *Hand-coloured lithograph from C.L. Blume's* Rumphia *(vol. 1, plate 55, Leiden 1837).*

These two Malay gardeners, named Mahomed and Gorung, were sent back to Sumatra with a consignment of plants including nutmeg, coffee, cinnamon and other fruit trees, as well as sixteen bundles of Chinese sugar-cane at the very end of the year.

Once William Roxburgh junior arrived to take over the Spice Plantations, a very much fuller review appears from a number of the India Office Record archives. These were prepared about the end of 1802 or early 1803 but contain earlier documents as well, most are to or from his father as Superintendent of the Calcutta Botanic Garden. What becomes evident is that Roxburgh was overseeing the development of the spice plantations to the east from Calcutta but was finding the inertia of many of the people hard to cope with. Thus, writing of William Hunter who was on Prince of Wales Island, Roxburgh stated that

> 26,000 nutmeg plants, now in the garden, only 600 have yet been planted out, consequently 22,000 remain in the nursery, and in the boxes in which they were imported, ... that it may be advisable to direct Mr Hunter to make every exertion in his power, to have the whole planted out either in the company's Plantation or distributed among such Gentlemen as are in possession of land fit for the growth of this Tree.[44]

A week later it was being suggested that Hunter should be put in charge of these plantations which extended to about 140 acres, with the rank of either Surgeon or Superintendent. This dire state of the plants was reinforced a fortnight later, when young Roxburgh had had time to report back to his father who was able to write that, unless something was done quickly, 'a very large proportion of the small plants, particularly of the nutmeg tree will be lost if suffered to remain untransplanted any longer.' He then continued to expound the fact that considerable expense had been laid out introducing them and unless this was to be wasted, the best answer would be to distribute them, partially as the ground was not yet ready for them with the commensurate damage done with repeated disturbance, and secondly by sending them to a variety of sites, they 'will of course be tried in various places over the island, by which it will sooner be determined by actual experiment, the soil and situations they succeed in best'. By this time there appear to have been nearly 20,000 nutmeg trees from one to seven years old and well over 6000 clove plants from two to seven years old growing on the Company's plantations on the Island with over 7,000 more growing on private plantations.

The wastage, however, does seem to have been heavy. Thus, out of 15,958 clove plants sent between March 1800 and September 1801, nearly 3,500 perished on the voyage and in July 1802 there were still 3,231 waiting to be planted out. The position for the nutmeg was similar, with 24,820 sent of which 1,330 died en route and 10,132 were awaiting planting out. This does not take into account the success rate when those still in the nurseries were planted out.

The position in Amboyna was similar, with Smith himself in charge. Here we see again the problems that Roxburgh had with Smith, for although he had received nearly 2,000 nutmeg plants before he left for the Cape, while he was away Smith had brought some thousands in from Amboyna but took many with him to Prince of Wales Island, but the 'particulars of these Transactions I cannot ascertain'. Between 1797 and January 1802, 5,163 nutmeg plants had been sent to Calcutta, of which 500 were planted in the Botanic Garden; 9,362 still surviving in February 1802, with heights ranging from three to six feet, 179 sent to Bencoolen, 3,000 to Chittagong, 1,300 to Tipperah, and the remaining 863 distributed among 112 native and European planters. The latter group probably had only about 10 per cent still alive.[45] The overall rates of plant survival for the year 1796 to 1800 are summed up in Table 23.

	Cloves	Nutmeg	Others	Total	Per cent
Not recorded	3674	715	502	4891	20.3
No information'	1563	208	5130	6901	28.7
Perished	2356	1706	3312	7374	30.7
Sickly	193	60	522	775	3.2
Survived	892	2987	234	4113	17.1
1 shipment all dead			24054		100

Table 23 Survival rates for plants sent by Christopher Smith between 1796 and 1800.

Some idea of the costs of plant collecting can be gleaned from Table 24, which covers William Roxburgh junior's expenses for the period starting at the beginning of June 1802 until the end of February 1803. These months produced a collection of 21,483 nutmeg plants and 6,910 clove plants plus 'many more were daily arising from the seeds he had sown in Boxes of Earth at Amboyna'.

Personal expenses	1223
Chests for plants	255
Boat and Coolie hire	104
Mr Beale's expenses	574
Mr Mollies expenses	100
Williams House expenses	657
Native gardeners	24
Incidental expenses	29
Total	2966

Table 24 Expenses of William Roxburgh junior during the nine months from June 1802. These costs are in Spanish dollars, and equate to about £631 15s. 2d. on the given basis of 213 rupees per 1,000 Spanish dollars.[46]

The problems that William junior faced and the extortion that some of his fellow Europeans exacted can be glimpsed from an episode in May 1803. William had loaded a consignment of spices at the Moluccas for Bencoolen on the *Transit*, captained by Francis Lynch. However, Lynch decided to sail up the east side of Sumatra rather than the west that would have taken him passed Bencoolen. To persuade Lynch to alter his route, William had to agree to pay him an extra 5,000 rupees, however it appears that Richard Wellesley, the Governor General did not approve this extortion, seeing it for what it was!

By March 1804, Roxburgh senior was asked to comment on the nutmeg and cloves that had been produced on the Company's spice plantations on Sumatra. As usual, he did not rely just on his own knowledge but 'applied to two eminent native spice Merchant Viz.ᵗ Tam Kistna Paul and Radda Kistna Huldar, who have pronounced both (the Nutmeg and Mace) to be of a good quality, and equal to what is generally imported into Calcutta from the Moluccas and elsewhere.'[47] This request for his comments must have accompanied a rather critical view of the nutmeg, for Roxburgh continued 'Mr Ewer observes in his report, that the Nutmegs are very small'. Roxburgh's familiarity with the growth and development of the plant then become evident, for he went on: 'But on comparing them with the general run of Nutmegs in the Bazar, and in the possession of the above mentioned Merchants, the difference in size was by no means so great. However Mr Roxburgh, who paid considerable attention to the culture and produce of the Nutmeg and Clove trees, while at the Moluccas not only observed, but was informed that the Nuts, the produce of young trees, for the first year or two of their bearing, were generally smaller than when the product of trees more advanced. This information is further strengthened by the same remark being made in the instructions which accompanied the Spice Plants, sent from Amboyna to Bencoolen in 1798.' His knowledge was further reinforced, as he then wrote:

> On the three Banda Isles, Vizt Lonthoir, or Great Banda, Neira, & Pulo-dy, where the Nutmeg tree is chiefly Cultivated, they begin to bear when five to eight years old. On the West coast of Sumatra, a few of the most forward trees began to bear when rather under five years. This favorable circumstance, together with the luxuriant healthy appearance of the trees, and the quality of both the Nutmeg and mace, which they have produced at so early a period as five years, give us every reason to flatter ourselves with the hope of a successful issue to those Plantations, provided the expence can be kept so moderate as not to exceed the value of the produce.

The possibility of criticism is strengthened by his comments on the cloves that he had received with this consignment, writing:

> The specimens of Cloves came preserved in Spirit, consequently no judgement can be formed of their quality. But on comparing them with numerous similar specimens from Amboyna, they exceed in the number of Cloves, or flower buds, (which compose what Botanists call a Corymbus) those of the Moluccas. This Spice is now successfully cultivated on the Island of Bourbon, in Cayenne, and in the West Indies, and in all the Cloves have been found to arrive at the heighest perfection, and not distinguishable from those of the Moluccas. We may therefore rest assured that the Cloves of Sumatra will prove equally good.

The fact that all was not well with Smith's work and attitude in the Moluccas is highlighted in a long letter a few months later, when Roxburgh went into some detail about the plants sent by Smith, the state they were in and the information with which Roxburgh was supplied by Smith. In the archives, this letter covers some 33 sides, consisting of more than 1,500 words and made up of 14 paragraphs. He is highly critical of Smith's work: 'it appears that above five sixths of the number [of plants] shipped had perished before 23ʳᵈ October 1802. Of this deficiency, so much greater than we had reason to expect, I had no intimation ... . For want of authentic reports I was obliged to trust to private informations.'[48] Notwithstanding these criticisms and shortcomings, Roxburgh, having given cogent reasons, felt 'that with due care and attention, the plantations at Prince of Wales's Island will be a valuable property to the Company'. However, to achieve this, he would need to have from Smith 'a full account of his progress in the cultivation, stating the number of plants, whether planted out where they are to stand or in nurseries. The number planted out monthly, age, size, appearance, period of blossoming, and ripening their fruit, with an account of the soil situation, aspect, and irrigation by which they thrive best; together with any other remarks which he may think necessary, to enable us the better, to rear those trees in other places to the greatest advantage. A memorandum respecting the state of the weather, ought to be added to his report'; in other words, Roxburgh did not trust Smith and hence asked for a very detailed report that would show Smith's care in nurturing these valuable plants.

The problems that these plantations were going through was further highlighted in a letter from Roxburgh junior to his father, in which he stated:

> During my residence in the former place [Prince of Wales Island], about two years since, I saw most of those, sent by Mr Smith from the Moluccas, amongst the Nutmeg first introduced in 1797, were a number of grown up, small trees, in a thriving condition – several of them had flowered, but they proved unfortunately male plants, excepting one, a fruit bearing Tree, which had some months before nearly ripened a single nutmeg, when it was accidently pluckt. We are therefore still ignorant of the quality of the spice, that may be produced on that Island.

These plantations were in an area previously forested, but young Roxburgh reckoned that they would have done better if 'several of the forest trees had been allowed to remain, to afford them a partial shade, against the powerful rays of the sun' and continued:

> On the three Banda Islands, Great Banda, Neira, and Pooloo Aij, where they are cultivated, they have large trees growing scattered amongst them, in the Parks (plantations) to break the direct intense rays of the Sun, and cause a more agreeable sunshine. This in situations, were the shade was exactly suited, and where the Nutmeg trees had space to expand their branches freely, they appeared beautiful, and were bearing fruit, of a very large size; which are generally better than those, from trees too much shaded, or entirely exposed to the influence of the Sun.

He went on to write about the water requirements of the nutmeg, which, again, were such that 'the cause of their succeeding so well, in the vicinity of Bencoolen where they are refreshed by frequent showers'. His letter ended with the statement: 'Of all the different countries where the Nutmeg has been tried, there is no place where they appear to have succeeded so well, as on that part of Sumatra, lying under the same parallel of Latitude, and having a climate, which would appear to be in every way similar to that of the Moluccas.'

The final chapter of the spice plantations on Prince of Wales Island is encapsulated in an enigmatic document that stated:

> In consequence of the Instructions of the Supreme Government for the reduction of the Expence of the Botanical Establishment, Mr Roxburgh, then in charge thereof, recommended the disposal of the Company's Spice Plantations at Prince of Wales Island, which being acceded to, they were sold on the 20th June 1805 by Public Sale (twelve days previous notice having been given for that purpose) for Spanish Dollars 9,656, a sum (which the new Government of Prince of Wales Island observe) not equal to the value of the ground they occupied.[49]

The date is during Roxburgh's visit to England so the Mr Roxburgh referred to was young William. The document, having continued with an exposition of the Company's attitude restricting servants of the Company farming lands, and ended with the paragraph that stated:

> In August 1800, agreeably to the suggestion of the Lieutenant Governor of Prince of Wales Island the quantity of Land to be granted to any individual was not to exceed four hundred Orlongs, equal to 500 Acres, and a clause was directed to be inserted in all Grants issued to those applying to clear Ground, that if not brought into a state of Cultivation within a fixed period the whole should revert to the Company.

This makes one wonder whether young William organised the sale, somewhat hurriedly and cheaply it would seem, with himself as the purchaser, but this is very much speculation for there is no evidence for this either then or at his death five years later.

For whatever reason, some of the plantations were sold and one can only surmise as to the causes. Certainly amongst them must be the high loss of plants during the

106 Mundulea sericea (Robinia suberosa). *This small tree with a corky bark occurs only in southern and western India (and Ceylon), and was therefore probably drawn by one of Roxburgh's artists before he moved to Calcutta in 1793, although the drawing probably dates from c.1798 (Roxburgh drawing 1275).*

Mundulea suberosa.

Robinia Suberosa. R.

Ulicato

development stages, and a growing impatience on the part of the Company to see a return on its investment: that the nutmeg took from five to seven years to bear fruit in an area where life expectancy of the crucial people was short would seem to bear this out. However, writing to William Mackenzie who was living at Amboyna, at the end of November 1811, Roxburgh wrote: 'this year I have got above 40 ripe nutmegs of a very good size. They are employed to rear plants from, because I think this offspring will succeed better here than the plants from the Moluccas.' This would suggest that there was still an interest in plants from further east.

Similarly, young William Roxburgh, writing in the middle of June 1806, when acting as Superintendent at Calcutta, stated:

> it gives me much pleasure to state, ... that five of the Nutmeg trees out of the 22,323, plants landed by me at Fort Marlbro' in May 1803 have already flowered. The nuts were sown in the latter part of the Month of January of the same year, at Amboyna, & the trees in March last, when they blossomed were thus scarce three years & two months old. Many more were in a forward State, & likely to flower in the course of the month. The accounts from Dr Campbell also make the most favorable mentioned of the growth of the clove trees.[50]

OVER-VIEW

Roxburgh started work on introducing spices once he arrived at Samulcottah but took time to become aware of the different species and the physiology of the various plants, particularly pepper. However, once at Calcutta, the success of some of these introductions was evident, especially cinnamon and pimento. With cloves and nutmegs, the problems were as much due to the people involved, with Christopher Smith posing real difficulties in the care that he took in the transportation of the young plants. The success of these has, therefore, been more difficult to determine but the long-term effect of the work is undeniable. Equally, Roxburgh must be given credit for following his remit, of developing potentially economic crops which could then be distributed to appropriate growers around the Company's lands throughout the Far East, and indeed further afield, to the Cape and West Indies, for example.

CHAPTER 12

Sugar

INTRODUCTION

The politically powerful West Indian plantation owners exercised an effective monopoly of sugar for the British market, whereby duty on sugar coming from elsewhere in the world was virtually double that imposed on that exported from the West Indies. There was, therefore, a strong incentive for the East India Company to try to break this monopoly. Roxburgh's involvement with this crop exemplifies a number of interesting aspects of his work and character. First of all, there was his humanitarian concern for the natives set within an awareness of the peasant farmer's intrinsic conservatism, which is a characteristic that also will be highlighted in the following case study in Chapter 13.

Having looked into the methods already used for the purification of cane sugar in India (*Saccharum officinarum*), Roxburgh then set about finding a suitable source of better varieties of cane. One such came from China and which he described as a new species (*S. sinense*). After studying the way that the Chinese extracted sugar from the cane, he was in a position to grow canes that could be distributed with a view to its commercial introduction.

INTRODUCTION OF SUGAR CANE

For many years the Company showed little interest in Indian sugar-cane (*Saccharum officinarum*), but the high price and poor quality of that produced in the West Indies in the 18th century prompted a petition to the Company to consider its importation. As a result, the first sugar factory was opened in Bengal in 1784. By 1792 there was sufficient competition with the West Indies for there to be a meeting in London of the 'Consumers and Traders in Sugar' which included the Chairman and his Deputy of the Company, to discuss whether it should remain a monopoly, and 'concluded with general reasoning upon the wisdom of opening the trade to India'.[1]

This was followed a couple of months later by a meeting of the General Court of the Proprietors of the Company at their office in Leadenhall Street, when a number of resolutions were passed, regarding the importation of sugar and its current high price. Quoting what amounted to the theory of supply and demand, they stated, 'the present enormous price of Sugar is owing to the annual importation of that Article being very unequal to the increasing consumption in Great Britain.'[2] They felt that they could 'speedily and permanently supply a considerable quantity of Sugar for the relief of Great Britain, provided they are placed on the same footing with respect to Duties and Drawbacks, as the West India planters.' This was followed by the statement that 'the present high Duty of £37..16s..3d per cwt on East India Sugars, while the West India pays only £15 per cwt was accidental', the explanation being that, when deciding the rates of duties for imports, sugar had not been of sufficient importance to be separately listed and was, therefore, included under the general head of 'Manufactures Goods non-enumerated', all of which were charged at the rate of £37 16s. 3d. These arguments very much follow those laid out by Adam Smith in his *Wealth of Nations* that was published in 1776, where he wrote that inequalities from policy came from 'restraining the competition in some employments to a smaller number than would otherwise be disposed to enter into them, ... [and] by obstructing the free circulation of labour and stock, both from employment to employment, and from place to place.'[3]

The initial reference to this commodity as far as Roxburgh is concerned, appeared in a Commercial Letter from Fort St George, regarding his pepper plantation, in which they suggested that it would be better to grow sugar-cane or paddy rather than to allow him to expand his as yet untried pepper plantations.[4] This idea for sugar rather than pepper was reiterated the following year by the inimitable Ross, 'Pray tell me how it is [the pepper plantation] and whether the Cultivation of Sugar might not be practicable on a large scale'.[5] The eventual outcome of this comment was an extensive report written by Roxburgh in June 1792, sent to the Court of Directors who forwarded a copy to Dalrymple who published it in his *Oriental Repertory*.[6]

Before setting out the system of cultivation, one of the few insights into Roxburgh's view of Indians and his philosophy regarding introductions of new concepts and practices, may be found:

> No pursuit is more pleasing to the benevolent mind than such as tend to add a new source of happiness to Man.
>
> Amongst the Natives of India the transition from one stage of improvement to another, are exceeding slow as scarce to deserve the name, except it be the few who have benefitted by the example of Europeans; They naturally possess a strong disinclination at departing from the beaten path, established from time immemorial; however when they see a certain prospect of gain, with little additional trouble, they have been frequently known to adopt our practices. We ourselves ought more generally to keep in view & to instill into their minds this maxim – That every new proposition, merely on account of its novelty must not be rejected, otherwise our knowledge would no longer be progressive & every kind of improvement must cease.

Similarly, referring to the use of the left-over cane residues as cattle fodder, compared to its conversion in the West Indies into rum, he stated: 'Let it continue to be so employed is my sincere wish, for the longer they are ignorant how to convert what is at present wholesome into a poison, the better it is for them; they have already too many ways of furnishing themselves with spirits, particularly near the residence of Europeans.' This typically late 18th-century view of Indians, whom he repeatedly terms as 'natives', is further reinforced when he calumniated the Zemindars in the same article: 'I observe that the Farmer would require to have the agreement he makes for the Rent of the Land strictly adhered to, because the Zemindar raises his demand, if the Crop is good, so that he will often in a favorable season make Farmers of all denominations pay probably a fourth more than the original agreement; such injustice they are obliged to put up with, as custom has rendered it common, & they have no idea of applying for redress; yet it no doubt damps the spirit of industry.' While on this topic, it is interesting that Roxburgh considered it worth mentioning the possibility of the planters in the West Indies emancipating their slaves, and the economic results of this action: with increased mechanisation and lower labour costs.

107 Saccharum officinarum, *the Indian sugar-cane. Drawn and engraved by James Sowerby, for William Woodville's* Medical Botany *(vol. 3, plate 196, London 1793).*

His interest in the costs of labour, and the general economics of such a project are also shown in a copy in Roxburgh's collection of a review of Abbé Rochon's *Voyage à Madagascar*, in which there is discussion on the advantages of introducing a steam engine: 'Here we have a history & description of this machine, from its invention, by the Marquis of Worcester, in 1663, to its present perfection, as made by Messrs Watt & Bolton. If this machine were adopted by the colonies in the mills employed in pressing the sugar-cane, how many thousand slaves devoted to excessive labour, would be relieved by it.'[7]

As is usual with Roxburgh, there is once more evidence of his reading round the subject, for he refers a number of times to 'Mr Beckford's history of Jamaica' which had been published in 1790. In the collection of his manuscript archives, there are also a number of further excerpts relating to sugar and sugar products:

From the Appendix to vol. VI of *Monthly Review*, 1 folio on the manufacture of sugar
From Woodfall's *Register*, of 31st January 1792, 24 folios on sugar
From *Calcutta Magazine*, of August 1792, 4 folios on distilling rum
From *London Chronicle*, of 17th March 1792, 4 folios on sugar duties from the East Indies[8]

Roxburgh's account of the 'Hindoo method of cultivating the Sugar Cane' considered most areas that were germane to persuading the Court to take it up.

The soil that suits the Cane best in this Climate is a rich vegetable earth, which on exposure to the air readily crumbles down into very fine mould; it is also necessary for it to be of such a level as allows of its being watered from the River by simply damming it up (which almost the whole of the Land adjoining to the River admits of) and yet so high as to be easily drained during heavy rains ... . The situation of all the Sugar Lands hereabout is exactly alike, being the middle of an extensive plain adjoining to the before mentioned River; the soil in all is also much alike, so that the produce is nearly equal in all, when no unfavorable circumstances happens.

Having then dealt with the annual farming cycle, he stressed the importance of improving the soil:

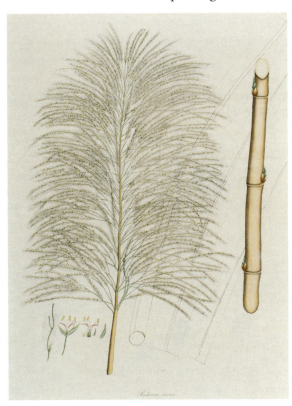

108 Saccharum sinense, *the Chinese sugar-cane described by Roxburgh. Hand-coloured engraving from a drawing by one of Roxburgh's Calcutta artists, c.1800 (Roxburgh drawing 1322), from* Plants of the Coast of Coromandel *(vol. 3, plate 232, 1815).*

the cane impoverishes it [the soil] so much that it must rest or be employed during the two or three intermediate years for the growth of such Plants as are found to improve the soil of which the Indian Farmer is a perfect judge, the leguminous tribe they find the best for that purpose ... having been well meliorated by various crops of leguminous Plants.

This awareness of good agricultural practice, of crop rotation and the use of legumes which are now known to have nitrogren-fixing bacteria in their root nodules, ties in with his correspondence with Sir John Sinclair and the knowledge of the aims of the Board of Agriculture.

Finally, as far as this account goes, Roxburgh took some pains to compare the economic benefits of importing Indian sugar over that from the West Indies. He started by emphasising the low costs: 'The method of cultivating the Cane & Manufacturing the Sugar by the Natives hereabouts is like all their other works, exceedingly simple; the whole apparatus, a few pair of Buffaloes or Bullocks excepted, does not amount to more than a few (15 or 20) Pagodas. As many thousand of pounds is necessary to set out the West Indian Planter.' He then went on to discuss the relative productivities of the two sources of sugar:

The half Vissum or one Acre of Sugar Cane, in a tolerable season yields about 10 Candy of the above mentioned Sugar, or rather more if made into Jagary; each Candy weighs 500 lb, & it is worth on the spot from 16 to 24 Rupees according to the demand; In the West Indies the Acre (so far as my information goes & it is chiefly from Mr Beckford's history of Jamaica) yields from 15 to 20 cwt of their Raw Sugar, worth on the Island from 15 to 20£ currency. Here the produce is more than double, but on account of the inferior quality, & the low price it bears on the spot, the produce of the Acre does not yield a great deal more Money than in the West Indies; however, as here labour is uncomparably cheaper, the Indian Planter must make much larger profits.

The detail into which Roxburgh went in this Account obviously surprised Ross, for it is presumably this to which he referred in his letter, stating with a tinge of irony that 'when I said I wished you to make your Observations, what Grounds might be fit for the cultivation of Sugar, it was not in mind, that you should Manufacture it, but only that there might be an apparent advantage in knowing it.'[9]

THE POLITICS AND ECONOMICS OF SUGAR IN INDIA

Roxburgh was well aware of the pressure that the West Indian interest could exert, and offered alternatives if this should prove insurmountable:

Should political Motives prevent the Importation of East India Sugars into England, it is even then of political import to the Companies Territories to have the qualities of their Sugars improved, so as to render unnecessary the importation of those of China & Batavia, large sums being annually thrown into those places for this commodity, while we at the same time possess every advantage for making this necessary article of the best quality to the full in as high a degree as either the Chinese or Dutch. Besides our own wants, we have every reason to imagine that we might soon be able to supply the Malabar Coast, Persia & Arabia with Sugars, whereas at present they are chiefly supplied from China & Batavia.

This appears to have been a wise attitude, for writing in August 1792, Ross said first: 'They [the Court of Directors] have lowered the freight of two more ships coming for Sugar to £8 – outward, £12 – homeward & the private freight to £15 – per person.'[10] However, three weeks later came the letter stating:

I had met with an English Paper of the 17th March giving an Account of what passed at a Meeting of East India Directors, West India Merchts &c on the Subject of lowering the Dutys on Sugar to be Imported from the East Indies – whereof I send you a Copy & by which you will observe that the Measure has not met with the support & encouragement which at the outset of Business there was every reason to expect tho' you will also see that the Matter seems still to remain in suspence.[11]

This takes us back to Corcondah, for towards the end of October, Ross wrote: 'I took the occasion 3 days ago to talk a little to my friend the new revenue Member – C. White about your idea of getting a Grant or long Lease – for the Cultivation of Sugar', and suggested that ground nearer where Roxburgh was, Samulcottah, might be considered more suitable.[12] Progress must have been made, for in March 1793, Ross was forwarding 'the Muster of Sugar' to the Directors.[13] The following year, Ross was sending samples of sugar from China, Manilla, Batavia and Bengal with their relative prices: Pagodas 21, 20, 19½ and 19 respectively, so that Roxburgh could compare them.[14] Discussions about the cost of manufacturing the sugar must have taken place at the highest level, for Ross wrote in September 1793,

Lord Cornwallis [the Governor General] is strongly of opinion, that the Manufacture of Sugar in this Country, by the Expensive Machinery of the West Indies, will by no means answer, as Mr Paterson knows to his cost. You will inform yourself of this important matter, at Bengal, & with certainty. His Lordship says, that its Manufacture by the Natives, is the only proper mode to obtain success.[15]

The earlier success of this approach, in June 1792, was reflected in a report of May 1793, for the Court: 'By Mr Roxburgh's accounts, the Cane of this Country yields more than double the Produce of that in the West India Islands. The price of Labour here, is much lower, and the Cane not so liable to Accidents, as in Jamaica. The great Sugar Crop also happens at a Season of the Year unfit for the Cultivation of Grain and the only Desideratum appears to be, the improvement in the Manufacture by the introduction of the Sugar Work used in the West Indies.'[16]

Again, the slowness of communication between Calcutta and London is shown by the fact that the Revenue Department's letter of May 1793 was not answered until July 1795, probably not arriving in Calcutta until the following summer as the main fleet usually sailed at the turn of the year. The Directors, however, were well pleased with Roxburgh's idea of developing a sugar industry:

> We have perused with much satisfaction, the very able Letter from Doctor Roxburgh ... and have shewn the Sample of Sugar which accompanied it to Persons possessing a competent judgment of the Commodity, who declare it to be of a most excellent Quality, and that any Quantity of it would at all times be highly acceptable at the British Market.
>
> It is therefore our earnest wish, and desire, that you will afford all due, and fitting encouragement to the Planters, and Manufacturers, for inducing them to extend the Cultivation and Manufacture of this Article ... by giving the assurances that the Company will readily become the Purchasers of any Quantity that may be produced, provided the needful attention is paid to it's Quality, and that it is afforded at a fair, and just price, not materially differing from that quoted by Dr Roxburgh.
>
> The circumstances we find noticed in the Doctor's letter, of the Zemindars exacting from the Farmers, in favorable Seasons, a larger Rent than is originally stipulated for, we consider as a Act of Oppression, which, in its consequences cannot fail of proving a check to Industry, and to improvement of the Soil; you will therefore particularly direct your attention to this point, and act therein as you shall judge the most effectual for remedying so injurious practices.[17]

This attitude of the Directors to the behaviour of the Zemindars can be easily explained as a case of self-interest, in that, as they point out, it would inevitably lead to a reduced motivation for the farmers to improve the quality of the crops and the soil. It is also interesting that, even though they may not have won a reduction of the duties paid on the importation of East Indian sugar, they still deemed it a potentially profitable commodity at this stage. This is confirmed in a detailed letter the following year, when it was still making some profit, but was also being used to replace saltpetre as a ballast:

> 107. Presuming that every exertion was used for effecting this contract at the lowest possible rates, we do not hesitate in sanctioning it with our approbation, although the prices materially exceed those quoted by Dr Roxburgh in his Letter of the 20th June 1792, referred to in the 40th Para. of your Commercial Letter of the 2nd May 1793. From the tenor of our former advices you will have perceived the importance in many points of view, of introducing this commodity into our Investments upon an extended scale; in addition to which, the necessity we shall be under of providing an increased quantity of ballastable commodities suitably proportioned to our enlarged orders for Piece Goods, is a consideration by no means to be overlooked. The article of Saltpetre, which has hitherto been principally resorted to in the arrangement of our Cargoes as an article of Dead Weight, will not, we are apprehensive, under any expected permanent scale of consumption, be hereafter equal to the exigencies of both settlements. We therefore direct, that you use every endeavor for extending your provisions as far as possible, so as to avoid occasioning any advance in the price by the sudden pressure of demand.
>
> 108. By the Account of Sales in the Margin [see below], you will perceive the prospect is encouraging. We therefore wish you to give every needful degree of encouragement

to the Riots that may have a tendency to promote an increase in the cultivation, which may, we conceive, be effected with little difficulty when the Natives find a ready market opened to them for the disposal of it, on fair and equitable terms. It will be necessary however that the utmost attention be paid to prevent any debasement of the quality, as upon that alone will depend its future success.

109.   We observe that the Charges of Packing in Chests tend very much to enhance the cost Rate. The mode of Packing in Bags as practised in Bengal, has been found fully competent to the purpose. We therefore direct, that this mode of packing be adopted in your future consignments.

Estimate Profit on Ganjam Sugar, sold 19 April 1796.

Chest, 43 for 1795 per Queen

Prime cost per cwt. Pagodas 8	£1	5	3				
Charges in India		11	7	£1	16	10	
Freight				1	4	5	
Custom							
Charges Merch[ants] in England				0	3	9	
				3	5	0	
			Profit	0	10	6	
Average Sale Price per cwt				£3	15	6	

From £3. 13. 0 to £3. 17 per cwt.[18]

This worry with price had arisen the previous year, when the Secretary of the Madras Commercial Department wrote: 'We are happy to inform you that the Commercial Resident at Vizagapatam has been able to reduce the prices of Ganjam Sugar ... 1st sort 25 lb 3 Rupees; 2nd sort 25 lb 2½ Rupees; and 3rd sort 25 lb 2 Rupees. These Terms being considerably lower than those of the late contract have been accepted.'[19] This worry about price and quality of the sugar cane, made Roxburgh consider getting the plants used by the Chinese, and obtained permission for this in March 1796 when Captain Carnegie 'of the Nancy Grab has been so obliging as to offer to bring any Plants procurable at China that I may wish for'.[20] By the end of the year, Roxburgh had made approaches 'to China, to the Coast, to Manilla and to Bencoolen, the War has hitherto deprived me of an opportunity of writing to Batavia, but hope Mr Smith, the Nursery Man, who is gone to the Moluccas, will be able to procure all the sorts known at Batavia among those Islands.'[21] He continued this letter to the Governor General: 'Some parcels of cane I have already received from Manilla, Fort Marlbro' and the Coast but from the length of the various voyages, scarce any arrived alive, and before they can be sufficiently multiplied to enable the necessary experiments to be made to determine their quality, and comparative value with those already cultivated in Bengal, will require a great length of time.' However, rather than delay the dissemination of the knowledge he had acquired until this research was complete, 'these Memorandums accompany this, together with a model of the Sugar Mill employed in the Rajahmundry Circar .... It is so simple and cheap, at the same time so efficacious that I think there will not be much difficulty in prevailing on the Natives of this Country to use it.' This reinforces the point he made about

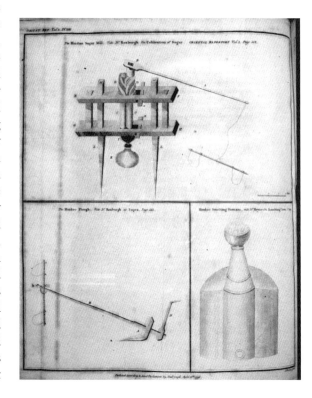

109 Sugar press and traditional Indian plough used in sugar cultivation and production (and also an iron smelting furnace). Engraving by J. Walker after a drawing by Roxburgh from Dalrymple's Oriental Repertory *(vol. 2, plate XIII, page 485, 1797).*

110 Saccharum arundinaceum (S. exaltatum*), a wild relative of the sugar-cane. Drawing by one of Roxburgh's Calcutta artists, c.1794 (Roxburgh drawing 774*).

the natives not being prepared to take on new ideas, so to offer something that was easy to implement and understand had a greater chance of acceptance.

The letter then went on to discuss the need for the sugar to be extracted quickly, as 'no cane, let it be ever so good, can yield good Sugar, if the juice is not quickly expressed and put over the fire before there is any tendency towards fermentation & consequent evolution of the acetous, and which can never be prevented while the slow working rude machines used about Calcutta and I believe even the adjoining Provinces, are employed.' The point that he made here was that, if fermentation takes place, either alcohol can be produced, which would have been transformed into rum, or there are conditions when it converts into acetic acid, which is the basis of vinegar. Although the diagrams for the machinery were published by Dalrymple (see Fig. 109), the plans for this sugar mill have not yet been found.

A month later, a number of plants arrived from Canton in good order, due to 'the uncommon great care that must have been taken of them by Mr Carnegy the Commander of the Nancy Grab during the passage'.[22] Amongst these plants were two chests of sugar cane which were already 'planted out, and promise perfect success', each chest containing a different type of cane. One was used for making sugar which was 'remarkably small, but uncommonly heavy, more so I think than any cane I have seen. The second, which the Chinese eat raw, or gently roasted, is much larger, and not unlike the Cane of this Country.' The first sort which he described as a new species (*Saccharum sinense*), published in 1813 in the second volume of the *Plants of the Coast of Coromandel*, continued to thrive, but 'the other sort does not thrive, however it is of less import, being only cultivated for eating raw'.[23] The new species continued to do sufficiently well for Roxburgh to write 'considerable quantities for planting have already been distributed according to the directions of the Board of Trade',[24] thus carrying out one of the requirements of the Botanic Garden.

Roxburgh's sojourn at the Cape did not interfere with his interest in the promotion of this commodity, as he was able to report on his return to Calcutta in December 1799: 'the China Sugar Cane has been cultivated with the utmost possible success. Many hundred thousand have been distributed over the Country amongst the cultivators of that article. It continues to resist on account of its hardiness, the attack of the white ant, and the Jackall, which is of infinite importance, as these animals annually destroy a large portion of the common Cane but the same quality renders it difficult to express the juice, by means of the Common Bengal Mill, which operates against a more general cultivation amongst the natives. I hope to obviate the inconvenience by introducing from the Coast of Coromandel, a very simple, at the same time powerful mill.'[25] The problems with the white ant were also causing havoc with Roxburgh's herbarium specimens and drawings, but this is the first time that jackals are mentioned as a nuisance eating crops, but when he described the Chinese plant as a new species, *Saccharum sinense*, he commented that the 'White Ant, and the teeth of the Jackall: two great enemies of our East Indian Sugar plantation.'[26] Two years later, in the last of the letters which mentioned it, the sugar cane was still being distributed to all those who wanted it.[27]

SUMMARY

Before he had left the Coromandel Coast, Roxburgh had investigated the taxonomy of the species of sugar-producing *Saccharum* and the conditions necessary for their growth and dissemination. At the same time, he proposed a system by which it could be developed using refining processes that were, as with indigo, preferable to the labour-intensive methods preferred in the non-mechanised Indian culture. Real success with sugar did not come, however, until the introduction of the Chinese species in 1796 which also allowed him to differentiate the various species.

Although he was not successful in assisting the break-down of the West Indian monopoly, he provided the Company with a profitable source of sugar that could be used for the distillation of spirits and its export to the emerging colonies in Australia, as well as for consumption in India itself. I was fascinated on my visit to India in early 2004 to see the vast areas of the Rajahmundry district still growing the crop on a commercial basis.

CHAPTER 13

Watering the Circars

INTRODUCTION

As has been shown in the previous case studies, Roxburgh was a scientist with a broad range of interests, and this is exemplified again in the work that he did in proposing the irrigation of the Circars, during and following the devastating famines during his last three years in the district, by canalising the River Godavary. Spurred by his humanitarian feelings in seeing thousands of Indians dying of starvation, he sought a solution that would overcome these comparatively frequent natural disasters.

He set about this problem from two distinct angles. First was the provision of crops that would be less susceptible to drought, so he looked for species that were grown in areas that had low rainfall or were used by the natives during famines. His second approach was to propose the use of the vast quantities of water that came down the Godavary River, by a system of canals and irrigation channels. For this, the first thing was to survey the area and, although a Company Surveyor had been appointed to the area, it was Roxburgh himself, with the aid of his friend Denton, and help from the engineer Beatson, who accomplished this. His prescience is shown by the fact that, once again, the proposals were not followed through for, in this case, another sixty years.

FAMINES

The interest of the Royal Society in meteorology was predated in India by a strong awareness of the frequency of extremes of cyclones and droughts. Thus, Love quotes for Madras nine cyclones occurring between 1640 and 1695, and their frequency was sufficiently great for him to be able to say that 'the usual months for such storms are May and November'.[1] It is interesting to compare these results with Roxburgh's own published meteorological results, but in the published diaries, the wind strength is never given greater than 3, although these do indeed occur more frequently during May and November.

During the same time, there were three periods of famine, one lasting for two years. The causes of the famines were easily explained by the failure of the monsoons. While there was usually grain enough to tide the farmers over one poor year, if this were followed by a second even partial failure of the monsoons, then famine followed. The sequence of events that led to such a state is well given by Philip Mason:

> If there is no rain, there is no harvest of rice and millet in September and the ground is too hard to sow the wheat and barley that ought to be cut in March. The peasant seldom had enough grain in hand to carry him more than a month or two beyond harvest-time. The grain-dealer of course has stocks, but prices rise and the peasant cannot buy without running into debt. There is scarcity, debt, hunger, and something near starvation. Then perhaps next year there is a poor crop and partial recovery, then another failure; the dealer's stocks are exhausted and there is no food in the area. This is famine.[2]

111 *Overflowing tank, showing the problem of variation in water supply between monsoon excess and scarcity during droughts. Hand-coloured engraving from Blagdon's* A Brief History of Ancient and Modern India *(plate 25, 1805).*

Famines had been endemic and there was certainly among the Mogul rulers a sense that they were beyond the power of human rulers to mitigate or prevent. This was partly due to the very slow speed of moving surplus grain from one area to another over land, and if this was to be done by bullock cart, the pair of bullocks would eat their cargo in a week. This meant that transport of much more than fifty miles was out of the question, as these carts only travelled about ten or twelve miles per day. There is also the fact that local politics would have precluded movement much further than this before the Company's rule.

Another contributory factor to famines and shortages was the wars that had been such a prevalent feature of the 18th century. Thus, from 1779 there had been scarcity in the Madras area 'which was due mainly to scanty rainfall, but aggravated latterly by Hyder's devastations. As early as March, '78, the situation was regarded as serious.'[3] Love went on to say that 'the records of 1783 contain few allusions to famine, but there is no doubt that scarcity continued until the conclusion of peace with Mysore in March 1784, when the bazars were suddenly flooded with foodstuffs, and prices fell 50 per cent.'[4] This attitude is also shown in the answer to one of a series of questions about watering the Circars. The question was, 'As you must often have experienced great distress from the want of Rain, by what means has it happened, that no Tanks of consequence have been formed, nor any steps taken, to avail yourselves of the water, that passes annually unemployed into the Sea?'[5] The answer was, 'Our Ancestors, being subject to the depredations of every neighbouring Poligar, who was inimical to his interest, had no encouragement for undertakings, that involved much expence, and only promised a distant and uncertain advantage; and even since we have enjoyed the Company's Protection, the nature of our Tenure of these Lands, and an opulence requisite, have been sufficient to frustrate any such designs.' There was also the fear that if the farmers improved the productivity of their land, the only effect was that their rents were increased.

The changing attitudes of the Court of Directors is shown by a reply of 1793 to a Public Letter sent from Madras in October 1791:

We approve of the measure that were taken in consequence of the famine in the Northern Circars, as mentioned in your Advices and proceedings received in the course of the past Season. The Governor General and Council have lately had under their consideration the steps that may be necessary to be taken in order to guard against the frequent return of the Calamities which have been so often experienced in Bengal from the scarcity of Grain, occasioned by the failure of the periodical rains; and have suggested the outline of a Plan admirably calculated, as far as we are able to judge, to prevent in future, or at least considerably to diminish, the miserable effects which have been so frequently produced by a failure in the Crops. Enclosed is a Copy of the Bengal Advices on this subject, with our reply thereto; and we direct that you refer to the Board of Revenue to take into their consideration the propriety of adopting a similar Plan for the countries subject to your Authority. Should you be in need of further information from Bengal on the subject, you must apply for it. It is stated in the papers you have transmitted to us, that the produce of one years good Crop in the Circars, is sufficient for two Years consumption by the Inhabitants. Surely under such a circumstance the consequences of a partial failure might be easily obviated, by causing a certain proportion of the Grain produced in a good year to be lodged in Magazines to answer the exigences of a bad one, and by the formation of an adequate number of Tanks and Reservoirs. Humanity and good policy require that this be made the object of your earliest deliberation.[6]

When Roxburgh arrived in Madras, he soon became aware of the likelihood of both these extremes of weather, and experienced both: the cyclone which destroyed much of Coringa in 1787, as stated in Chapter 2, was of sufficient strength to destroy his house, belongings and nearly his family as well as killing about 15,000 people and possibly as many as 100,000 cattle.[7] The five years 1788 to 1792 produced one of the serious famines, on a par with the well-publicised famines of Bihar of 1866 and Bengal of 1770, when it was estimated that between a third to a quarter of the populations died. There had also been a serious famine in the Northern Circars

betwixt the Latitude of 16° and 18° on the Coast, there so little rain fell, during the years 1764, 1765 and 1766, that the Country was desolated by Famine …. The same thing has now happened again, in the same Country, in so much, that I am credibly informed, One half of the Inhabitants are no more! and the remainder so feeble and weak, that, on the report of Rice coming from the Malabar Coast, by order of the Governor General, 5,000 poor people left Rajamundry, and very few of them reached the Sea-side, although the Distance is only 50 miles.[8]

In the case of the famine that Roxburgh experienced, we have the rainfall records which he kept (see Table 25), so that the train of events can be followed in some detail, as well as his descriptions: his interest in meteorology has already been considered in Chapter 7. Further evidence of the calamitous nature of this famine is obtained from a letter from Andrew Ross of 1793:

the dreadful effects of the famine have not only far exceeded any Description from us, but we are persuaded far exceed all belief; and it may be impossible for us to impress that conviction on the minds of others, which we ourselves feel. In many places where populous villages formerly stood, there is at present neither vestige of Man or Beast, & not a village in the County, which does not exhibit the most melancholy marks of depopulation & decay.[9]

Roxburgh had been able to talk to the Rajah of Pittapore's family Brahmin, who

informs me that he finds among the records of his Grand-Father and Father an account of a most dreadful Famine which prevailed over the northern Provinces during the years 1685, 86 and 87. During the first year, grain was not so scarce, and sold at about 20 Seer of Paddy, or 10 of Rice, to the Rupee. During the second it became more scarce, and sold at double that rate; but the last year there was only one shower fell, so that every thing was most completely burnt up. There was scarce any kind of grain to be had. The price rose to be at the rate of one seer of rice the rupee.[10]

In spite of the comment above about the movement of grain during famines, the Madras Government did make efforts to lessen the effects of the famine in 1792:

> the scarcity of the Grain in consequence [of the failure of the rains] was most severely felt by the poor classes of Inhabitants. Every measure has been taken that occurred to us as likely to afford relief. We suspended all duties on Grain and every kind of provision, we wrote to Bengal in the most pressing terms for supplies, and we sent directions to the several Chiefs and Councils to search suspected places for Grain that might be concealed by dealers with a view to take advantage of the increased scarcity. As it was possible that some of the Countries bordering on the Circars might have escaped the calamity, or suffered it only in a partial degree, we wrote to the Maratta Chiefs at Cuttack and the Resident at the Nizam's Court to obtain such assistance as those countries could afford for the preservation of the Inhabitants of the Circars.[11]

This Public Letter continued with the statement that 12,200 bags of rice had been sent to Ganjam, that the middle and southern Circars had had plentiful rain, so surplus grain was being sent from these areas, and that the poor people of the Northern Circars were selling their children as they could not afford to feed them. The problems, however, in this area were exacerbated by the fact that Bengal was in a similar position and had also placed an embargo on the export of all grain; even so 15,000 bags of rice had been sent to the Circars to help. All this bulk movement of grain, however, was almost certainly by sea.

112 *Grains from Gujarat (a dry area of western India). Top: chena* (Panicum miliaceum)*; left: codra* (Paspalum scrobiculatum)*; right: buntee* (Echinichloa frumentacea)*; bottom: natchee* (Eleusine coracana)*. Hand-coloured engraving by William Hooker after a drawing made by James Forbes in 1780, from Forbes's* Oriental Memoirs (vol. 2, following page 405, 1813).

Year	Rainfall (inches)
1788	$75\ ^{5}/_{12}$
1789	$46\ ^{10}/_{12}$
1790	$17\ ^{4}/_{12}$
1791	$26\ ^{11}/_{12}$
1792	$43\ ^{6}/_{12}$

Table 25 Annual rainfall for the five years, 1788-1792, of severe famine.

One of the comments that Roxburgh made about the famine was to Banks at the end of August 1791:

> The Famine of these provinces begins to rage with double Violence, owing to a failure of our usual rains, a continuance of such distressing misery constantly before my Eyes, is almost the only thing that renders our Residence in these countries during such times distressing. There ought to have fallen about 40 inches from the usual time of commencement of the Rains in May or June till the end of August, and there has fallen only 13½.[12]

This confirms that these famines could take two years of failed or partially failed rains before becoming really devastating. It also demonstrates how enjoyable Roxburgh felt his time at Samulcottah to have been. Roxburgh used this disaster to some advantage, for he continued, 'these distresses have been a means of bringing to my knowledge many indigenous Vegetables that the poor in great measure live on'.

This letter to Banks also highlights a major difference between the Moguls and the Company as rulers: whilst the Moguls had a fatalistic attitude, in that it was

113 Grains from Gujarat (a dry area of western India). Top left: juarree (Sorghum bicolor); top right: bahkeree (Peunisetum typhoides); bottom: batty or rice (Oryza sativa). Hand-coloured engraving by William Hooker after a drawing made by James Forbes in 1780, from Forbes's Oriental Memoirs *(vol. 2, facing page 406, 1813).*

in the hands of Allah, the Europeans, not being used to seeing such mass starvation, took a Christian view, to try to obviate suffering, but in this were often hampered by the grain merchants hoarding stocks and waiting for prices to rise. Amongst Roxburgh's manuscript collection is found an example of a scheme to help poor rural populations in a quite different part of the world. These take the form of excerpts from Newte's *Prospects and Observations*, in which the work of the British Society had set out to help those in poor regions and included the quotation, 'Since the publication of the first Edition of this Tour, I have been informed of a small fishing station erected on Loch Torridon, an excellent harbour upon the west coast of Rossshire, opposite to the Isle of Skye.'[13] In fact, it would seem that this refers to the development of Ullapool on Loch Broom, but the idea of developing a 'fishing station' for the locals as a source of employment would have appealed to Roxburgh as a transferable concept to be applied to the starving Indians. This idea of the importance of the population as a source of wealth was described in Newte's first *Tour*:

> soldiers, sailors, merchants, physicians, and others, in whose imaginations Scotland has been uppermost amidst all their peregrinations and all the vicissitudes of life, returning home with the earnings of industry and the favours of fortune, add to the general wealth of the nation. Scotland, though barren in many things, is yet *ferox virorum*; and men undoubtedly are the most important articles in any country.[14]

The local zemindars, however, took a very different approach to the problems of famine and depopulation, as shown in a letter from Ross, who had been told by Haliburton that 'no reliance whatever ought to be placed on the Statements delivered by the Zemindars, or upon any information of any kind obtained under the influence of their authority.'[15] This was even though Ross had used 'Such Natives as were believed to be best qualified, & least under the influence of the Zemindars were deputed to examine into the accounts of different Zemindar.' It appeared that these people that Ross used had taken the interest of the Zemindars, to alter the taxes that were due to the Company, and Ross himself was sympathetic to their attitude, continuing,

> that an uncommon degree of indulgence & forbearance, towards the remaining Inhabitants, is absolutely necessary, to admit of their recovering themselves in some degree & enabling them to procure the Cattle & implements necessary for cultivation & other purposes whatever loss (if it can be considered a loss) may at first be sustained by a lenient Conduct of this kind, will be amply made up for in the end, as an impoverished & depopulated Country whatever may be the fertility of its soil, can never be expected to pay a Revenue.

Roxburgh, as a botanist, realised that there must be plants that could withstand periods of drought and provide sustenance; and again, as a botanist, he could see that the root cause of the famines was lack of water, yet flowing comparatively nearby were two enormous rivers, the Godavary and Kistna. If these could be canalised and the water spread for irrigation, this was the second prong in reducing the effects of famine. There was a third, of which he was aware, and that was to import from other areas of India the surpluses, but the problem, as mentioned above, was transporting them. It is interesting to note that it took the devastating famine of 1866 in Orissa before the solution to these problems was resolved, requiring the rapid transport offered by the railways which had been developed by then.

One inevitable effect of this reduced population was that large areas did fall out of cultivation. Another consequent problem was the strains it put on the tenancies and the well-being of the zemindars. This aspect was repeatedly used by Roxburgh and Ross when they were negotiating the cowle for Corcondah, for instance: 'I have recd a letter from him [Roxburgh] wherein he gives me to understand that he considers such valuation to be so much beyond what the Country may be expected to produce in the distressed condition to which it is reduced from the melancholy effects of the late Famine, & particularly in the great decrease of its Inhabitants'.[16]

As a result of this awareness of the causes and consequent problems of famines, it is not surprising that Roxburgh set out to become involved in the two ways that he felt able to help: in recommending alternative plants as sources of food, and, as a practical scientist who knew the area well, the use of the waters of the Godavary and Kistna for irrigation.

ALTERNATIVE FOOD SUPPLIES

As early as 1791, Roxburgh was suggesting two plants immediately which could be used as alternatives for food: *Phoenix pusilla* and *Caryota urens*.[17] The latter he described as 'being an immense large Tree, yields a large quantity of farina, or Sago. It is a native of the Mountainous parts of this Country only. There most is sold in their public Bazaars.' Samples of these two plants were sent to Banks at the end of the year, by Lieutenant Uzield who

> has some of the Pithy part of the Tree & some of the Meal of the first [in this case *Caryota urens*] for you to examine, it has saved many Lives during the dreadful Famine which has raged hereabout for some time past, & is still raging with Violence, & will, no doubt, continue to be a means of subsisting many. It is esteemed much more wholesome than the other. The Tree grows to an Immense Size, being by far the Largest Palm I ever saw, & yields a Large proportion of Pith which gives two thirds of its Weight of the finest sifted meal, such as I now send you. The Ripe Fruit are always eat by the Natives, in short it deserves even more praise than Thunberg has bestowed on his Cycas Revoluta of Japan, for besides yielding pleasant wholesome Fruit, & much good nourishing Farina, it yields a Large Quantity of Rettia Toddy (palm Wine) than any other Palm in India.[18]

The fullest account of Roxburgh's ideas about alternative food supplies was written at the end of the famine, in January 1793, but were also reproduced in a letter of November 1799:

> I will further presume to suggest to the Right Hon'ble the Governor General in Council the planting of Teak and other timber trees as well as those which yield adventitious articles of diet to the natives in the time of scarcity on all such corners, hedgerows, embankments, borders of Canals &c as are at present covered with filth and bushes which serve only to harbour wild beasts, contaminate the air, promote disease, and the production of innumerable Reptiles and Insects which devour or otherwise injure the products of the earth. Such spots afford, at the present no profit, either to the owner or occupier, consequently may be considered a public loss of no small consideration, as such places taken collectively over Bengal alone, will amount to an immense extent certainly to many hundred thousand Biggahs. Add that the soil of these spots, where filth and Rubbish has been collecting for years, will always be found rich, yet from their situation fit for little else than the growth of Trees .... In consequence of the dreadful famine which carried off so many of the Inhabitants of the Northern Circars, during the years 1791, 2 & 3, I was induced to offer some suggestions to the Madras Government on the means I thought most likely to lessen the Effects of such direful Visitations in future. Probably that paper may not have reached the Supreme Board, I have therefore taken the Liberty of sending a Copy of it, with this Letter, because it is connected with the foregoing part of it relative to the Culture of such Trees as yield Sustenance to the poorer Class of the Inhabitants.[19]

The cause of the famine once a drought has started was:

The greatest part of the Circars depend chiefly upon the Paddy crop consequently is the most important article of Consumption, this being a water plant requires to be constantly flooded, or nearly so from the time it is transplanted till within a few weeks of its being cut, a period of about four months; of course when the usual periodical rains fail, the Paddy Crop also fails, more or less as the rains have been withheld even should they prove plentiful during the early part of the season and hold off during the latter, it still suffers because there are no reservoirs of any note over these provinces to retain the superfluous waters that fall. This most precious gift is therefore suffered to run off into the sea, nor are there any aqueducts or mode of conveying the waters of large rivers to the Paddy lands.[20]

Having then stated: 'Yet after all it would certainly be judicious, to have as many resources for the poorest and of course the most numerous and most useful Classes of the Inhabitants, as can possibly be found', Roxburgh then went on to comment: 'On Ceylon, a well peopled Island, little Paddy grows, yet I do not remember to have heard that they ever suffer from Famines in the dreadful manner the more fertile Provinces on the Continent do, owing to their having a greater abundance of various kinds of Fruit Trees which yield them sustenance.' He thus showed that part of the problem for the poor of the Circars was their over-reliance on a single crop, rice, which depended on a plentiful supply of water during its crucial growing season; and that other areas, such as Ceylon, which had a similar climate but, because the people used a greater variety of plants for their food, were not so liable to famine.

Roxburgh was then in a position to start offering alternative plants, and as was his custom, using economic arguments as well as purely humanitarian. Thus the first plant mentioned was the coconut (*Cocos nucifera*), and the presentation took in all aspects of why it should be used as well as how it should be grown and the mechanism within the social context that it could be managed: a large part of the letter is given below, as it covers the various plants that Roxburgh suggested, to give a full idea of his attitude and approach.

Cocoanuts, the Oil thereof and the Coir, of which our best Indian Cordage is made, are very considerable Branches of Commerce from thence, yet when a Scarcity of Grain takes place, the exportation of Cocanuts is prohibited. This took place during the time I was Surgeon at Nagore between 1778 and 1780, those nuts yield much Nourishment one or two of them with the water they contain, will support a Man for 24 hours. We are told that the Inhabitants of many of the South Sea Islands, a stout Race, live almost entirely on them and the Bread Fruit. This Tree (the Cocoanut) is exceeding scarce throughout the inland part of the Circars and but to be found in very few parts on the Coast. I therefore conclude the cultivation thereof ought to be attended to, for I conceive it to be the most important of all the Articles of

114 The sugar palm, Arenga pinnata (Saguerus rumphii), *one of the species that Roxburgh recommended as a food crop. Engraving from Rumphius's* Herbarium Amboinense *(vol 1 t. 13).*

Food which we may call adventitious and not much affected by Drought. The Banks of all the Aquiducts, Banks of Tanks, Water Courses, Streets of Towns, Villages &c ought to be planted with them, particularly, where the Soil is not too stiff, for it delights in a light Soil, they grow high and give so little Shade, when in a single Row, that almost every other Vegetable will grow readily under them. The Headman of each Village may be bound down under a certain penalty to plant a certain Number according to the Extent &c of his Lands or at least made to rear and keep up the Number should they be for the first time planted to his hand. The Coir would soon render our Shipping independent of other Countries for their Cordage and Cables which is also an Object, tho' greatly smaller than finding Sustenance for the Ryots or labouring Inhabitants and Manufacturers during times of Scarcity – to enumerate the various uses of this Valuable Tree would require a Volume.

Since I have begun with the Palms, I must beg leave to mention some more of them, that are certainly well deserving attention. I deem the next in rank to the Coca-nut that large indigenous mountain palm which Linneus called Caryota Urens, and the Hindoos of the Circars Jeerooga. It yields a larger quantity of Toddy (Palm Wine) than any other that I know, even as far as 100 pints. I am told the best will give every 24 hours during the best period of the Season, and the whole of the pith of old trees makes excellent Sago.

Cycas circinalis and Revoluta are two sorts of the real Sago palm (more properly Ferns) the next deserves to be introduced: Seeds and plants may be had readily from the Malay Islands and the Travancore Country, where the first grows naturally, and forms a very considerable part of the diet of the Natives.

Date Tree (Phoenix dactylifera Linn.) may be easily reared from the seeds found amongst the freshest Dates from Arabia.

There are still two Sorts of the Sago palm natives of the Malay Islands which deserve the greatest attention. These are the Sagus, or Palma farinaria, and Saguerus[21a] of Rumphius. The first of these is the real Sago Tree which we have not yet got on the Continent; the second also yields Sago, much Palm wine, and a kind of black fibrous substance called Ejoo used for Cordage by the Malays.

The last Palm I shall mention is the Palmaira Tree (Borassus flabelliformis Linn) considered as yielding an Article of Food it holds a place, tho' in an inferior Degree.

Next to the Palms I must place the Plantain Tree, but it ranked according to its Utility, it would claim the first place. This Tree, altho' well known to be one of the most valuable in India, is but little cultivated for sale in these parts (the Northern Circars) they will grow luxuriantly on the inside of Banks of Water courses, where their Roots are kept constantly moist, while the cocoanut Trees ought to occupy the Top or Outside of the Banks.

The Jack, Bread Fruit of the South Sea Islands &c various sorts of Fruit Trees of a hardy Nature, claim the next place and after them, those seemingly (to the unobserving eye) insignificant wild plants which the Natives have recourse to at all times, but particularly during those of Scarcity or Famine. They are numerous, I will only just name a few of them placing such as I think most useful at the head of the List.

	Botanical Names	Tellinga and Hindoo Names	English Names & Remarks
1	Chenopodium esculentum	Ella Kura T	Indian Samphire, a Native of Salt Marshes near the Sea, the leaves and tender Shoots are eat
2	Canthium parviflorum	Balusoo Kura T	No name, is a common thorny Shrub, the leaves eat in Curries, &c
3	Guilendina Moringa	Mollunga T Sajama H	Marunga, the leaves, flowers and unripe fruit are universally eat
4	Æschynomene Grandiflora	Avice T Bascana H	Agatty – The leaves are eat in Curries &c
5	Nympheas	Caalwas T Sundihala H	Water Lilies of several sorts, the tuberous roots and seeds are eat in Stews &c
6	Padma	Gul carninal H	The Lotus or Egyptian Milufer, the tuberous roots and tender Joints as well as the Seeds are greedily sought after by all Hindoos

7	Arums	Ishamas T Cutches H	Eddoes or Water yams of various Sorts, both Roots and Leaves are eat by the Natives of almost all hot climates
8	Cucumis latissimus	Doss cay T Kakrie H	Milem Cucumber in some parts, particularly the Guntoor Circar this plant is much Cultivated, the ripe and unripe Fruit form a considerable article of Diet. The seeds are ground into meal and baked into Cakes. A sweet useful Oil is also expressed from them, which is used in diet and to burn
9	Dioscoreas	Wangoradoo T Hura T Alloos H	Yams of various kinds, they are well known

It is unnecessary to mention more of the small and for the most part wild Plants, however I should blame myself, was I to omit mentioning in a more particular Manner, the large Fruit such as the Tamarind Mango.

Wood Apple
Bassia latifolia, Illippi of the Tamuls, Ipie of the Talengas, Madhuca or Mahwah Tree of the Asiatic Researches Vol 1. This Tree is of a slow Growth but of infinite use, it grows wild among the Circar Mountains. The fleshy flowers are eat fresh or dried like Raisins – an ardent Spirit is also distilled from them, the Pulp of the Fruit is also eatable and from the Seed an Oil expressed.

Not only the Leaves of the Tamarind Tree and Pulp of the Fruit are Articles of Diet but also the Seeds, they are powdered and boil'd into thick Conjee which the poor Natives of the Hill Countries chiefly eat.

The Seeds of the Mango when boiled in the Steam they also eat.

The Opuntia plants, particularly Cactus Cochenellepher, contains much Nourishment in very little Bulk and altho' the Natives do not immediately like to eat it, yet there is no doubt, but they will as soon as it becomes familiar and they get reconciled to it. It grows readily upon almost any Soil with little or no trouble and certainly ought to be

115 The fish-tail palm, Caryota urens – Roxburgh noted that the pith of this species was used as a famine food. From Rheede's Hortus Malabaricus *(vol 1 t. 11).*

made as common as possible with a View to help to mitigate the distress of the poor dureing times of Scarcity.

Should Government approve and wish that Attention be paid to multiply such Resources as I have pointed out, I should imagine that it would be advisable to order some Sack of ripe coconuts for Seed to be brought from Colombo where the best Sorts grow, so as to have the young plants ready to plant on and along the Banks of any Aqueducts, Tanks &c that may be constructed or mended.

Roots of the Bread Fruit Tree (the best sort is propagated from piece of the Root, as that sort bears no seed), may be had from various parts of India.

Seeds and plants of the Sago Trees may be had from Sumatra &c Malay Islands. The rest of the plants pointed out are natives of this country. But after all I fear that no great deal of good can be done, while the present annual system of renting the lands of these provinces prevail, where the sower scarce knows whether or not he will be permitted to reap the Crop he sowed &c if he mends the Bank of a water course or digs a well, he knows not but it may be for the immediate benefit of another; such being the situation of the cultivation throughout the lands under the management of the hereditary Zemindars. Those called Havelly, under the immediate management of Government, are, I hope, better managed; it is therefore on the Havelly Lands that we may soonest expect to see resources for the poor, hitherto unknown in these parts springing up.[22]

Roxburgh's friend, the Rev. John, had advised him on the growing conditions of the coconut:

116 Hardwickia binata, named by Roxburgh after Colonel Hardwicke who supplied him with a number of specimens when he was in Calcutta. Drawing by one of Roxburgh's Calcutta artists, c.1802 (Roxburgh drawing 1432).

The Coconuts grow here best in clay ground at least in a mixed one. Those in sandy ground are much exposed to a most pernicious small insect, like the meal worms, they call here Wercarayan or root carayan, which destroy yearly some trees in our mission garden, & against which I have not yet found out a proper remedy effective, in every case, especially when this dangerous insect has got already two or three feet high. If I discover their presence near the root by the gum & meal, they throw out through very small wholes, I cut off a part of the bark which is most affected, and beat the place where I observe them with a wooden hammer several days to disturb them in their nest and destructive work. This Tree can suffer to be deprived of a great deal of his bark, without being hurt in its growth.[23]

John continued this letter by saying, 'near the paddy fields & on both sides of the channels running through them, & upon the high ground that divide them into several parts, are often planted with Coconut trees, where they grow very fast if they are only watered three or four years', and added that he could get Roxburgh the sort of coconut palms that grew in Ceylon and Jaffna 'which are of better sort than here.'

This lead to the successful adoption of the coconuts and 'such Plants as would yield them [the Natives].'[24] All this hard work on the part of the Madras Government was appreciated by the Court of Directors, who wrote that 'Too much praise cannot be given for your endeavors to obviate the effects of scarcity in the Northern Circars.'[25] Roxburgh also took the opportunity to point out that if his ideas of using all the waste ground was used, 'innumerable clumps of fine stately trees will make their appearance in every part of the Country, adding greatly to its beauty and wealth, as well as to the convenience of the Natives by affording them and their Cattle, at least, a delightful shade, without obstructing a free circulation of air, provided they are kept clear of underwood.'[26]

As part of his work in providing plants for distribution to those interested or in need of them, Roxburgh, writing in 1802, stated that 'Saguerus Rumphii, Gomotoo, or Sago palm continues to thrive here as well as any indigenous tree of the Country. Many thousand plants have been distributed from this Garden, and many thousands (say about a Lack) remain to be disposed of.'[27] The rate of distribution and the problems with devoting such a large area to these plants was highlighted a year later, when he wrote

> Altho' many thousand Saguerus plants have been distributed, and numbers planted out in the Botanic Garden during the last twelve months, yet a great many (say about 50,000) still remains. The approach of the Rains, and the encreasing size of the plants, induces me to beg you will make those circumstances known to His Lordship in Council, and also that there is a piece of ground adjoining to the East Side of the Botanic Garden, which is at present almost unoccupied, having been deeply cut in various directions for making Bricks for many years past, which might with little expense be made to hold the greatest part, if not the whole of the remaining plants.[28]

As was usual, the cautious Council asked for more details, which Roxburgh sent ten days later, stating that the area was

117 The East Indian butter tree, Diploknema butyracea (Bassia butyracea). Hand-coloured engraving, doubtless based on a drawing by one of Roxburgh's Calcutta artists, c.1803 (Roxburgh drawing 1568), from his article on the species in Asiatick Researches *(vol. 8, 1805).*

about Sixty Biggahs nearly one half of which has been, for the present, rendered unfit for cultivation by making Bricks. To bring into order, and plant the other half with Saguerus, would be attended with an expense of about ten Rupees the Biggah: no additional charge need be incurred, as the People already appointed to this Garden could attend to the Plantation.[29]

Roxburgh could not give an indication of the cost of the land, as 'the present proprietor, Peetembur Cundoo a salt Merchant who resides in the Village of Moyeri, a little to the Westward of Calcutta, declines giving me any satisfactory answer'. This application appears to have been unsuccessful then, for William Roxburgh junior, during his time acting for his father who was in Britain, wrote that the ground was still necessary, and this time it was ordered 'that the Board of Revenue be authorized to conclude the purchase of the ground on account of Government, on terms recommended by the Collector of Buddwan'.[30] The piece of ground was defined rather more clearly in June 1806, 'extending from the first assigned piece to Sir John Royds's, and measuring one hundred and sixty Biggahs, which was lately delivered over to me by the Collector of Budwan.'[31]

Towards the end of 1803, there was an example of the way that decisions as to where these plants should be distributed.

> In the accompanying Letters from the Collectors of Bengal, Behar, and Benares, we have now the honor of submitting the result of our enquiries made in obedience to the orders of your Lordship in Council of the 30th September last, in regard to the best means of distributing through the provinces the plants of the Teak Tree, and Goomoota or Sago Palm and for promoting the cultivation of those useful trees.
>
> Your Lordship will observe, that the Collectors in general, express themselves desirous of receiving a supply of the plants in question and that they entertain the prospect of their being propagated to a considerable extent, without any further expense to Government than may arise from the conveyance of them from the Botanical Gardens to the different Provincial Stations.
>
> We would therefore beg leave to recommend, that the accompanying letters be communicated to Doctor Roxburgh, and that he be desired to furnish the plant applied

for at such times and in such quantities as he may deem conducive to the object in view, charging the expense of conveyance to Government, that Doctor Roxburgh be further requested to correspond on this subject with the Collectors themselves, and to furnish them with such general instructions, in regard to soil, situation, and Management, as may appear to him Necessary for the culture of the plants.

It does not appear to us necessary that Government should incur any further expense in this pursuit, than for the conveyance above mentioned, and we would therefore propose to withhold authority for any expense beyond it, unless in very particular Cases which may be previously explained to your Lordship in Council.[32]

A fortnight later, Roxburgh was able to report to Philpot that

Already the following gentlemen have had small supplies sent to them:

Mr Enst, Collector at Midnapore	1620 Teak plants, &	1090 Sago
Mr Middleton, Judge at Jessore	100 Ditto	70 Do
Mr Elliot, Judge at Tipperah	100 Ditto	100 Do
Mr Massie, Collector at Patna	473 Ditto	150 Do
Mr Barton, Collector at Benares	500 Ditto	147 Do
Mr Crommelin, Commercial Resident at Gazypore	326 Ditto	143 Do
Mr Arbuthnot, Judge at Benares	100 Ditto	50 Do
Total	3219 Teak, and	1760 Sago

These Gentlemen have also been furnished with full directions for the Management of the plants, and young trees as they grow up.[33]

Roxburgh continued to grow and distribute these, together with a variety of other economically useful trees, reporting to this effect in June 1804.[34]

IRRIGATION

Amongst the manuscript collection of Roxburgh's in the India Office Library, which has been referred to before, there is an interesting set of excerpts that give some indication of his interests in canals and irrigation. The first of these is from Thomas Newte's *Prospects and Observation; on a Tour in England and Scotland*.[35] The quotation that is of particularly relevance states:

On a general view of the natural face of Scotland, it must occur to any man who has been a witness of the aqueducts in other parts of the globe, how practicable it is, in that mountainous country, to enrich a great portion of the land by means of water; and to form canals that might serve the double purpose of manuring the land, and of water carriage. The former of these improvements naturally leads to the latter. The farmer begins by leading the springs and rills from the sides of the mountains towards his waste, or, as they are called in Scotland, his out-fields grounds, upon an uniform descent of a foot, or a little more, every hundred yards. By and by, he unites several small streams into one, and finds that half of that fall, executed with judgement, will answer his purpose. Thus the fall or descent of the drains and aqueducts being lessened in a reciprocal ratio of the quantity of the running fluid, the attentive husbandman will at last stumble, as it were, on the idea of water carriage, without ever suspecting that his little and partial endeavours, when united with others of the same kind, would produce so grand an effect.[36]

This quotation is followed by one which appears earlier in Newte's book (page 179), on the use of the national debt as bounties, but the next is pertinent here, for it discussed the Union Canal near Carron: 'This canal is forty miles long, and near fifty feet broad, which is a very unnecessary width, as boats of fifty tons are quite large enough for carrying on commerce by canals, and will answer every purpose better than larger vessels .... Vessels come from Glasgow to the sea on this canal, in ten hours.'[37]

We do not know the date when the above extracts were written, but they are echoed in an extract from Roxburgh's 'Meteorological Diary with some consequent remarks', quoted in a letter dated 23 July 1791:

118 The Godavary estuary. Engraving by J. Walker, after a drawing by W. C. Lennon, from Dalrymple's Oriental Repertory *(vol 2, following page 170, 1797).*

Last night the (Elyseram) river rose for the first time this year. I find upon inquiry that the Godavary is also just come down, and as the rise of this river constantly corresponds with the swelling of the Godavary, and as also there has not been any rain near this, nor appearance of any having fallen elsewhere within the limits of our horizon, I must believe the report that prevails of this rivers having a communication with the Godavary far up to the North or NW. Be that as it may, it now brings a most beneficial & timely supply of water to enable the Farmers to begin the cultivation, for by dams across the river, and its various subdivisions, the waters are raised sufficiently high to overflow the adjoining paddy lands. Its natural banks as well as additional embankments being considerably above their level, & the descent of those lands from the hills from whence the river issues is so great as to admit of a dam every half mile, or mile & a half. It is but a small river, the breadth is very various & so shallow, that a tall man standing in its bed will in most places see over its banks. Consequently those dams are easily made, or repaired, with some sticks, straw & Earth, which easily gives way when rises to any considerable height during heavy falls of rain a thing required, that the water [next seven words lost on crease] its banks, & by them we may also nearly calculate the whole descent from the hills to the sea. The higher up the more it is, as the dams are nearer one another than towards the sea, (indeed there are none after you come within about 3 miles of the sea) so that upon the whole I may safely venture to say it cannot be less than two feet to the mile in a direct line, or in other words, that the level of the paddy lands next to the hills is 40 feet above that of those bordering on the sea at Cockanada, where the south end of my direct line terminates & the north end of the same line I consider to be at the hills on the west side of the Talapella pergunnah.[38]

In the remarks that follow, he referred back to the 'melancholy proof that the descent from this place to the sea is full what I have stated it at on the 21st May 1787'. He continued that 'in consequence of the favorable level & descent of lands through which it passes, we clearly see the infinite benefit that must arise from the waters of large rivers when a method of making them subject to the will of man is affected.' His final comment here was on the practicability of making the necessary measurements:

I conceive the descent of the lands adjoining to the Kistna & Godavary as well as the lands of these rivers could be ascertained in a similar manner, during the months of April & May when there is very little water in them. The natives understand only this mode of levelling, it is probably the best, & cheapest and most expeditious that can be taken when there is only a small stream of water to dam up, however it is also probably my ignorance of the art makes me form such an idea.

Here, for almost the first time, is direct evidence of Roxburgh using information that he had gathered from a book for promoting a project in India. A few excerpts later, there is a further Memorandum, undated, but which started with a quotation from Newte: 'Providence never fails to be kind, but very often leaves something to our own industry to deserve.' This is followed by a detailed description of the possible benefits of making two tanks: 'I will mention only two not only from my own observations, but from the authority of the most respectable native Farmers up there. At a village called Shankarlapoody one might be made, that would water a very large extent of country that has hitherto been but partly cultivated for dry grain, those natives who are better acquainted with the expence of such works think.' The cost of this would be 5,000 Pagodas and would affect the ground belonging to eight villages:

Shereporam	annual value in general	1000	would then be worth 2000
Gerra-waram		1200	2600
Pedenapilla		800	1600
Sullanka		800	1600
Patta-pore & Gaindaporam		330	600
Ellore		300	600
Rayarum		400	800

At Peddashankarlapoody a second Tank may be made, it will be but small, and may cost 1600 pagodas. It will water three villages, vizt

Rant-pallam	now worth 200	would then be worth 500
Maraty-massarum	200	500
Peddashankkarlapoody	300	1000
	700 annually	2000

To support his argument, he continued that 'At Lingunput, a village four miles west from the above mentioned places one Rameauze a man of note, made a Tank about ten years ago at his own expence which is said to have cost 5 or 6 thousand pagodas. It waters the land belonging to six villages, they were reckoned worth about 4 or 500 pagodas yearly before the tank was made, & since they say they have yielded more than twice that sum.' As with the case of the Scottish farmers collecting the rain from small streams and rills, so with the two villages mentioned above,

> a small Brook issues out from the hills, during the driest weather it has, the natives say, always a small run of water in it up there, which is supplied from various springs up amongst the mountains, but that almost as soon as it passes the skirts of the hills it gets into loose soil & is there absorbed except during the rainy season, so that were the forementioned Tanks to be made their banks would run across this Brook, so that they would be filled with a very moderate portion of rain even during seasons when the low lands are parched up with drought, for it is a well known fact that hills & mountains attract the clouds & that in such countries there is much more rain than in those that are flat & low.

This, again, supports the argument that Roxburgh was knowledgeable about the causes of weather, in this case, what used to be called Consequential rain.

This Memorandum, after mentioning another two possible rivers for creating further tanks, ended with the uses to which the water could be put:

> The waters of these or as much thereof as may be necessary & of many more such small Rivers, Brooks, & Springs might certainly be preserved in Tanks for the purpose of fertilizing the adjoining paddy lands during a scarcity of Rains. The cultivation of the Sugar cane, Ginger, & Turmeric might then with much ease be carried on over a great extent of country, which is chiefly confined, at present, to the lands bordering to the Elyseram River which is a little Nile & to trace its source would also require another Bruce. The benefit arising from those reservoirs would be immediate even while making,

119 Map showing the stretch of the Golconda River entering the Circars. From an engraving by W. Harrison after a drawing by Lieutenant Alexander Denton in Dalrymple's Oriental Repertory *(vol. 2, following page 168, 1797).*

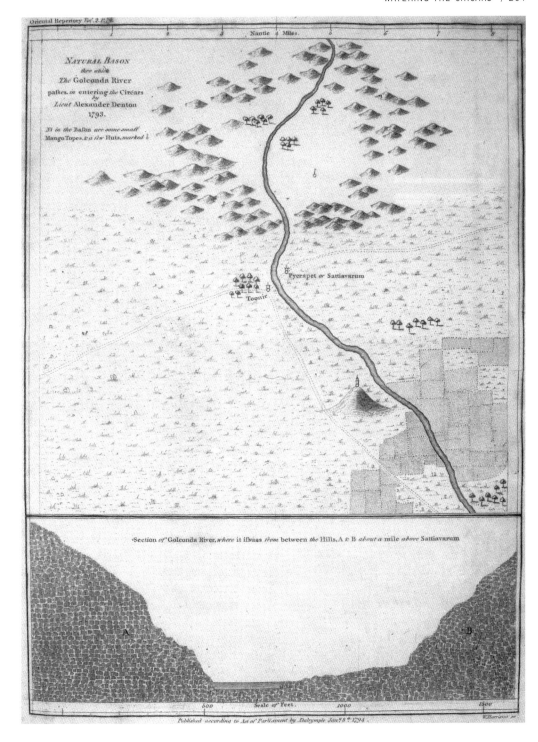

because the labouring people would find employment during the present dry season, which is without equal, else many must, I fear, still perish, for the price of grain, when compared with the price of their labour in dules, at the low exchange they pass, is by no means adequate to their real wants however few. Breadfruit, Coconut &ca other useful Trees before mentioned might <u>at the same time, with little additional expence,</u> be planted along the Banks & on such adjoining spots as may be deemed proper.

James Bruce (1730-94) was a great African traveller, who reached Egypt in 1768 and sailed up the Nile. From Aswan he crossed to the Red Sea and landed at the port of Masuah in Abyssinia in September 1769 from where he explored the sources of the Blue Nile, returning to Aswan in November 1772. His *Travels* were published in 1790.

The first indication that a survey was taking place in the Circars, was the appoint-ment of Mr Topping who was to be the 'Company's Astronomer & Surveyor' and the receipt by the Court of Directors of his 'Survey of the Bay of Coringa'.[39] The following paragraph stipulated that

> Mr Topping, till he can set about the intended admeasurement, may be employed in surveying the Circars, particularly the Rajahmundry Circar, as it would be of very great consequence to have the Company's former Orders for a Compleat survey of the Lands, carried into execution. But this should be a <u>mere Land Survey</u> expressing the kind of Land, without any reference to the value, which might raise Jealousy and discontent; such a survey would at once shew, not only the position and nature of the Lands present in Cultivation, and with what cultivated, but would shew also such as are not in cultivation, and what Improvements might be made. ~~By turning the Stream of the Rivers Godavary and Kistna, to good effect.~~ [This sentence, as shown, was crossed out.]

The final paragraph relevant to this issue, stated that

> One consideration of much moment is, the easy communication with the Sea, or Water Carriage; because, for example, if the transport is remote from the pepper Plantations to the Sea, the expence of carriage would very much enhance the price of the pepper, which the Chief and Council of Masulipatam did not attend to when they recommended the Pepper to be cultivated in the distant Hills.

This would seem to refer to the pepper plantations that Roxburgh had instigated and the importance that the Court put on their future, that have been referred to before. Roxburgh, in a letter to Colonel Kyd, stressed the need for something to be done:

> I have often been astonished that in no one place of the Circars, have I ever met with, or heard of, the least trace of any work, ancient or modern, for retaining, or conveying the Water to fertilize their Paddy Lands: the Cultivators here depend entirely on the rains, when they fail, a famine is, and must ever be, the consequence, till some method is taken to supply their Fields with water in case of a failure of rain.[40]

In a note accompanying this extract, Dalrymple added, 'This can relate only to that part of the Circars which is watered by the Kistna and Godavery.'

This letter of Roxburgh's is quoted at some length by Dalrymple, as it gave suggestions as to which rivers of the Kistna, Godavary and their deltas could be canalised. At the end of the extract, there is the comment, 'This last is what Dr Roxburgh refers to; *vide* the whole of this River, in the Hydrographical Map from the Kistna to Coringa, by AD, 1783.' The implication here is that Dalrymple was involved with the mapping of the Kistna in 1783, by which time Roxburgh was at Coringa, giving the suggestion that they may well have met, hence the inclusion of so much of Roxburgh's material in the *Oriental Repertory*.

The problem that is referred to again and again over the ensuing pages of Dalrymple, was that the sources of the Kistna and Godavary were not known. When Lieutenant Denton, 'who is stationed here, was Commanding Officer, of an advance Party, to explore the roads, &c for Colonel Cockrell's Troops, his account corresponds exactly with Capt. Beatson's remark: but still no information from whence that Nulla comes.'[41]

A letter at the end of 1792, from Andrew Ross to Roxburgh, gave much more information on the development of this plan, to bring water to the Circars. 'I am now to acquaint you with a subject of which you will see the importance on the perusal of the three first of the inclosed papers which I have put into the hands of Sir Chas Oakeley, on the utility & practicability of affording every advantage of a supply of Water to the Northern Circars – by means of the Rivers Guadavary & Kistnah & the Leakes & Streams which are already connected to them. The eligibility of this great Object is too self evident to require any observations from me; nor will it escape

*120 Map of the Northern
Circars. From an engraving by
W. Harrison after a drawing
by Lieutenant Alexander
Denton in Dalrymple's*
Oriental Repertory *(vol 2,
following page 168, 1797).*

your recollection, that I have sometimes spoke of it to you, & that you expressed the strongest wishes that it could be accomplished.'[42] Sadly, the three enclosed papers do not appear to survive, but some indication of their contents can be gathered, for Andrew Ross continued:

> when I received from Doctor Roxburgh the inclosed Copy of what he had written to Coll. Kyd upon the Subject, I asked my friend George Baker to give me his opinion of it and when he did so, & gave me what you receive here inclosed – I asked Dr Anderson for the inclosed Copy of what I knew that he also had written to Coll. Kyd – and I determined to make the best use of all this excellent Matter with Sir Chas Oakeley – who on reading my Address to him (also sent herewith) & those papers which accompanied it, expressed an intire satisfaction, and assured me that he would give the Object every attention.

Andrew Ross suggested that Major Beatson was the 'most proper person to consult, as well from his knowledge of the subject as of a great part of the Country' but 'that Mr Topping is to be employed as the Chief Surveyor & Conductor of the Plan, and it is fortunate that so able & honest a Man is at hand to execute so great a Work – & which the Major [Beatson] tells me, that he is highly pleased & happy at being employed in.'

Whilst Beatson was in favour of building dams on the Kistna and Godavary, which was strongly disapproved of by Dalrymple, Roxburgh was more interested in developing tanks from the overflow of these rivers.

Here Capt. Beatson's zeal has prevailed over his judgement! for, not to mention that it is by no means certain, any Dam, is necessary for conveying a sufficiency of water from the Kistna, into the Guntoor Circar, or into that part of the Masulapatam District, to which it can be conveyed; it is obvious, that turning the Kistna out of its present Channel, would ruin the Country to the Westward, and to the southward, of Masulapatam.[43]

While Roxburgh was quoted as writing that

the point to be attended to, immediately, seems to me to be enlarging those Channels and making Tanks; for it is far too advanced in the Season, to begin any grand work, such as a Dam across the Kistna: but the sooner Tanks, &c on a small scale, are begun, the sooner will the remaining part of the Poor be enabled to live.[44]

Roxburgh concluded this letter by taking the longer view, of the problems of over-production:

should the succeeding Seasons be as favourable for the Crops, as there is a right to expect, what is to be done with the Overplus Produce? For there are not Inhabitants left, sufficient to eat half a good Crop: it cannot be exported, on account of the exorbitant inland duties; for only between the Hills and the Sea, near this, they come to about 30 per Cent on Grain.

Following the discovery of a natural basin by Denton,

when on his trip towards Golconda; many such are to be found all along the skirts of these Hills; and, in my opinion, they are the places where abundant supplies of Water can be retained for every purpose; throwing Dams across (between two Hills) is a thing that can be done at a small expence, when compared with an attempt to dam up either of the two large Rivers. Golconda River, is not near so large as the Elyseram River; yet the latter may easily be banked up, between those Hills that it passes on its way into the Circars.[45]

After a brief comment,

We continue to have a dreadful hot weather. The Thermometer has been as high as 109 in the Shade; it is by far the hottest April I have ever felt in India', Roxburgh continued that such basins 'are, to my knowledge, very numerous; and with small Rivers, Brooks, or some such supply of water, running through them, they have a number of advantages over small Tanks; for a Bason of such capacity occupies infinitely less land. Than as many small Tanks as would contain an equal quantity of water; consequently they expose a smaller surface for evaporation; the waste of water will therefore be the less; and Banks confined between two hills, must be constructed at less expence, than when carried on, over Land that is void of hills, more flat, and every way less favorable for forming reservoirs.

The fact that digging out the tanks would provide extra occupation was another approach that Roxburgh had used, for in a note, Dalrymple stated 'Occupation is urged by Dr Roxburgh as an argument in favour of the proposition, from the consideration that it would afford occupation to multitudes of People, and I remember, in the war of 1756, multitudes of Natives were employed on the Works at Fort St George.'[46]

The ideas that Roxburgh had gleaned from Newte were used when he wrote: 'there may be less trouble in forming Aqueducts than was at first imagined; particularly as there is as much descent in 45 miles, through the Valley of Peddapore to the Sea, as mentioned in my Yesterday's letter; as there is in 65 miles down the courses of the Godavery.'[47]

By the middle of April, Roxburgh was determined to set out to have a look at the lie of the land, with Denton:

I shall examine well the Country, and every other circumstance, as soon as Mr Denton and myself can go out. It will be a dreadful hot trip; yet it is the only time of the year that we can venture in amongst the hills; and there is no prospect of doing good to the

Country, by means of Tanks, Canals, &c without going in there, where, by the economy of nature, we are to look for, and expect, similar elevated situations, to that sent you the other day [Denton's report].[48]

The driving force behind this project, other than Dalrymple in England, was Sir Charles Oakeley, of whom Ross had stated, 'this is what naturaly pleases the Bart most, & recommends you chiefly to him, as it will redound greatly to his Credit!'[49] The reason for Roxburgh's involvement is evidenced from a letter a few days later, in which Beatson wrote: 'I think your friend Dr Roxburgh might be enabled to bring us very satisfactory information were he to take <u>Bezoara & Bampettah</u> in his intended Journey to the Presid[cy] and upon his natural disposition to promote useful undertakings, & this especialy – which he has certainly the merit of suggesting.'[50] Ross, being based in Madras, was able to apply pressure on Beatson to provide

> a sketch of a Map which contains all the Districts in which the Kistna runs that are connected with the places & situations of which advantage may be taken to procure the expected assistance of Water from that River – with a Letter of explanation … and both these, accompany this – franked by him on the Service – the Map rolled up in Wax cloth.[51]

This letter emphasises the closeness with which Roxburgh worked with Beatson on this project, laying the foundation for the time they had together in St Helena twenty years later. Also in this letter, Ross passed on the approval that Sir Charles Oakeley had given to Roxburgh becoming involved: 'I have also told Sir Charles Oakeley that it is your intention to pay him a Visit & that I would make it my earnest (& indeed absolute) request, that you would come by land, for the purpose of exploring the Country – for the pursuit of the great Object in view of which he much approved.' Indeed, Sir Charles had heard from Lord Cornwallis, the Governor General, that he 'earnestly recommends the speedy execution of it' rather than the 'slow process of Surveys & Levels', which explains why Roxburgh was requested to do the surveying rather than leave it to Mr Topping. Roxburgh, however, was obviously worried that he might

> give offence to the people of Masulipatam or to Major Gents or Mr Topping (who is pleased at it). All that has been done with the knowledge and consent of the Government & I have sent it all home to Mr David Scott – Govnr Hornby, Mr Dalrymple – So for Gods sake be easy on that score & let Mr Denton & yourself go on & do every thing that is possible & every where.[52]

In planning this survey, Roxburgh had been greatly hampered by his medical garrison duties and the Medical Board not granting him an Assistant to take over his duties while away. However, in mid-February Ross wrote: 'I have leave of Sir Charles Oakeley to acquaint the Medical Board & their acquiescence to say to you that you may immediately apply for an Assistant – on the grounds of the various other employments which occupy your time so much And it will be immediately granted.'[53] The next problem was to get some instruments with which to achieve this object, but Ross wrote:

> I am sensible that you will be under some disadvantage from the want of those Articles in the Surveying way which you desired me to procure from Major Beatson … but I am sorry to say without effect as he has told me that he has not any of them, nor can procure them … he has however, put into my hands all Matters Mathematicks, that I might extract what is said on the Subject of leveling which I send you herein, & he says that there can be not better Tube than what you can make of Bamboo, according to the Instructions in the paper … I now send you a Pocket Compass borrowed of Dr Ball, to whom I have promised another in place of it, in case you do not return it soon. A Perambulator & a Leveling Instrument I should suppose you might procure from Major Gent either immediately or thro' Dr Binny.[54]

The following day, the tube was despatched with instructions as to its use which Beatson had said were 'so plain that you cannot mistake them'.[55] The tube was a simple sort of theodolite, used for measuring height.

Towards the end of March, Ross was 'getting ready a Perambulator Wheel & I have ready a Theodolit to send you'.[56] These, together with the compass that Ross had already procured would have supplied Roxburgh with the three basic instruments necessary for his surveying work: compass for directions, theodolite for levels, and perambulator for distances, which together with the necessary mathematical knowledge would allow Roxburgh to work by the accepted method of triangulation. A few days later, Ross was able to send

121 *The pikota, a traditional way of raising water from wells into water channels, a method that would have been familiar to Roxburgh. Photograph by the author's mother taken in Guntur District in the 1950s.*

> The Mahogany Box contains a Theodolite which with its Frame or Stand I bought at Coll Ross Sale for Pgs 7 after Major Beatson had valued it at from 15 to 20 & thought it best (notwithstanding) to purchase it altho' he seemed to think that it is not what you had an idea of when you asked for a Leveling Instrument, but this he thinks, will answer the purpose tollerably. In my last I told you that I was getting a Perambulator made – but it is not yet finished – & will cost some trouble but the use of it will be material in your Examination of the Country on which you now proceed without delay or interruption – as your Asst Surgeon goes by the Minish.[57]

The perambulator was finally sent on 5 May, at a cost of Pgs 20: 40: 20 (see Table 26).[58]

6ft of Brass for the Plates	Ps	1	0
3ft of Tranquemalay Wood for the Wheel		1	33
1ft of Navel Wood for Do			15
Cooley for fetching the above Articles			6
2 Tiles			16
2 doz of screw Nails			6
Painting the Wheel three times			36
Pd the European Carpenters for the carpenters Work		4	0
Pd the European Brazier for casting 2 Brass Plates		1	0
The European Smith for the Iron work that fixes the Brass Plates to the Wheels		1	0
A Bar of Iron for the stroke of the wheel		1	0
Pd Mr Gordon the Watch Maker for Cutting the teeth of 2 Plates		5	0
for Engraving the 2 Plates		3	8
Straw for covering the Wheel			12	20
Fraight pd Capt Wheadon in advance		1	0
Pags		20	40	20
Pd a tin man for making a Tin Tube sent before			24
Errors excepted Pags		21	19	20

Table 26 Charges for making a Perambulator Wheel for Dr Roxburgh.

The story of the manufacture of this piece of equipment gives a good example of the problems confronted by Ross:

> It was altogether a chance, that I had the opportunity of procuring it to be made. None of the Public Offices, (Engineer or Artillery), had it. In the List of the Effects of Coll. Ross, which were left for Sale, I saw it mentioned, & determined to buy it at the Outcry; but then learnt, that the Coll. Had given it to Lieut. McKenzie of his Corps, & that probably it was the only one to be found upon this Coast, at present; & that it was more accident than industry or attention, that had put it into the Colls hands. Pringle, at his death, having left it to Captn Jennings, who bequeathed it to the Coll., &

he having no further occasion for it, determined to convert it into Money; & it might then have fallen into hands, that would use it for fire wood, but for the Accident of Mr Mackenzie's seeing it in the List for Sale, & prevailing upon the Coll. to part with it.

But this is not all. When Mr Mackenzie did me the favor to put it in my hands, that I might get another made by it, I applied to Major Hall, (of the Artillery), to assist me in procuring workmen & materials; & indeed I found, that he was the only person that could do so. It was fortunate, also, that he had the best disposition to do so. He told me, that there was but one man, a sad drunken Invalid Sergeant, (who could not be trusted with the money to buy the brass, or for the work, until it was finished) that knew how to cast the plates, & that it was the same with the workman for the wheel; but by the Majors influence & care, they were employed & performed their parts. It was then necessary to get the assistance of an able & expert watch, or rather, Clock, Maker to cut the teeth on the edges, & the figures on the face of the plates, which is a very nice piece of work, as the smallest deviation or Mistake, renders the Whole of the Measurement abortive. This was done, By Mr Gordon, who arrived from England in the last ships, & has a high Professional Character; all these person had Mr Mackenzie's model before them. When the Whole was finished, & painted, Mr Mackenzie did me the favor to try it, twice, upon the road, for 1 or 2 miles, & I have notes from him approving of it, so that you may rely upon it best, to make no mention of it to Mr Topping; but from late communications with him, I am in doubt whether he has any Perambulators, as he has applied for it at Bengal, & (with several other articles in the surveying way), I saw that he has Indented for two from the Court of Directors; desiring that they might be made with a double Wheel, but this, tho' of some advantage, Mr Mackenzie says, is not of much consequence, nor can I explain it. Major Hall has now had a Wheel made for himself; & Mr Mackenzie has set out with his, to join the Nizams Detachment. He is an excellent ingenuous Man.[59]

This was not the final hurdle, for the captain of the ship taking it would not accept it on board until the freight, of one Pagoda, had been paid in advance.

There is no evidence either way as to whether Roxburgh and Denton actually did this survey, for none of the letters from Ross state Roxburgh's address and nor do those quoted by Dalrymple, so it has been impossible to trace Roxburgh's whereabouts during this critical period. However, it must be remembered that Roxburgh was aware by this time that he would be moving to Calcutta. In fact, the monsoons returned to their usual level in 1793, and the need for watering the Circars was no longer immediate. The urgency thus disappeared and the project was not taken up again for nearly 60 years, for Sir John Kaye wrote that 'Similar works [canal building] for the Godavery and Kistnah rivers, in the northern part of the Madras Presidency, have been sanctioned, and are in the progress of execution ... in 1850. £150,000 [was sanctioned] for the latter [Godavary and Kistna].'[60] It is probable that the many tanks (artificial reservoirs that are used for irrigation as well as storage of drinking water) were built at this later time. Roxburgh's involvement with the planning of an earlier scheme gives an insight into his care for ensuring that all the details were in place, that he was aware of up-to-date thinking and of the practices that would be used, that he chose the time that was right for the job rather than what might have been the easiest in less severe conditions, and that he was aware of the feelings of others, both to improve the lot of the natives as well as those of his fellow countrymen.

122 The Godavary Barrier below Rajahmundry as it is today. It was first constructed by Sir Arthur Cotton in 1847–52.

123 *Fishing on the Coromandel Coast. Photograph taken by the author's mother in the 1950s, but showing a scene that must have been familiar to Roxburgh.*

SUMMARY

Crop failures are frequent events in areas that experience monsoons and these, if they last for more than two seasons, can cause famines on a major scale. Before the introduction of modern methods of transporting large quantities of cheap food and grain, these famines caused the death of thousands of the local inhabitants. Roxburgh sought to alleviate this suffering, first by seeking alternative crops that would withstand failure of the rains, and a system of reliance on a wider spectrum of food crops that has been followed ever since. Once more showing him to be at the forefront of humanitarian and economic innovations: over and over again in his letters he refers to the starving and underemployed natives who need occupation, hence his work on silk, sugar and alternative crops.

His other alternative was to suggest the use of rivers, in this case the Godavary, which has its source sufficiently far away that even in the area affected by the failure of the monsoon, sufficient rain still falls there for the flow of water to be plentiful. This required surveying which itself produced further problems, of relevant equipment some of which he had to have specially made, plus the time and energies of experienced people. It is a reflection of the regard in which he was held, that he was deemed to be the person to carry this out in spite of there being a Company Surveyor in the area.

Travelling round the Rajahmundry area when I visited it, and from my memories of my childhood, the importance of the canals for the rice was everywhere very evident. The tanks were also much used fifty years ago, and still are, with their importance being as much for the fish and waterfowl that live in them as a source of water.

CHAPTER 14

Roxburgh at the Cape of Good Hope

BACKGROUND TO THE JOURNEY

From his first voyage to the east, Roxburgh became familiar with the Cape, one of the main revictualling and watering places for these journeys. The pivotal position of the Cape is treated at some length by Harlow, who stated that

> in the long sea-passage between Europe and India the only port in British possession was St Helena. Like Mauritius this Island was barely self-supporting, so that the needs of the garrison and of the Royal and Company ships that were constantly calling there were chiefly supplied from the Cape .... If ever the Cape were controlled by an enemy, St Helena would be crippled as a port of supply and would also be in danger of attack from the Cape itself. The British trade-route to the East, upon which the Empire in India depended, was thus in the weak position of relying upon a South African port owned by a Foreign Power .... The Directors of the East India Company were scarcely exaggerating when they declared that whichever of these powers [France or Britain] shall possess the Cape, the same may govern India.[1]

For long years, the Cape was governed by the Dutch East Company who remained neutral between France and Britain. From the early 1780s its position was more equivocal after an abortive attempt to capture it, and then in the winter of 1794/95 France overran the United Provinces (modern Holland). As a result two British squadrons set sail from England in April 1795 with a total force of 6,600 troops and successfully accomplished the Cape's occupation. Thus, when Roxburgh arrived there in 1798, it was still recently within the British sphere of influence.

It is, indeed, possible that Roxburgh met the plant collector Francis Masson (1741-1806) at the Cape on one of the latter's collecting trips for Banks between 1772 to 1774, and this connection may have continued, for there is an enigmatic list of plant shipments headed 'Masson & c.'[2] Among this shipment of 22 listed consignments are three that include Roxburgh's name, numbers 13 to 15:

13. Capt Gerard, Deptford, Indiaman. Seeds Madras Dr Roxburgh
14. Capt Moffat, Phoenix, Indiaman. Live Plants, Madras Dr Roxburgh
15. Dᵒ Dᵒ Dᵒ Dᵒ Dᵒ Mr Dundas & Dᵒ

The importance of the Cape botanically, and Francis Masson in particular, is shown by the number of plants that he is claimed to be responsible for introducing into England by his death in 1806: 183 species of *Erica*, 175 of *Mesembryanthemum*, 102 of *Pelargonium*, 57 of *Oxalis* and 42 of *Stapelia*.[3] Access to the colony became easier after the mid-1780s when it came under British influence, and led to Masson's second period there, from 1786 to 1795. In September 1795, six months after Masson left the Cape, Britain finally took possession of the Cape for the first time.[4] It was at this point that Roxburgh made his suggestion that since we have been in possession of the Cape of Good Hope, it has often occurred to me that a small Garden established there would be of infinite use. In the first instance as a resting place for plants to recoup at from India to Europe & the West Indies; & from those countries again to our colonies to the Eastward of the Cape, & for forwarding the Plants of the Cape itself, as well as for introducing others into that Colony should we retain it; or even should it be restored to the Dutch at the close of the [Napoleonic] War.[5]

Masson had, in fact, developed a small garden at the Cape where he had carefully cultivated the plants that he had collected before forwarding them to Banks.[6]

Roxburgh was aware of the botanical wealth of the Cape, and in February 1796 had asked Lt Owen to 'make a collection of any curious plants, Flowers, Bulbs or seeds I might meet with during my residence at the Cape of Good Hope, and forward them' to Sir John Shore.[7]

As early as April 1795, Roxburgh was complaining that his health had deteriorated since leaving the Coromandel Coast: 'I am now from home on account of a bad state of health which I have been more or less afflicted with ever since I left the Coast but more so for these last two months.'[8] By the end of the year, he was saying 'my health has suffered so much for these last nine or ten months, that I fear I shall be obliged to try some other climate, which is the thing, of all others, I wish most to avoide.'[9] Six weeks later, he wrote to Colin Shakespear 'my constitution has suffered so much during these last twelve months that I am under the necessity of soliciting the Hon'ble the Governor General in Council permission to proceed on a voyage to sea for the recovery of my health should it be found adviseable so to do', which was agreed.[10] This demonstrates a major change of attitude for which the draw of the Cape could perhaps provide an answer.

124 The Dutch East India Company's arsenal at Stellenbosch.

The study of Roxburgh's health has been considered in Chapter 5, but it is of some relevance here, in that his letter of 10 December 1796 to Banks he continued 'my health has been so long on the decline that a longer residence in India without a little renovation in a cooler climate may be fatal'. He must have given this some thought, for his proposal is well developed even if far fetched and unlikely to have been approved:

I could therefore wish to be allowed to remain some time at the Cape to take care of the Garden. The extensive correspondence I am favored with in the Botanical way would soon enable me to make it one of the first Gardens in the World; & if the company is allowed to defray the expense, I can engage to reduce that of the Garden here so much as to defray all that would be necessary for the Cape Garden. I could go backward & forward between India & the Cape without rendering it necessary to have any superintendant for the Bengal Garden other than Mr Smith during my absence, & for the Cape such another man as Mr S – when I may be in India, and I fortunately have a very excellent one here, who lost his employment as Nurseryman, on Mr Smiths arrival.

This presumably refers to Douglas.

Roxburgh completed this section of the letter with a plea to Banks to 'endeavor to have it put in execution' as he had no doubt that it would meet with Sir John Shore's approval as he was 'always ready & willing to grant his aid when any thing which promises benefit to mankind is going on'. There is no mention in any letter from Banks furthering this idea, but he had written to H. Macleod, the Secretary to the Council, requesting that Roxburgh should be allowed to proceed to the Cape for his health, which was again agreed.[11] The same day that he wrote to Banks, he also corresponded with Dr J. E. Smith, indicating that he was on his way to 'the Coast of Coromandel, & will proceed to the Cape of Good Hope, should I not derive benefit from the coast air'.[12] It is not certain whether he went to the Coromandel Coast, but he was repeating his intention of embarking for there in January 1797.[13] However, he was again requesting for permission to go to the Cape in November, when he excused the repetition by saying that 'various circumstances intervened to prevent my benefiting by the indulgence granted me, ... in December last, to go to sea for the benefits of my health, which still Continues to suffer. I have therefore to solicit the same indulgence may again be granted for going as far as the Cape of Good Hope.'[14] It was agreed that Thomas Douglas would sign the monthly bills until Dr Fleming arrived to take over as Acting Superintendent but Roxburgh was still in Calcutta in the middle of January.[15]

Roxburgh had arrived at the Cape before the end of April, after a voyage on which his 'health suffered while on Board ship, but find myself better since I came on shore, so that I hope to derive benefit by the trip. I mean to remain here six or eight months, or till William, and his sister join me', and ended this letter to Dr James E. Smith with the request, 'if you have any commands that I can execute for you here write me soon'.[16] The next day he started a letter to Banks, and it is interesting to note this priority, that Smith was written to before Banks, suggesting once more that Roxburgh was beginning to look to other people for advancement in England, even though in his letter to Banks he asked him to procure a Writership for William junior.[17]

PLANT COLLECTING AT THE CAPE

However ill Roxburgh might have been on the journey to the Cape, it did not stop him working, for in the letter to Banks he commented, 'on the passage from Bengal I completed an account of the Tasseh silk worm', and included copies of the drawings which are in the Linnean Society. These, together with various specimens of plants including a 'Tea plant I have found in one Garden growing luxuriantly', were sent under the care of Dr Fontana of the Bengal establishment on the Danish ship *Dannenberg*, which Roxburgh had himself travelled on from the Bay. In an addition to this letter, he wrote:

> I have seen Lord Macartney [Governor at the Cape and who had headed the embassy to China five years previously and from whom Roxburgh had obtained tea plants on that occasion] find him much inclined to encourage establishing a Garden here, but as he means, I understand, to leave this place in the course of three or four months, he will not likely be able to do much.

This was reinforced by the fact that he had brought with him 'many fine plants in the highest order on shore with me here. Amongst them are the grafted mango, China fruits, &c.' Two further comments are of interest from this letter. The first was that, even though Banks may 'not much want Cape seeds, yet I will collect & send you some as soon as I have a little time' and later he compared the costs of labourers: at the Cape they received four shillings a day as against three pence in Bengal.

The replies that Roxburgh received from Banks, particularly regarding the search for new plants, are illuminating. As one of the gardeners accompanying Governor King, on his way to New South Wales, was George Caley who, of the three gardeners on board, 'has more Genius than the others' and Banks had

> directed Caley to attend principaly to the plants on that Peninsula between False Bay and Table Bay which I believe has not been so much examined as the more distant parts our Botanists used to travel 100 miles at least before they thought it worth while to begin their collections. Groote Vaaders Bosch a few miles from Schwellendam is I am told one of the most productive spots for a Botanist in beautifull Trees & Shrubs.[18]

This area that Banks mentions is either just south of Cape Town (the peninsula between False Bay and Table Bay) or only about 120 miles due east of Cape Town (the modern Swellendam), compared with distances of nearly 200 miles north of the Cape that Roxburgh visited (Zwartberg).

By March 1799, Roxburgh had had a successful time plant collecting, for he sent Banks numerous seeds amongst which

> was a parcel marked "probably a Daphne from Gardens Bay". This I now find is Lachnea conglomerata, a plant, I am told, much wanted in England. I have therefore sent you another parcel of seeds gathered in Simons Bay. Also parcels of other seeds collected since my last despatch, as also the seeds of 27 species of Protea, almost 50 of Erica.[19]

He complained that the lack of books meant that 'names I have often been obliged to give them myself, tho I do not imagine they can all be new'. This complaint was expanded upon three months later to Smith who was thanked for

> your volume of tracts on Natural history, for Jussieu, & for your own papers from the third & fourth vols of your transactions. Jussieu I have found of real use in helping me out with the Genera, where otherwise at a loss, and here I want assistance much, having brought only Gmelins Edit[n] of the Systema with me, which is but poor help.[20]

In this letter to Smith, he mentioned that he had sent to Aylmer Lambert

> specimens of 63 proteas, & some seeds. The latter I requested he would divide with you. I have found in all about 80 species of Protea, 162 of Erica, and about twice as many of Brunia, Phylia, Diorma &ca exclusive African Genera as even Gmelin enumerates. But want of Books to help me out with the species, & paper to preserve the specimens in I have felt severely. I had little thought of staying for a few weeks when I left India, otherwise I had been better provided.

His mention of 'paper to preserve the specimens' does reinforce the theory that he collected a substantial herbarium which has since disappeared, or was relabelled and distributed by Wallich when he took over the Calcutta garden. His interest was broader than just flowering plants, for amongst the packages that Smith had sent was his essay on 'Dorsiferous Ferns' and Roxburgh had collected many while in the Cape, 'to amuse myself arranging on my return to India, when, I think, I shall find it an easy task, with your paper before me'.

125 Protea pulchella, *a member of the family* Proteaceae *found by Roxburgh at the Cape, and discussed in a paper by Robert Brown. Hand-coloured engraving by H. Andrews for* The Botanist's Repository *(vol 4, plate 270, 1802).*

A month later, Roxburgh's son, William junior, and daughter had arrived and he was making plans to depart for India, leaving his natural son, John, to collect 'all the seeds and specimens that can be found at and about Graatvaaders Bosch', having himself 'examined the greatest part of the peninsula from the point of the Cape to Table mountain, for I have lived on it during half of the summer & autumn'.[21] It is worth noting that he refers to 'Mr Burrow, the Gentleman in charge of the Botanic Garden here', but I have as yet not been able to find out whether this was a new project or whether Roxburgh discovered the remains of Masson's garden. By his comment to Darell at the beginning of December,

> it gives me infinite satisfaction to find that it is the intention of Sir George Young [he had been Governor of St Vincent and had corresponded with Roxburgh from there], the Governor of the Cape to restore and improve the Botanic Garden there. It is what I strongly recommended in my private letters to Sir Joseph Banks ever since we have been in possession of that Colony, and personally to Lord Macartney when I saw him at the Cape, and had the pleasure to make a beginning at his Lordships request. I sent by various conveyances before I went to that place myself, a great variety of the most useful plants and seeds that I could think of, and also carried many with me, several of which I had the satisfaction to leave in a thriving state.[22]

Pl. 15. *E. longifolia* Ait.

126 Erica longifolia, *one of Roxburgh's Cape finds discussed by Richard Salisbury, who noted that Roxburgh's manuscript name for it was 'Erica pinifolia'. Colour print from a painting by Fay Anderson in Baker and Oliver's* Ericas of Southern Africa *(plate 15, Cape Town and Johannesburg, 1967).*

By the middle of October, Roxburgh was reporting back for work at Calcutta.[23] He continued to have a strong affection with the Cape and the development of its garden.

The continuing interest is further reflected in the 'two Chests of Growing Plants, and a Box of Seeds' he sent to the Cape in August 1800,[24] though sadly there is no mention of his staying there on either of his journeys to England in 1805 or 1813.

IMPORTANCE OF ROXBURGH AT THE CAPE

The importance of Roxburgh's 18 months at the Cape can be gleaned from a number of contemporary and later sources. He arrived there probably shortly before James Niven and the two of them almost certainly went on joint plant collecting journeys, as can be seen from a study of the distribution of their collecting localities.[25] Niven had been sent out by George Hibbert, a powerful West Indian merchant and MP, who had to await the opening up of the Cape before being able to expand his collection of South African plants. Like Roxburgh, Niven had studied at the Edinburgh Botanic Garden. They were certainly acquainted, for in Roxburgh's instructions to his son John, whom he left there until 1804, he wrote, 'you & Mr Niven know well how they [grafted fruit trees for India] are to be sent'.[26]

There are two near contemporary publications which highlight Roxburgh's important role in the discovery of the Cape flora. Richard Salisbury published an extensive article in 1802 on the 'Species of Erica' and Robert Brown wrote an even longer article, 'On the Proteaceae of Jussieu' in 1811.[27] Salisbury credited Roxburgh with 21 species of *Erica*, and Brown stated that 'Numerous observations on the same subject [separation of sexes in *Leucodendron*] have also more recently been made by Dr Roxburgh and Mr Niven, who have bestowed much pains on ascertaining its limits'. Ted Oliver, from Kirstenbosch Botanic Garden, has supplied me with a list of Niven's herbarium specimens of *Erica* which contain references to Roxburgh: these extend to 34 species, and a further 17 herbarium specimens which were Roxburgh's and appear to have followed Niven's herbarium collection to Dublin. Of these some are at Trinity College, and the rest at Glasnevin.[28] As far as Roxburgh's work with the Proteaceae is concerned, in addition to the comment by Brown quoted above, Brown also made extensive reference to 'Rox MSS', as did Salisbury. Sadly these manuscript notes have disappeared, but Salisbury gave due credit to Roxburgh:

> I have to acknowledge the great assistance I have derived from the extensive collection presented to this [Linnean] Society by my friend Dr Roxburgh, who during his short residence at the Cape appears to have paid particular attention to this tribe of plants, and who, besides the many new species discovered by him, has given a greater value to his Herbarium by numerous observations on the sexes, the size, and the place of growth, which I have everywhere inserted on his authority.

Over the years, Roxburgh built up a considerable herbarium collection of Cape specimens at Calcutta, for he wrote in 1812,

> I must have near 200 sp of Erica, & about 100 of Protea, but Insects destroy them so fast, & so certainly in India, notwithstanding my utmost care, that I have no pleasure in them. I believe I will send the whole home. I wish Mr Salisbury may have seen the Synoptical Catalogue of Erica which I made out while at the Cape, & sent Sir J. Banks from thence. Did you observe the two species I thought new & named after you & Dr Smith.[29]

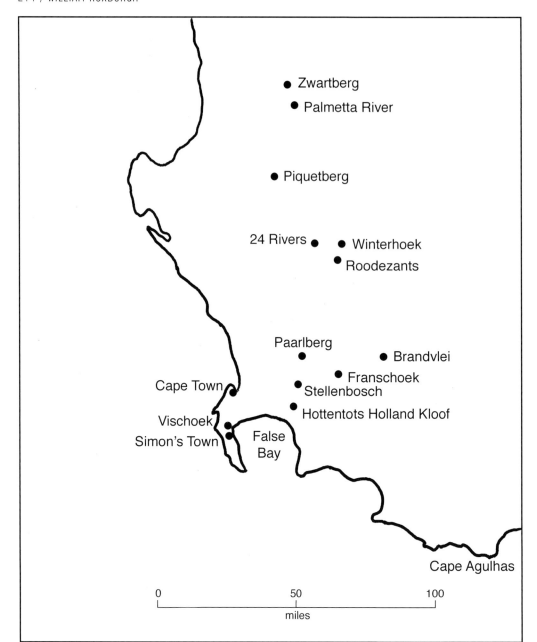

Sadly, as with so much of the Roxburgh material, this list no longer seems to be in existence.

A glance at the map (Fig. 127) of his plant collecting localities, taken with his statement above about the time he spent between Cape Point and Table Mountain, demonstrates the enormous energy of Roxburgh. In one of his obituaries, it states that 'on more than one occasion he was known to have wandered over 40 miles in a morning, over the immense mountains of the Cape of Good Hope.'[30] A certain amount of exaggeration must have been built into this description, for Viscount Valentia used two modes of transport when he visited the Cape, either on horseback, or 'covered waggons, each drawn by eight horses, with a Hottentot for a driver, and a slave to assist him .... These waggons are the only machines adapted to the roughness of the roads, as they have the advantage of strength, and difficulty of being overset. The Dutch ladies use them constantly.'[31] Certainly to have covered 40 miles in a morning in those days would have been almost impossible.

SUMMARY

This description of Roxburgh's time at the Cape gives support to the fact that he must have kept an extensive set of notes and that he also collected herbarium specimens. The latter must have been dispersed after his death and Appendix 2 list those institutions which now hold some of his herbarium specimens; and the collection of notes may well have been left with John Roxburgh at Calcutta as was originally planned for his drawings before Colebrooke borrowed them, or went to William Carey at Serampore. In either case, their disappearance and that of his Synoptical Catalogue are a major loss to the botany of both India and South Africa.

The ground he covered was enormous, bearing in mind his comment about spending half of the summer and autumn of 1798/99 between the Cape and Table Mountain, and the time taken collecting plants, writing up notes, and pressing specimens. All this suggests very strongly that Roxburgh went to the Cape with the intention of becoming involved in what must have been one of the most exciting under-explored botanical areas that was opening up during his lifetime.

Roxburghia gloriosoides

CHAPTER 15

The Founding Father of Indian Botany

When one attempts to assess Roxburgh's achievements, there is a mass of evidence, both contemporary and later. Obituaries are, almost by definition, laudatory as they are rarely written to undermine or criticise the recently departed, but the eulogies that were bestowed on Roxburgh indicate his perceived standing in India and Europe: 'the hand of death ... deprived the world of a most scientific and zealous man, who would have adorned even the chair of Linnaeus, and have added new lights, had he lived, to European learning';[1] 'in the extent and profundity of his knowledge on botanical subjects he was unrivalled, being esteemed the first botanist since the time of Linnaeus.'[2]

Describing his taxonomic work, Thomson wrote, 'The descriptions, which are remarkable for their accuracy', whilst Sir Joseph Hooker said of his fieldwork, 'Old Roxburgh stood alone', and more recently Cronk combining these two areas referred to 'the importance of Roxburgh's list [of the St Helena flora] lies in the new species described and the notes given on certain of the endemics'.[3] A more specific view was given by William Roscoe, reviewing Roxburgh's 'Descriptions of the Monandrous Plants of India', in which he wrote, 'it was, indeed, reasonably to be expected that the observations of so experienced a Botanist, founded on actual inspection of the living plants, in their native climate, must be highly valuable; and in this, his readers will not be disappointed.'[4] More recent accolades have been that 'through his labours on Flora Indica and the development of the Botanic Garden under his stewardship [Roxburgh] earned himself the appellation "the Linnaeus of India".'[5] Roxburgh also earned the sobriquet, 'The Father of Indian Botany'.[6]

A short review of the Calcutta Botanic Garden, written in 1971, gives a further confirmation of Roxburgh's role in the history of Indian botany: 'Dr William Roxburgh was one of the greatest botanists of his time, and during his term converted the garden in its character from its original economic purpose to the service of Scientific botany .... William Roxburgh made the rest of the world interested in India's immensely rich and varied flora – then new to horticulture.'[7] Viscount Valentia, writing in 1803, was also highly complimentary, stating that the Botanic Garden 'affords a wonderful display of the vegetable world, infinitely surpassing anything I have ever before beheld ... at present it is a complete centre, where the productions of every clime are assembled.'[8]

He was elected to the Royal Society of Edinburgh in 1791 and the Linnean Society in 1794, but a more telling indicator of how he was viewed in his own lifetime for his work on economic botany were his awards of the Gold Medal by the Royal Society for Arts twice, first in 1805, 'for his valuable Communications on East-India Products', and again in 1814, for 'the different Products of the East Indies, and their several Applications to the Arts, Manufactures, and Commerce of the United Empire.'[9] This

128 Stemona tuberosa (*Roxburghia gloriosoides*). *Hand-coloured engraving from a drawing by one of Roxburgh's Indian artists,* c.*1791 (Roxburgh drawing 208). From* Plants of the Coast of Coromandel (*vol 1, plate 32, 1795*).

second Medal was awarded by the Society's President, the Duke of Norfolk himself, and 'was prefaced by an eloquent and appropriate speech, in the presence of an elegant and crowded assembly'.[10]

His botanical work forms his most substantial legacy in the wealth of his publications, and the collections of botanical drawings. A selection of the latter was published in his *Plants of the Coast of Coromandel*, under the editorship of Banks. He also collected a herbarium that is now unfortunately spread over a number of institutions: the collection that he left at Calcutta was partially destroyed over the years by the ravages of damp and insects, and Nathaniel Wallich dispersed most of what still remained there in 1832. With the delay in publication of Roxburgh's *Flora Indica*, many of the species that he first recognised and described were taken up by other authors and so do not bear the name that he gave, and his work on them is therefore not recognised.

Roxburgh's collection of 2,572 drawings with their associated written plant descriptions is a monument of art, science and industry. The drawings are on folio sheets, including both plant habit done to natural size and enlarged dissections. The Indian artists themselves were considered one of the attractions of the garden, and a number of visitors had copies of the drawings made. These are now widely spread and explain the existence of such collections as those at the Royal Botanic Garden Edinburgh and the Public Library at Reading.[11] Although a full study of the drawings at Kew and Calcutta has been made, the existence of some of the Edinburgh drawings with late 18th-century and early 19th-century watermarks, and containing handwriting contemporary with some of the Kew drawings, shows that more work requires to be done on these collections. There is, for instance, the problem of the location of the 619 drawings done for Dr Hare, the temporary Superintendent in 1816-18.[12]

The botanical achievements of William Roxburgh have been described in some detail, but his lasting influence in other fields is less easy to unravel. Richard Grove assigned him an important role in the development of ideas on climate theory and its implications for famine control, based on the meteorological diaries and the article on the land-winds, an area on which Grove is still working, with particular relevance to El Niño.[13]

As a scientist, he based his methodology very much on that taught by Dr John Hope at Edinburgh and descended from the great Leiden academic, Boerhaave: study of past literature, carefully controlled experiments, and rigorous analysis of data. As a result, his published work, for instance on possible alternatives for hemp, were developed over such a long period of time that he was sometimes overtaken by events, in this case the end of the Napoleonic Wars. His vision, however, was such that in a number of cases his suggestions were ahead of his time: a particular example was his work on tea, which he suggested be tried in north-east of Bengal, 'approaching Assam'. His practical ideas were largely based either on the economic benefits that could accrue to the East India Company, such as his studies on dyes, sugar, pepper, spices, timber and fibres; or on the benefit they could provide for the indigenous inhabitants. Of these latter, his work on silk, on sugar as a mode of employment, on alternative food crops to overcome famines caused by failure of the monsoons, and the concomitant use of the River Godavary as a source of irrigation, can all be quoted.

Like any educated scientist of his time, he did not devote himself solely to botany. Thus his first publications were on meteorology and this long-lasting interest led to his theories on the origins of the land winds that he published in 1810.[14] This showed his awareness of contemporary thinking, particularly that of French physiocrats such as Poivre and Céré, and their work on Mauritius, but he would have been equally aware of the work on St Helena. This introduces the subject of his place in the scientific world, where he occupied significant positions both in Britain and India. He was a correspondent of such pivotal people as Sir Joseph Banks, President of the Royal

Society, with Sir James E. Smith, President and founder of the Linnean Society, as well as many of the Directors of the East India Company, and other botanists such as William Roscoe in Liverpool and various Edinburgh scientists, such as Dr William Wright, a Fellow of the Royal Society of Edinburgh. In India, especially from the time he reached Calcutta, he was closely involved with the Asiatic Society, being on its Publishing Committee for ten years, and thus in close contact with such important scientists as Sir William Jones and Henry Colebrooke, and would have been aware of the important research done by such people as Wilkins.

Much has been written about the pivotal position of Sir Joseph Banks in the spread of science during the latter part of the 18th and early 19th centuries, and Roxburgh must be seen as part of this. However, as time went on, so his reliance on Banks did decrease as he published more through *The Asiatick Researches* and the Linnean Society. This does question the central position of Banks, and this doubt about Banks's importance is underlined in the way that he used other people's discoveries as his own: as far as Roxburgh is concerned, this is very evident with the lack of recognition of the vast quantity of new plants that he sent to England, and the fact that Banks was not involved in Roxburgh's appointment to Calcutta.

Of Roxburgh's zoological works, the most significant are his descriptions of the lac insect, *Chermes lacca*, that of the Ganges dolphin, *Platanista gangetica*, and that of the Indian bison, *Bos gaurus*. And the fascinating story of the elephant for Philadelphia must not be forgotten!

When considering how he viewed the natives of India, his attitude is more difficult to sort out. During his time in Edinburgh and under the influences of the Enlightenment, there were three ways that variations in humans were approached. The first was climatic: this proposed, since the time of Hippocrates, that environmental conditions caused these differences and tied them in with the four humours – tropical people were scorched by the sun, became dark and lethargic. One of the main contemporary proponents for this idea was the Comte de Buffon, and Roxburgh did have his study of natural history in his library. John Locke, however, proposed a four-stage subsistence theory which became popular in Scotland during the Enlightenment, so Roxburgh would have been aware of this from his time in Edinburgh. The third attitude was based on Linnaeus's ideas of classification, whereby people were grouped by their appearance and behaviour, but this idea, although the basis of the modern ideas of race, was criticised by many contemporaries, including Buffon. Roxburgh usually referred to them as 'natives' but this does not seem to have any derogatory overtone, as he seems to have taken the humanitarian view to try to improve their lot as best he could, by introducing crops and manufactures that would ease their hardships.

Roxburgh, married for 15 years to the daughter of a Lutheran missionary and with strong connections with the Dutch missionaries at Tranquebar, does appear to have been guided by Christian and humanitarian ideals. His role as a surgeon and physician would also appear to give added weight to this idea. This personal philosophy would seem to be demonstrated by the extensive work that he did, almost from his first days in India, to alleviate famine and suffering. Furthermore, Roxburgh was a friend of the Baptist missionary William Carey who had arrived in India in 1793. Additional evidence for Roxburgh's Christian ideals is more indirect and posthumous, in that his children asked Bishop Heber to write the epitaph for the monument raised to his memory in the Calcutta Botanic Garden in 1823. But it should be noted that Heber arrived in India only in 1823, and the erection of the commemorative urn must therefore have been one of his first acts. The information in the inscription must therefore have been based on second-hand information, no doubt provided by contemporaries who had known Roxburgh personally. From the inscription, Heber seems to have empathised with Roxburgh's care for the sick and poor, describing the 'care of the sick coolies in his entourage with touching zeal'.[15] The plinth bears a

129 The Roxburgh coat of arms, as granted to Roxburgh's son Bruce in 1854, and including two leaves of Roxburghia gloriosoides.

eulogistic Latin inscription recording Roxburgh's botanical work as was appropriate for its setting in the Botanic Garden.

He was a highly energetic person, shown by the distances he covered while in South Africa. He appears to have been a thoroughly honourable man, and Sealy wrote: 'I have never found an instance of Roxburgh using any other person's work without acknowledging the fact and giving a reference. He seems to have been most scrupulous.'[16] Although he suffered from ill health for much of his life, particularly after reaching Calcutta in 1793, to survive for nearly forty years in India at that time indicates an unusually strong constitution, which was commented on by a number of his contemporaries.

As a man of business, he appears to have been somewhat hesitant, but, with the assistance of Andrew Ross in Madras, highly successful. Thus, to leave an estate of in the region of £50,000 was a considerable achievement. This led to 'gentrification' of his family and the granting of a coat of arms to his son. His family was extensive, and reflects the *mores* of the time. Thus his first son appears to have had a native mother, for he was educated in India and never achieved any degree of success other than as a nurseryman in the Calcutta Botanic Garden and plant collector in both South Africa and the East Indies. This is in contrast to Roxburgh's other children who were all well provided for in his will as were his step-siblings and their children. His own children were educated in Edinburgh and other than William junior, who predeceased his father, his sons were either commissioned in the Indian army, the Royal Navy or, in the case of the youngest William, trained as a doctor. The daughters who married seem to have been well set up, with husbands who were in the employ of the East India Company.

Roxburgh's comments on the Arrow palm, *Saguerus rumphii*, sum up many of his attitudes: the appreciation of plants; the need to research earlier works where available; the use of plant materials, in this case for making cordage; his slightly puritanical views, in that it can be used to make 'ardent spirits', but that the pith can be used as the basis of sago, thus helping the natives in times of famine; in view of its usefulness, the distribution of thousands of plants free of charge; and the description and drawings of the plants sent to the Court of Directors.[17]

PUBLICATIONS OF DR WILLIAM ROXBURGH

1. 'A meteorological diary, Ec. Kept at Fort St. George in the East Indies. Communicated by Sir John Pringle, Bart. P.R.S. Read January 29, 1777', *Philosophical Transactions of the Royal Society*, 68 (1778), 180-95.

2. 'A continuation of a meteorological diary, kept at Fort St. George, on the coast of Coromandel. Communicated by Joseph Banks. Read January 20, 1780', *Philosophical Transactions of the Royal Society of London*, 70 (1780), 246-71.

3. 'Process for making Indigo on the Coast near Ingeram', in Charpentier de Cossigny de Palma, J. F., *Memoir, containing an Abridged Treatise on the Cultivation and Manufacture of Indigo* (Calcutta, 1789), p. 157.

4. 'On the Lácshà, or Lac, Insect (*Coccus lacca*)', *Asiatick Researches*, 2 (1790), 361-4; 1799, Tilloch, *Philosophical Magazine*, 3 (1790), 367-9. Includes one illustration.

5. Unpublished article. Remarks made about the year 1790 by W. Roxburgh, on the effect of sunlight on different coloured materials used by soldiers. Paper for the Medical Society (1790).

6. '*Chermes Lacca*. Read May 19, 1791. Communicated by Patrick Russell, M.D. F.R.S.', *Philosophical Transactions of the Royal Society*, 81 (1791), 228-35.

7. 'A description of the plant *Butea*', *Asiatick Researches*, 3 (1792), 469-74.

8. 'A Botanical Description, and Drawing of a new Species of Nerium (Rose-Bay) with the Process of extracting, from it's Leaves, a very beautiful Indigo', in Dalrymple, *Oriental Repertory*, vol. 1 (1793), 39-44. Includes one drawing.

9. *A Botanical Description of a new species of Swietenia (Mahogany), with Experiments and Observations on the Bark thereof, in order to determine and compare its Powers with those of Peruvian Bark, for which it is proposed as a Substitute, etc.* (London, 1793), pp. 24.

10. 'A description of the *Jonesia*', *Asiatick Researches*, 4 (1795), 355-8. Includes three illustrations.

11. '*Prosopis aculeata*, Koenig. Tshamie of the Hindus in the Northern Circars', *Asiatick Researches*, 4 (1795), 404-8. Includes one illustration.

12. 'Botanical observations on the Spikenard of the ancients, intended as a supplement to the late Sir William Jones's papers on that plant', *Asiatick Researches*, 4 (1795), 433-6. Includes one illustration.

13. 'Watering the Circars', in Dalrymple, *Oriental Repertory*, vol. 2 (1797), 33-84.

14. 'An account of the Hindu method of cultivating the sugar cane and manufacturing the sugar and jagary in the Rajamundry Circar', in Dalrymple, *Oriental Repertory*, vol. 2 (1797), 497-514. Tilloch, *Philosophical Magazine*, 21 (1805), 264-75.

15. 'A botanical description of *Urceola elastica*, or Caout-chouc Vine of Sumatra and Pullo Pinang, with an account of the properties of its inspissated juice, compared with those of the American Caout-chouc', *Asiatick Researches*, 5 (1798), 167-75; Nicholson, *Journal*, 3 (1800), 435-40; Tilloch, *Philosophical Magazine*, 6 (1800), 154-61. Includes one illustration.

16. *Plants of the Coast of Coromandel, selected from drawings and descriptions presented to the Hon. Court of Directors of the East India Company, Published by their Order, under the Direction of Sir Joseph Banks*, vol. 1 (London, 1798).

17. 'An account of a new species of *Delphinus*, an inhabitant of the Ganges', *Asiatick Researches*, 7 (1801), 170-4; *Edinburgh Review*, 9 (1807), 283-7. Includes one illustration.

18. 'An account of the Tusseh and Arrindy silk-worms of Bengal'. *Transactions of the Linnean Society*, 7 (1804), 33-48. Includes one illustration.

19. 'Communication on the Culture, Properties, and comparative Strength of Hemp, and other Vegetable Fibres, the Growth of the East Indies', *Transactions of the Society for the Encouragement of Arts, Manufactures, and Commerce*, 22 (1804), 363-96; Nicholson, *Journal*, 11 (1805), 32-47; *Gill. Tech. Report*, 6 (1824), 184-94, 240-4.

20. 'A botanical and economical account of *Bassia butyracea* or East India Butter tree', *Asiatick Researches*, 8 (1805), 477-86; Nicholson, *Journal* 19 (1808), 372-9; Gilbert, *Annals*, 40 (1812), 334-40. Includes one illustration.

21. *Plants of the Coast of Coromandel, selected from drawings and descriptions presented to the Hon. Court of Directors of the East India Company, Published by their Order, under the Direction of Sir Joseph Banks*, vol. 2 (London, 1805).

22. Letter to Dr C. Taylor, 'Concerning the Aldacay, or Caducay Galls, with which the Yellow Colour in the Indian Chintzes is formed', *Transactions of the Society for the Encouragement of Arts, Manufactures, and Commerce*, 23 (1805), 407-14.

23. 'Observations of the Culture, Properties and comparative Strength of Hemp and other vegetable fibres, the Growth of the East Indies', *Transactions of the Society for the Encouragement of Arts, Manufactures, and Commerce*, 24 (1806), 143-53; Nicholson, *Journal*, 15 (1810), 32-47.

24. 'On the growth of trees in the Botanic Garden at Calcutta in Bengal', *Transactions of the Society for the Encouragement of Arts, Manufactures, and Commerce*, 24 (1806), 154-6; Nicholson, *Journal*, 17 (1807), 110-11.

25. Letter to Dr Charles Taylor, 'Concerning the extract of Tannin from *Embryopteris glutinifera*', *Transactions of the Society for the Encouragement of Arts, Manufactures, and Commerce*, 26 (1808), 241-5.

26. Letters to Dr Charles Taylor, 'Concerning extraction of an amber-like resin, and supplies of Fever Bark', *Transactions of the Society for the*

Encouragement of Arts, Manufactures, and Commerce, 27 (1809), 233-7.

27. 'Directions for taking care of growing Plants at Sea', *Transactions of the Society for the Encouragement of Arts, Manufactures, and Commerce,* 27 (1809), 237-8; Nicholson, *Journal,* 27 (1810), 69-76.

28. 'Description of several of the monandrous plants of India, belonging to the natural order called *Scitamineae* by Linnaeus, *Canna* by Jussieu, and *Drimyrhizae* by Ventenat', *Asiatick Researches,* 11 (1810), 318-62; Sprengel, *Jahrb.,* 1 (1820), 64-110. Includes six drawings.

29. 'Account of a new Species of Nerium, the leaves of which yield Indigo', *Transactions of the Society for the Encouragement of Arts, Manufactures, and Commerce,* 28 (1810), 250-72.

30. 'A brief Account of the Result of various Experiments made with a view to throw additional Light on the Theory of this Artificial Production', *Transactions of the Society for the Encouragement of Arts, Manufactures, and Commerce,* 28 (1810), 272-94.

31. 'Description of a second newly-discovered Indigo Plant', *Transactions of the Society for the Encouragement of Arts, Manufactures, and Commerce,* 28 (1810), 294-301.

32. 'Description of a third Indigo Plant, viz. Asclepias tinctoria, Roxb.', *Transactions of the Society for the Encouragement of Arts, Manufactures, and Commerce,* 28 (1810), 301-4.

33. 'Description of a New Species of Asclepia, which is said to yield a green dye, viz. Asclepias Tingens Rxb.', *Transactions of the Society for the Encouragement of Arts, Manufactures, and Commerce,* 28 (1810), 304-7.

34. Letters to Dr Charles Taylor, 'Concerning Fever Bark, and Resin', *Transactions of the Society for the Encouragement of Arts, Manufactures, and Commerce,* 28 (1810), 308-13.

35. 'Remarks on the Land Winds and their causes', *Transactions of the Medical Society of London,* 1 (1810), 189-211; Tilloch, *Philosophical Magazine,* 36 (1810), 243-53.

36. *Further Extracts of Correspondence [between William Roxburgh and R. C. Plowden, Acting Secretary to the Board of Trade and others] relating to Indian Hemp* (London, 1811), pp. 16.

37. 'Some account of the Teak tree of the East Indies', Nicholson, *Journal,* 33 (1812), 348-54.

38. Letters to Dr Charles Taylor on 'India-rubber', *Transactions of the Society for the Encouragement of Arts, Manufactures, and Commerce,* 30 (1812), 191-2.

39. Letters to Dr Charles Taylor on 'Transporting plants', *Transactions of the Society for the Encouragement of Arts, Manufactures, and Commerce,* 30 (1812), 193-203.

40. Letters to Dr Charles Taylor on 'Teak', *Transactions of the Society for the Encouragement of Arts, Manufactures, and Commerce,* 30 (1812), 203-11.

41. *Hortus Bengalensis, or a catalogue of the plants growing in the Honourable East India Company's Garden at Calcutta* (edited by William Carey, Serampore, India, 1813-14).

42. 'On the growth of trees in the Botanic Garden at Calcutta, in Bengal', *Transactions of the Society for the Encouragement of Arts, Manufactures, and Commerce,* 32 (1814), 208-10.

43. 'Remarks on and further Descriptions of the Calooee Hemp, or Urtica Tenacissima', *Transactions of the Society for the Encouragement of Arts, Manufactures, and Commerce,* 33 (1815), 188-92.

44. 'Posthumous Observations on Substitutes for Hemp and Flax', referred to in letter from J. Dart, Secretary's Department at Fort St George, dated 10 February 1816, British Library, OIOC IOR E/4/916, ff851-2.

45. 'Description by Dr Roxburgh, of the most remarkable of the plants indigenous to St Helena' in Alexander Beatson, *Tracts relative to the island of St Helena; written during a residence of five years* (London, 1816).

46. *Plants of the Coast of Coromandel, selected from drawings and descriptions presented to the Hon. Court of Directors of the East India Company, Published by their Order, under the Direction of Sir Joseph Banks,* vol. 3 (London, 1820).

47. *Flora Indica; or Descriptions of Indian Plants, to which are added Descriptions of Plants Recently Discovered by Nathaniel Wallich, M.D., F.L.S.* (edited by William Carey, vol. 1, Serampore, India, 1820).

48. *Flora Indica; or Descriptions of Indian Plants, to which are added Descriptions of Plants Recently Discovered by Nathaniel Wallich, M.D., F.L.S.* (edited by William Carey, vol. 2, Serampore, India, 1824).

49. *Flora Indica; or, Description of Indian Plants* (2nd edition, 3 vols., edited by William Carey, Serampore, India, 1832) [a publication of Roxburgh's complete manuscript without Wallich's additions].

50. *Flora Indica, Volume 4.* Reprinted from 'Cryptogamous Plants' (ed. by W. Griffith). *Calcutta Journal of Natural History,* 4 (1844), 463-520.

51. 'On the genus *Aquilaria,* with remarks by the late H. T. Colebrooke', *Transactions of the Linnean Society,* 21 (1855), 199-206, *Proceedings of the Linnean Society,* 2 (1855), 123-5.

52. *Flora Indica; or, Descriptions of Indian plants* (single volume, including Volume 4, Cryptogamous Plants, edited by C. B. Clarke, Calcutta, 1874).

WILLIAM ROXBURGH'S HERBARIUM COLLECTIONS

The main collections are at:
1. Botany Department, Natural History Museum, London [BM]
2. Botanic Garden, Brussels [BR]
3. Royal Botanic Garden Edinburgh [E]
4. Delessert Herbarium, Geneva, one of the largest [G]
5. Royal Botanic Gardens, Kew [K]
6. Botany Department, Liverpool Museums [LIV]

There are smaller collections at:
1. Arnold Arboretum, Cambridge, Mass. [A]
2. Herbarium Willdenow, Botanische Garten und Botanisches Museum, Berlin [B]
3. Botanical Museum and Herbarium, Copenhagen [C]
4. National Botanic Institute, Kirstenbosch [FC]
5. Herbarium Universitatis Florentinae, Florence [FI]
6. National Botanic Garden, Glasnevin, Dublin [GLAS]
7. Linnean Society, London [LINN]
8. New York Botanical Garden [NY]
9. Fielding Herbarium, Botany Department, Oxford University [OXF]
10. Muséum National d'Histoire Naturelle, Paris [P]
11. Botany Department, Academy of Natural Sciences, Philadelphia [PH]
12. Botany Department, Trinity College Dublin, Dublin [TCD]
13. The Herbarium, Institute of Systematic Botany, Uppsala [UPS]

The main source for the above lists was Stafleu and Cowan's *Taxonomic Literature*,[*] but additional collections have been found through various personal communications, such as Dr E. Oliver at Kirstenbosch who advised me about the collections at Glasnevin and Kirstenbosch.

---

[*] Frans A. Stafleu and Richard S. Cowan, *Taxonomic Literature. A selective guide to botanical publications and collections with dates, commentaries and types*, 2nd edition, 7 vols. (Utrecht, 1976-88). Vol. 4, pp. 954-6.

WILLIAM ROXBURGH'S DRAWING COLLECTIONS

1. Royal Botanic Gardens, Kew [K]
2. Botanic Garden, Calcutta [CAL]
3. British Library (India Office Records), London [BL]
4. Natural History Museum, London [BM]
5. Linnean Society, London [LINN]
6. Royal Botanic Garden Edinburgh [E]

PLANTS INTRODUCED TO KEW
BY WILLIAM ROXBURGH

The following list of plants has been abstracted from the following Kew archives: Record Book 1793-1809, Record Book 1804-26 and Inwards Book 1805-09. During the years covered by these records, most of the plant material was sent by Roxburgh to Sir Joseph Banks who then passed it on to the Head Gardener of Kew − initially William Aiton. On his death in 1793, he was succeeded by his son, William Townsend Aiton who held the post until his retirement in 1841.

Only those plants specified as coming from Roxburgh are included in this list. Thus not included are: a collection of 110 species brought together by Christopher Smith from the Calcutta Botanic Garden in April 1795 and sent via Peter Good who also brought back on the same trip a few trees from China, but many of these plants were given either native names or just generic names; a collection of nineteen species sent by Peter Good in 1800 from the Cape of Good Hope for forwarding to Edinburgh; and a collection of twenty-one species for the African Institution sent in 1811. However, many of these species brought by Smith and Good were sent by Roxburgh at other times. Interestingly Roxburgh appears to have sent no plants to Kew from the Cape. A number of the plants, such as *Brunfelsia americana*, *Carolinaea insignis* and *Pitcairnia angustifolia* would appear to be plants that Roxburgh had received from places as far afield as America and a number from further east including China.

Year	Number sent
1793	1
1796	96
1797	9
1798	36
1800	50
1803	43
1805	342
1806	303
1807	450
1808	213
1809	160
1810	55
1811	240
1812	146
1814	183

Table 1 Number of plants sent each year by Roxburgh to Kew.

The names have been checked against those in his published *Flora Indica* and where they do not appear in that, they have been checked against the index of Hooker's *Flora of British India*.* The names have been corrected to these spellings except for a few cases where the plants could not be identified from either source and these remain doubtful and have been marked with '?'. The list includes 1033 species, four having different colour forms or geographical varieties: *Abrus precatorius* with black and red varieties plus one where the colour was not specified, *Canna indica* (in fact from South America) with yellow and red forms, and *Carissa caranda* from Bengal and China: a total of 1037 entries altogether. Over the fourteen years for which there are figures, virtually half (499 out of 1037) were sent only once while a few were sent frequently: *Nyctanthes arbor-tristis* and *Sophora tomentosa* nine times each, and *Costus speciosus* and *Leea macrophylla* ten times each. Some, such as *Epygea cordata* and *Exygea cordata*, are probably the same species and have been incorrectly transcribed. Others such as *Lupinaus orubiginosus* and *Sooroogada glabra*, have been ascribed to Roxburgh and will need to be checked against the various manuscript versions of his *Flora Indica* to ascertain their correct name.

Significantly the three years when the greatest number reached Kew, are those, 1805-7, in which Roxburgh was actually in Britain. It must, therefore be assumed that the plants were being sent by William Roxburgh junior under his father's instructions, or brought home by Roxburgh senior.

* Sir Joseph Dalton Hooker, *Flora of British India*, 7 vols. (London, 1872-97).

Two aspects have not yet been studied, and both are important in the history of plant introductions to Britain. The first question is: how many of these plants germinated and/or survived? The second follows from this: how many should now be attributed to Roxburgh as his introductions? Certainly the number of species in this list suggests that Roxburgh should be viewed as a major introducer of plants in his own right. It should also be remembered that he was himself at the centre of a web of collectors throughout India and further east, as well as receiving plants from North and South America, and Africa, as can be seen from the material detailed in *Hortus Bengalensis*.

Plant introduced	No	Plant introduced	No	Plant introduced	No
Abroma augusta	4	Amyris punctata	1	Barleria cristata	2
Abrus precatorius black	2	Amyris subtriphylla	1	Barleria dichotoma R	7
Abrus precatorius red	2	Andersonia altissima	1	Barleria prionitis	1
Abrus precatorius	1	Andersonia rohitoka	1	Barleria triandra R	2
Acanthus ilicifolius	1	Andrachne apetala	1	Barleria variegata	1
Achras bulloh	1	Andropogon aromaticum	2	Bassia latifolia	3
Achyranthes aspera	2	Andropogon arundinaceus	1	Bauhinia acuminata	6
Achryanthes incana	2	Andropogon juarancusa	1	Bauhinia candida	2
Achyranthes lanata	3	Andropogon miliaceum	1	Bauhinia emarginata	4
Achyranthes nodiflora	2	Andropogon saccharatus	1	Bauhinia parviflora	1
Achyranthes lappacea	1	Anethum sowa R	3	Bauhinia purpurea	3
Acrostium emarginatum	1	Anneslia spinosa	4	Bauhinia spicata	4
Adansonia digitata	6	Anona altissima	1	Bauhinia tomentosa	7
Adenanthera pavonina	5	Anthistiria arundinacea	1	Bauhinia triandra	4
Aeschynomene aquatica	1	Anthistiria cymbaria	1	Bauhinia variegata	2
Aeschynomene cannabina	2	Antidesma napauliana	1	Bauhinia zeylanica	1
Aeschynomene diffusa	1	Antidesma olexiteria	2	Belamcanda chinensis	2
Aeschynomene grandiflora	3	Antidesma paniculata	1	Bentinckia condapanna	1
Aeschynomene indica	2	Antyllis cuneata	3	Bergera koenigii	2
Aeschynomene paludosa	1	Aporetica ternata	2	Berrya ammonilla	1
Aeschynomene sesban	3	Aquillaria agallocha	1	Bidens bifurcata	1
Aeschynomene spinulosa	6	Arachis fruticosa	2	Bidens bipinnata	4
Ageratum aquaticum	1	Aralia agitatrec	1	Bignonia chelonoides	5
Ajuga disticha R	3	Aralia digitata	3	Bignonia crispa	3
Ajuga fruticosa R	3	Ardisia solanacea	7	Bignonia indica	1
Alanguim hexapetalum	1	Ardisia umbellata R	7	Bignonia solanum	1
?Alcis fruticosa	1	Areca catechu	3	Bignonia tuberculata	1
Aleurites triloba	1	Arethusa plicata	1	Bixa orellana	3
Allium tuberosum	4	Aristolochia indica	3	Bombax pentandrum	2
Alpinia malaccensis	1	Artimisia indica	5	Bombax salmala	1
Alpinia mutica	1	Artimisia parviflora	2	Borago zeylanica	7
Alstroemeria salsilla	1	Artocaprus Lakoocha R	2	Borassus flabelliformis	1
Amaranthus blitum	1	Arum bulbiferum	1	Boswellia thurifera	1
Amaranthus caudatus	1	Arum campanulatum	1	Briedelia scandens	3
Amaranthus curuca	1	Arum curvatum	1	Briedelia spinosa	1
Amaranthus pendulus	1	Arum esculentum	1	Brucea sumatrana	5
Amaranthus retroflexus	1	Arum macrophyllum	1	Brunfelsia americana	2
Amaranthus viridis	1	Arum manchiandum	3	Buchanania latifolia	2
Ambrosina cachlara	1	Arum orixensis	3	Buddleia neemda R	5
Ambrosina ciliata	2	Arum ramosum	1	Buddleia racemosa	1
Ambrosina lotus	1	Arum sessiliflorum	1	Butea frondosa	6
Ambrosina rubra	1	Asclepias gigantea	4	Butea parviflora	1
Ammannia octandra	3	Asclepias odoratissima	1	Byttneria catalpifolia	1
Ammannia vesicatoria R	2	Asclepias parasitica	3	Cacalia bicolor	1
Amomum aculeatum	1	Asclepias tenacissima	6	Cacalia tricolor	1
Amomum angustifolium	1	Asclepias tinctoria	1	Cactus chinensis	1
Amomum campanulatum	1	Asclepias tunicata	1	Caesalpina bonduccella	1
Amomum comosa	1	Asparagus acerosus	2	Caesalpinia cuneiformis	1
Amomum echinatum	1	Asparagus racemosus	6	Caesalpinia paniculata	3
Amomum galanga	1	Asphodelus clavatus R	2	Caesalpinia sappan	3
Amomum ligulatum	1	Athanasia indica	6	Caesalpinia sepiaria R	2
Amomum maximum	6	Atragene zeylanica	3	Caesalpinia tortuosa	4
Amomum nutans	1	Aubletia tibourou	1	Caesulia axillaris	3
Amomum pseudo zingiber	2	Baeobotrys indica	1	Calamus rotang	1
Amomum subulatum	1	Baeobotrys ramentacea	1	Callicarpa cana R	7
Amomum taraca	6	Bambusa arundinacea	1	Callicarpa dentata R	4
Amomum xanthorhiza	1	Bambusa nana	1	Callicarpa incana	1
Amyris agallocha	1	Barleria acuminata	1	Callicarpa lanceolaria	1
Amyris cisparis	1	Barleria caerulea R	4	Callicarpa macrophylla	5
Amyris heptaphylla	3	Barleria candida	1	Callicarpa villosa	1
Amyris protium	1	Barleria ciliata	2	Calophyllum inophyllum	3

Plant introduced	No
Canna glauca	1
Canna indica floribus lutens	2
Canna indica flore sanguineo	2
Cannabis sativa	2
Capparis orientalis	1
Cardiospermum halicacabum	2
Carduus lanatus R	2
Carduus radicans R	2
Carissa carandas (Bengal variety)	4
Carissa carandas (China sort)	4
Carolinaea insignis	2
Caryota urens	1
Cassia arcturus	1
Cassia bacillus	2
Cassia chamaechrista	4
Cassia coromandeliana	1
Cassia dimidiata	6
Cassia denudata	1
Cassia fistula	8
Cassia foetida R	3
Cassia glauca	1
Cassia grandis	1
Cassia herpataoides	1
Cassia marginata	2
Cassia marilandica	1
Cassia nodosa	1
Cassia pulcherrima	1
Cassia purpurea	2
Cassia sophora	3
Cassia speciosa	4
Cassia sumatrana	2
Cassia tora	3
Cassia toroides R	2
Castanea indica	1
?Casulia axillaris	2
Callicarpa cana	1
Cauclea cordifolia	1
Cerbera dichotoma	1
Cedrela toona	5
Celastrus emarginatus	1
Celastrus montana	6
Celastrus monterruy R	1
Celastrus verticillata	2
?Celobius paniculata	1
Celosia argentea	1
Celosia baccata	3
Celosia cernua R	6
Celosia comosa	1
Celosia coromandeliana	2
Celosia cristata	3
Cerbera manghas	2
Chionanthus axillaris	2
Chionanthus ramiflora	1
Chionanthus terminalis	2
Chloris barbata	1
Chrysophyllum acuminatum	1
Cinchona thyrsiflora R	5
Cissampelos hexandra	2
?Cissura paludosa	1
Cissus pedata	2
Clematis gouriana R	4
Cleome chelidonii	2
Cleome icosandra	1
Cleome pentaphylla	1
Cleome viscosa	2
Clerodendrum infortunatum	1
Clitorea ternatea flor coerulea	5
Cluytia collina R	2
Coffea arabica	2
Coffea bengalensis	3
Coix aquatica	3
Coix lacryma	1

Plant introduced	No
Coix natans	1
Colebrookia bulbifera R	2
Colebrookia oppositifolia	5
Colebrookia ternifolia	4
Combretum decandrum	3
Combretum pilosum	1
Commersonia echinata	4
Conocarpus acuminatus	1
Conocarpus latifolius	1
Convolvulus argenteus	4
Convolvulus bicolor	2
Convolvulus calycinus	1
Convolvulus digitatus	3
Convolvulus dissectus	5
Convolvulus fastigiatus	1
Convolvulus filiformis R	2
Convolvulus flagelliformis	3
Convolvulus gangeticus	3
Convolvulus hirsutus	2
Convolvulus macrorhizos	1
Convolvulus mechoacum	1
Convolvulus medium	1
Convolvulus paniculatus	7
Convolvulus parviflorus	4
Convolvulus pentagonus	1
Convolvulus pentaphyllus	1
Convolvulus purpureus	2
Convolvulus repens	1
Convolvulus semidigynus	1
Convolvulus setosus	1
Convolvulus sphaerocephalus	2
Convolvulus stipulaceus R	2
Convolvulus tridentatus	2
Convolvulus trilobatus	2
Convolvulus turpethum	2
Convolvulus umbellatus	1
Conyza balsamifera	2
Conyza pinnatifida	1
Corchorus capsularis	2
Corchorus fuscus R	2
Corchorus olitorius	1
Cordia angustifolia	2
Cordia gerascanthes	1
Cordia latifolia R	3
Cordia myxa	3
Cordia nigra	1
Cordia polygama	1
Corypha taliera	1
Costus speciosus	10
Cotyledon rhizophylla	1
Crataeva marmelos	1
Crinum latifolium	1
Crinum longifolium	1
Crinum sumatranum	1
Crinum zeylanicum	1
Crotalaria angulosa	1
Crotalaria bialata	1
Crotalaria bracteata R	7
Crotalaria chinensis	1
Crotalaria cytisoides	2
Crotalaria fulva R	7
Crotalaria juncea	1
Crotalaria laburnifolia	4
Crotalaria montana	5
Crotalaria paniculata	1
Crotalaria pulchella	1
Crotalaria pulcherrima R	5
Crotalaria quinquefolia	2
Crotalaria ramosissima R	5
Crotalaria retusa	2
Crotalaria sericea	5
Crotalaria stipulacea	3

Plant introduced	No
Crotalaria tenuifolia	1
Crotalaria tetragona R	4
Crotalaria verrucosa	7
Crotalaria virgata	4
Croton aromaticus	1
Croton boragatch	1
Croton drupaceus	1
Croton lacciferus	1
Croton tapa	1
Croton tiglium	5
Cucumis madraspatanus	1
Curculigo orchioides	1
Curcuma longa	2
Curcuma mutica	1
Curcuma pentandra	1
Curcuma toman	1
Curcuma zedoaria	3
Curcuma zerumbet	2
Cycas circinalis	2
Cycas sphaerica	2
Cylista scariosa	2
Cynanchium viminale	1
Cynoglossum racemosum R	2
Cyperus inundata	1
Cyperus iria	1
Cytisus Cajan	1
Dalbergia emarginata	2
Dalbergia frondosa	1
Dalbergia sissoo	3
Datura fastuosa	2
Datura metel	2
Diadelpha crinita	2
Diadelphia decandria	1
Dianthera ganderussa	1
Dillenia speciosa	2
Dioscorea aculeascora	1
Dioscorea aculeata	1
Dioscorea alata	2
Dioscorea anfecupa	1
Dioscorea anguina	1
Dioscorea glabra	2
Dioscorea pentaphylla	1
Dioscorea rubella	1
Diospyros chloroxylon	2
Diospyros cordifolia	6
Diospyros ebenum	1
Diospyros glutinosus	1
Diospyros ramiflora	1
Dolichos bulbosus	3
Dolichos ensiformoides	1
Dolichos fabaeformis	4
Dolichos flammeus	1
Dolichos gangeticus R	3
Dolichos gladiatus	3
Dolichos glutinosus R	4
Dolichos lignosus	1
Dolichos minimus	1
Dolichos scabaeoides	1
Dolichos soja	2
Dolichos strictus	1
Dolichos tuberosus	1
Dracaena angustifolia	1
Dracaena nervosa	1
Echinops spinosa	1
Echites antidysenteria	4
Echites frutescens	3
Echites scholaris	1
Echites venenata	2
Ehretia laevis	1
Elaeagnus conferta	1
Elaeagnus inferus	2
Elaeocarpus monogynus	1

Plant introduced	No	Plant introduced	No	Plant introduced	No
Elephantopus scaber	2	Gelonium bifarium	3	Hibiscus esculentus	5
Eleusine aegygptaca	1	Gelonium fasciculatum	2	Hibiscus ficulneus	2
Eleusine coracana	2	Glechoma erecta	3	Hibiscus gangeticus R	4
Epidendron sesselloides	1	Globba marantina	4	Hibiscus heptaphyllus	2
?Epygea cordata	1	Gloriosa superba	6	Hibiscus lampas	6
Eranthemum montanum	1	Glycine debilis	1	Hibiscus malabaricus	3
Eranthemum pulchellum	2	Gmelina arborea	3	Hibiscus moschatus R	1
Erythrina alba	2	Gmelina asiatica	1	Hibiscus mutabilis white	7
Erythrina incana	4	Gmelina parviflora	4	Hibiscus mutabilis red	3
Erythrina suberecta	3	Gmelina villosa	1	Hibiscus phoeniceus	3
Erythrina villosa	1	Gossipium barbadense	1	Hibiscus pilosus R	5
Eugenia acutangula	1	Gossipium vitifolium	1	Hibiscus populneoides	2
Eugenia alba	1	Gouania tiliaefolia	3	Hibiscus pruriens R	4
Eugenia fruticosa	1	Grewia asiatica	5	Hibiscus quinquefolius	3
Eugenia jambolifera	3	Grewia aspera R	4	Hibiscus radiatus	5
Eugenia jambos	2	Grewia carpinifolia	1	Hibiscus rigidus	3
Eugenia malaccensis	1	Grewia dioicea	1	Hibiscus rosa-sinensis	1
Eugenia zeylanica	5	Grewia orientalis	3	Hibiscus sabdariffa	5
Eupatorium divergens	6	Grewia pilosa R	3	Hibiscus simplex	2
Euphorbia antiquorum	3	Grewia pinifolia	1	Hibiscus solandra	3
Euphorbia arborescens	1	Grewia polygama	1	Hibiscus strictus R	3
Euphorbia lactea	2	Grewia sepiaria	2	Hibiscus surattensis	1
Euphorbia nervifolia	1	Grewia ulmifolia	1	Hibiscus tetraphyllus R	5
Euphorbia trigona	2	Grislea tomentosa	7	Hibiscus tiliaceus	6
Exacum tetragonum	1	Guarea binectarifera	1	Hibiscus tortuosus	1
Excoecaria agallocha	1	Guga disticha	1	Hibiscus tubulosus	1
?Exygea cordata	1	Hamiltonia suaveolens	2	Hibiscus vitifolius	7
Ferreola buxifolia	2	Hastingia coccinea	4	Hippocratea arborea R	2
Ficus conglomerata R	1	Hedysarum alatum	5	?Hispoerea arborea	1
Ficus cordifolia	1	Hedysarum articulatum	1	?Hoorea paniculata	1
Ficus cunia	2	Hedysarum lagopoides	1	Hyperanthera moringa	1
Ficus elasticus	1	Hedysarum bupleurifolium	1	Hypericum pomiserum	1
Ficus heterophylla	1	Hedysarum canescens	1	Hypericum guyanaceum	1
Ficus indicus	1	Hedysarum capitatum	2	Indigofera atropurpurea	6
Ficus macrophylla	2	Hedysarum cephalotes R	5	Indigofera coerulea R	2
Ficus racemiferus	1	Hedysarum collinum	3	Indigofera enneaphylla	1
Ficus religiosa	1	Hedysarum crinitum	1	Indigofera glandulosa	2
Ficus repens	1	Hedysarum esculentus	1	Indigofera glutinosa	1
Ficus tjiela	1	Hedysarum gangeticum	7	Indigofera heterophylla R	3
Ficus venosa	1	Hedysarum gramineum	4	Indigofera hirsuta	5
Flacourtia cataphracta	2	Hedysarum gyrans	8	Indigofera linifolia	1
Flacourtia inermis	5	Hedysarum gyroides	1	Indigofera tetraphylla	1
Flacourtia ramontchi	1	Hedysarum heterophyllum	1	Indigofera tinctoria	3
Flacourtia sapida	1	Hedysarum junceum	5	Indigofera trita	2
Flemingia congesta	2	Hedysarum lagenarium	2	Indigofera viscosa	2
Flemingia decussata	2	Hedysarum lagopoides	2	Ipomea candmax	1
Flemingia grandiflora	1	Hedysarum latifolium R	4	Ipomea coerulea	1
Flemingia virgata	1	Hedysarum neli tali	1	Ipomea grandiflora	3
Gaertnera esculenta	1	Hedysarum obliquum	3	Ipomea muricata	4
Gaertnera racemosa	4	Hedysarum pictum	7	Ipomea pes tigridis	1
Galedupa arborea	3	Hedysarum pulchellum R	8	Ipomea phoenicea R	5
Galedupa indica	2	Hedysarum purpureum R	6	Ipomea quamoclit	4
Galedupa uliginosa R	2	Hedysarum sennoides	2	Ipomea sagittata	2
Galega heyniana	4	Hedysarum strobiliferum	6	Ipomea undulata R	1
Galega incana R	2	Hedysarum styracifolium	6	Ixora alba	2
Galega lanceaefolia	3	Hedysarum triquetrum	7	Ixora arborea	1
Galega pentaphylla R	3	Hedysarum tuberosum	1	Ixora barbata	2
Galega pinnata	4	Hedysarum umbellatum	1	Ixora coccinea	2
Galega purpurea	3	Hedysarum vaginale	3	Ixora incongesta	1
Galega villosa	1	Hedysarum violitans	1	Ixora parviflora	3
Garcinia celebica	1	Hedysarum viscidum	3	Ixora pavetta	1
Garcinia cowa	2	Hedysarum vispertilionis	6	Ixora polysperma	1
Garcinia polyadelpha	1	Helicteres isora	4	Ixora undulata	4
Gardenia campanulata	3	Heliotropium indicum	1	Jasmium auriculatum	1
Gardenia florida	5	Heynea spinosa	1	Jasmium multiflorum	2
Gardenia integrifolia	1	Heynea trijuga	1	Jasmium zambac	2
Gardenia latifolia	1	Hibiscus abelmoschus	3	Jatropha curcas	1
Gardenia longispina R	3	Hibiscus armatus	1	Jatropha glauca	1
Gardenia lucida	5	Hibiscus bifurcatus	3	Jatropha multifida	1
Gardenia pavetta	2	Hibiscus calyculatus	1	Johnia bengalensis	1
Gardenia sandia	1	Hibiscus cannabinus	1	Johnia coromandeliana	2
?Gaylalia verticillata	1	Hibiscus collinus	2	Johnia salacioides R	2

Plant introduced	No	Plant introduced	No	Plant introduced	No
Jonesia asoca R	5	Mentha verticillata	1	Ophioxylon serpentinum	2
Jussieua exaltata R	5	Menyanthes cristata	2	Orchis dependens	2
Justicia betonica	7	Mespilus japonica	2	Orchis viridiflora	1
Justicia bicalyculata	1	Mesua ferrea	1	Ornithotrophe serrata	4
Justicia ecbolium	1	Michelia champaca	4	Ovieda verticillata R	6
Justicia echioides	1	Michelia procumbens	1	Oxalis senistiva	1
Justicia glabra	1	Millingtonia linearis	2	Panax digitatum	1
Justicia monanthera R	2	Millingtonia pinnata	1	Panax integrifolium	1
Justicia paniculata	2	Millingtonia semialata R	4	Pancratium speciosum	1
Justicia ramosissima	2	Millingtonia trinerva	2	Pancratium triflorum	1
Justicia speciosa	2	Mimosa abstergens R	2	Pancratium zeylanicum	1
Justicia stricta	1	Mimosa adenanthera	6	Pandanus odoratissimus	3
Justicia thyrsiflora	5	Mimosa arabica	5	Panicum barbatum	1
Kaempferia angustifolia	3	Mimosa aspera	5	Panicum brizoides	1
Kaempferia galanga	3	Mimosa catechu	5	Panicum costatum R	2
Kaempferia rotunda	2	Mimosa catechuoides	3	Panicum dactylon	3
Kaempferia speciosa	1	Mimosa dulcis R	3	Panicum setigerum	1
Kleinhovea hospita	7	Mimosa elata	2	Panicum hispidulum	1
Lagerstroemia grandiflora	2	Mimosa elengi	1	Panicum holcoides	1
Lagerstroemia parviflora	4	Mimosa farnesiana	7	Panicum italicum	2
Lagerstroemia regina	7	Mimosa ferruginea	4	Panicum miliaceum	1
Lantana indica	2	Mimosa glauca	6	Panicum miliare	1
Lantana trifolia	3	Mimosa grandiflora	1	Panicum plicatum	1
Laurus cinnamomum	2	Mimosa lebeck	2	Panicum spicatum	1
Lawsonia inermis	4	Mimosa lucida	1	Parkinsonia aculeata	1
Leea aequata	1	Mimosa natans	3	?Parona paniculata	1
Leea crispa	7	Mimosa neemdra	1	Pavonia odorata	4
Leea hirta	1	Mimosa octandra R	5	Pentapetes phoenicea	5
Leea macrophylla R	10	Mimosa odoratissima	3	Physalis flexuosa	1
Leea staphylea	2	Mimosa plena	1	Phaseolus aconitifolius	1
Leechee chinensis	1	Mimosa pudica	5	Phaseolus alatus	2
Leonurus indicus	3	Mimosa serusa	3	Phaseolus aureus	1
Lettsomia bona nox	4	Mimosa sirissa	4	Phaseolus calcaratus	1
Lettsomia nervosa	5	Mimosa smithiana	1	Phaseolus dolichoides	3
Ligusticum ajouan	1	Mimosa stipulacea	6	Phaseolus flexuosus	1
Ligusticum diffusum	1	Mimosa triquetra	1	Phaseolus trilobus	1
Limonia crenulata	1	Mimosa tortuosa	1	Phillyrea bracteata	2
Limonia pentaphylla	5	Mimosa virgata	7	Phillyrea paniculata	1
Limonia trifoliata	2	Mimusops elengi	6	Phlomis biflora	2
Limodorum nutans	2	Momordica dioica	4	Phlomis cephalotes	1
Limodorum pierardi	1	Momordica graveolens	1	Phlomis nepetifolia	4
Limodorum virens	1	Momordica mixta	4	Phlomis zeylanica	2
Linum trigynum R	2	Morinda bracteata	1	?Phonicum pulcherrima	1
Liriodendron grandiflora	2	Morinda citrifolia	1	Phrynium capitatum	1
Liriodendron odoratissima	1	Morus paniculata	1	Phyllanthus bacciformis	1
Luffa luffa	1	Murraya exotica	7	Phyllanthus cheremila	1
?Lupinaus orubiginosus R	1	Musa ornata	2	Phyllanthus leucopyros	1
Lysianthus chleroides	1	Myrtus littoralis	3	Phyllanthus longifolius	1
Malva americana	1	?Nacibea coccinea	1	Phyllanthus lucidus	1
Malva spicata	1	Nauclea parvifolia	1	Phyllanthus multiflorus	1
Manisuris granularis	2	Nelumbium speciosum	1	Phyllanthus obcordatus	1
Mangifera indica	2	Nepeta malabarica	1	Phyllanthus scandens	3
Maranta grandiflora	1	Nerium odorum	7	Phyllanthus sepiarus	1
Martynia diandra	3	Nerium tinctorum R	5	Phyllanthus setosus	1
Melaleuca cajuputi	1	Nyctanthes arbor-tristis	9	Phyllanthus tetrandrus	1
Melaleuca leucodendron	1	Nymphaea caerulea	3	Phyllanthus virosus	2
Melastoma ornatum	2	Nymphaea cyanea	1	Phyllanthus vitis idaea	3
Melastoma savanens	1	Nymphaea lotus	3	Phytolacca acinosa R	1
Melia azadirachta	1	Nymphaea rubra	3	Pinus longifolius	1
Melia azedarach	6	Ochroma lagopus	2	Piper betle	2
Melia latifolia	1	Ocimum cristatum	1	Piper longum	2
Melia robusta	2	Ocimum goolutulose	1	Piper nigrum	1
Melia superba	1	Ocimum gratissimum	2	Pistia stratiotes	3
Melochia corchorifolia	1	Ocimum pilosum R	2	Pitcairnia angustifolia	2
Melodinus monogynus	2	Ocimum ram-tulan	1	Plantago isphagula	1
Memecylon capitellatum	1	Ocimum robustum R	2	Plectranthus monadelphus	1
Menispermum cocculus	2	Ocimum sanctum	2	Plectranthus scutellarioides	2
Menispermum heteroclitum	1	Ocimum staminecum	2	Plectranthus strobiliferus R	5
Menispermum tricuspidatum	1	Ocimum thyrsiflorum	2	Poa amabilis	1
Menispermum verrucosum	1	Odina wodier	1	Poa uliginosa	1
Mentha quadrifolia	4	Olea robusta	1	Poinciana pulcherrima	6
Mentha secunda	4	?Opalatoa aublet	1	Polygala prostrata	1

Plant introduced	No	Plant introduced	No	Plant introduced	No
Polygala telephioides	1	Scaevola koenegii	1	Strychnos potatorum	2
Polygonum perfoliatum	5	Scaevola taccata	4	?Sussodia oppositifolia	1
Polygonum pilosum	1	Scilla indica	1	Swietenia chloroxylon	2
Polypodium quercifolium	1	Scirpus complanatus	1	Swietenia febrifuga	2
Pontederia hastata	1	Scirpus kysoor	1	Swietenia trifoliata	1
Pontederia vaginatus	1	Scirpus pentagonus	1	Tabernaemontana coronaria	5
Porana paniculata	3	Scytalia verticillata	3	Tabernaemontana crispa	1
Porana racemosa	1	Securidaca scandens	1	Tabernaemontana dichotoma	1
Posoqueria dumetorum	4	Semicarpus anacardium	1	Tacca pinnatifida	2
Posoqueria spinosa	1	Senna alata	6	Tagetes patula	1
Posoqueria uliginosa	3	Senna arborescens	6	Tamarix dioica R	3
Pothos pinnata	1	Senna auriculata	8	Tamarix indica	1
Premna longifolia	1	Senna bicapsularis	7	Tectona grandis	2
Psoralea corylifolia	3	Serratula anthelmintica R	4	Terminalia alata	1
Pterocarpus dalbergioides	1	Serratula carthamoides	2	Terminalia balerica	1
Pterocarpus indicus	1	Serratula cinerea	1	Terminalia catappa	2
Pterocarpus marsupium	1	Sesamum indicum	1	Terminalia procera	2
Pterocarpus sisoo	2	Sesamum orientale	1	Theobroma guazuma	1
Pterospermum asperifolium	3	Sida asiatica	1	Thomea bona nox	1
Pterospermum canescens	1	Sida cristata	5	Thomea muricata	1
Pterospermum maximus	2	Sida graveolens	6	Thomea ornata	1
Pterospermum semisagittatum	3	Sida humilis	1	Thunbergia fragrans	7
Pterospermum suberifolium	1	Sida indica	5	Thunbergia grandiflora	2
Pterospermum uliginosus	2	Sida lanceolata	2	Tradescantia discolor	1
Pyrethrum indicum	1	Sida montana	2	Trapa bispinosa	1
Randia racemosa	1	Sida nudiflora	2	Trewia nudiflora	2
Rhamnus nitidus	1	Sida opulifolia	1	Tribulus terrestris	3
Rheedia montana	1	Sida periplocifolia	6	Trichosanthes palmata	1
Rhus succedaneum	1	Sida polyandra	5	Trifolium indicum	3
Ricinius dioicus	2	Sida populifolia	2	Triphasia aurantiola	2
Rivina laevis	4	Sida rhombifolia	2	Triumfetta bartramia	2
Robinia candida R	7	Sida tomentosa	6	Triumfetta rotundifolia	2
Robinia cannabina	3	Siegesbeckia brachiata R	2	Triumfetta trilocularis	1
Robinia grandiflora	3	Sinapis dichotoma	1	Trixis vervacea	1
Robinia paludosa	1	Sinapis juncea	1	Trophis aspera	1
Robinia segmenta	1	Sinapis ramosa	2	Unona odoratissima	2
Robinia sesban	1	Siphonanthus indica	2	Urena catechu	1
Rondeletia stricta	1	Smilax sagittata	2	Urena conferta	3
Rosa involucrata	3	Smithia aspera	1	Urena lobata	2
Rottboellia exaltata	2	Smithia sensitiva	8	Urena repanda	3
Rottlera paniculata	1	Solanum auriculatum	3	Urena speciosa	3
Rottlera peltata	1	Solanum decemdentatum	2	Urtica ferocissima	1
Rottlera tinctoria	1	Solanum hirsutum R	2	Urtica pentandra	4
Ruellia cernua	1	Solanum indicum	2	Urtica tenacissima	1
Ruellia dependens	2	Solanum jacquinii	2	Uvaria grandiflora	2
Ruellia fasciculata	1	Solanum pentapetaloides	1	Uvaria longifolia	2
Ruellia imbricata	2	Solanum torvum	1	Uvaria odoratissima R	2
Ruellia obovata	1	Solanum trilobatum	2	Uvaria villosa	4
Ruellia ringens	1	Sonneratia apetala	1	Vangueria spinosa	1
Ruellia secunda	2	?Sooroogada bilocularis	1	Verbena bonariensis	2
Rumex acutus	2	?Sooroogada glabra R	2	Verbena jamaicensis	1
Saccharum diandrum	1	Sophora tomentosa	9	Verbesina lavenia	1
Saccharum fuscum R	2	Spermacoce scabra	2	Vitex alata	5
Saccharum munja	1	Spermacoce teres R	3	Vitex arborea	4
Saccharum officinarum	3	Spermacoce sumatrensis	3	Vitex incisa	4
Saccharum procerum	1	Spilanthus amella	2	Vitex leucoxylon	1
Saccharum semidecumbens	1	Spilanthus articularis	1	Vitis candida	1
Saccharum spontaneum	1	Spilanthus oleracea	6	Volkameria infortunata	1
Sagittaria obtusifolia	3	Stapelia umbellata	1	Volkameria odorata	3
Saguerus rumphii	1	Sterculia argentia	1	Volkameria odoratissima	2
Salix tetrasperma	2	Sterculia balanghas	1	Vulna lobata	1
Salvia brachiata	1	Sterculia colorata	3	Wampas chinensis	1
Sandoricum indicum	2	Sterculia pterosperma	1	Xanthochymus tinctorius	1
Santalum album	6	Sterculia urens	2	Xanthochymus pictorius	2
Sapindus fruticosus	1	Sterculia villosa	4	Xyris indica	1
Sapindus laurifolius	1	Stilago bunius	2	Zingiber cassumunar	1
Sapindus rubiginosus	1	Stratiotes alismoides	2	Ziziphus trinervia	1
Sapindus saponaria	4	Streptium asperum R	3		
Sapium sebiferum	4	Strychnos colubrina	1		

BOTANICAL CORRESPONDENTS

William Roxburgh was in correspondence with a large number of people throughout the world, and a study of his *Hortus Bengalensis* gives the names of more than 140 who are listed below. Against each person is the number of plants that they are credited with sending to the Calcutta Botanic Garden. Three of these may be the same person: Dr I. Fleming and Dr J. Fleming; Captain Hardwicke and Colonel Hardwicke; and Sir G. Young and G. Young. It is worth noting that there are some Indian plant collectors mentioned, for instance, Ghosal and Ram-dun Choudri whose situations cannot be identified further that being in India, while Null-mandu and Palla-nundu were both from the Coromandel Coast area. Equally, there are a number of Frenchmen, such as M. Céré, M. de Cossigny, Dr Jacquin, M. Jannet and F. Peirard. Roxburgh also appears to have relied heavily on medical men and those serving in the army, with 12 of the former and 27 of the latter, although some may have been captains of ships rather than military men. His two sons whom he employed are also mentioned, with John introducing 92 species and William junior 48, while there is also an enigmatic entry of a G. Roxburgh who sent one species from Hardwar, about 120 miles north-east of Delhi, in 1801.

In terms of the origins of the plants, they came from most of the known world: North America, the West Indies, South America, Europe, Arabia, South Africa, India, Bhutan, Nepal, East Indies, China, Japan and New South Wales. People such as Dr J. Anderson and William Hamilton were major contributors from the Americas, while Roxburgh himself was responsible for introducing 38 from the Cape of Good Hope, and William Kerr 58 from China and one species from Japan. What is also interesting is that 47 species were introduced accidentally, presumably as seeds in the soil of intentional introductions.

Mr Abraham	2	H. T. Colebrooke	113
J. Addison Esq	1	M. de Cossigny	4
Dr J. Anderson	8	R. W. Cox Esq	2
Capt Anderson	2	H. Creighton	7
W. Aiton (Kew)	9	Capt Denton	4
Mr Arbuthnott	2	R. K. Dick	5
Mr Atkinson	1	Capt Dickinson	5
Lady Ann Barnard	1	G. Dowedeswell Esq	1
Col. Beatson	1	Col Dyer	1
J. Bentley	4	Samuel Dyer Esq	1
Dr A. Berry	67	Mr W. Dyer	1
Capt B. Blake	5	W. Egerton Esq	2
Mrs Blaxland	3	Mr Ewer	4
Sir C. Blunt	1	Dr I. Fleming	1
C. Borram Esq	2	Dr J. Fleming	3
B. Boswell Esq	7	Mr Forsyth	1
Mr Bowie	1	Capt Frazer	1
Mr P. Bowil	1	Capt Garnault	2
T. Brooke Esq	4	Col Garstin	4
Mr M. Brown	6	Rev. Gericke	1
Hon. C. A. Bruce	3	Ghosal	1
Hon. N. C. Bruce	1	Mr Gibbons	2
Dr F. Buchanan	176	I. Glass Esq	3
Mr Burchell	2	Goodlad	9
Dr C. Campbell	45	Mr Gott	4
Mr F. Carey	15	Alexander Gordon Esq	1
Dr W. Carey	96	T. Graham Esq	1
M. Céré	9	Hon C. Greville	8
Mr Chapman	3	Mr Griffiths	1
Lady Clive	2	Mr W. Gwilt	1

W. Hamilton	185		Lt McKenzie	3
Capt Hardwicke	6		Lord Minto	1
Col. Hardwicke	36		Mr M'Mahon	18
J. Harington Esq	1		Mr Moddem	2
H. Harris	1		Capt Murray	1
S. Harris	22		Col P. Murray	5
Sir A. Hesleridge	1		Nulla-mandu	40
Mr P. Heshusius	14		Palla-nundu	1
Capt Hewitt	1		Col Patterson	2
Mr B. Heyne	22		F. Peirard	7
A. Hogue Esq	1		Mr Pleydell	1
Mr Honeycomb	1		Ram-dun Choudri	1
Mrs Honeycomb	2		Sir J. Rhodes	1
H. Hope	3		Capt Richardson	1
Dr Hornman	10		A. Roberts	1
F. Horsley	1		Robertson Esq	3
Sir A. Hume	11		Mr Rottler	17
Dr W. Hunter	8		G. Roxburgh	1
Dr Jacquin	5		John Roxburgh	92
Jandi-mali	1		William Roxburgh junior	48
Mon. Jannet	10		Sir J. Royds	3
Dr John	5		Mr H. Russell	7
Mr Peter Julian	2		Dr P. Russell	1
W. Kerr	59		Mr Salisbury	1
Kerum	1		J. Scott	1
King of Nepal	3		Capt Simson	1
Dr Klein	113		B. Smith	4
Kurrim-khan	10		C. Smith	33
Col. A. Kyd	7		M. R. Smith	204
Mr R. Kyd	7		Mr Snodgrass	4
A. B. Lambert	25		Capt Wright	1
Capt Learmonth	1		Lt Stokoe	1
Mr W. Lockhead	12		S. Swinton Esq	1
Loddiges & Son	1		Capt Tennent	10
Dr Lumsden	1		Capt Thomas	2
Mr Maclew	1		Mr Turnbull	2
Mr W. Madden	7		Viscount Valentine	2
Maj Gen Malcolm	3		Dr Wallich	4
Gov. of Manilla	1		Mr White	1
T. Marriot	1		Capt Wilford	1
Gen Martin	11		Sir G. Young	4
Mr Meyne	2		G. Young	4
Gen H. M'Dowall	14			

LIST OF BOOKS IN WILLIAM ROXBURGH'S
LIBRARY

This Appendix lists those books and periodicals which Roxburgh appears to have had in his library, and are referred to in his *Flora Indica* and numerous letters in which he mentions books, etc. that he has received. In most cases, the title is given only briefly, such as 'The new edition of the Hortus Kewensis.'* This has made it difficult on a number of occasions to identify either the exact edition, especially in the case of such authors as Linnaeus, and some have not been traced at all, such as *Culture of Rhubarb*. To identify these publications, the *British Museum Catalogue* has been used, supplemented by the American *Library of Congress Catalogue*.†

Arnoldus Nicolaus Aasheim, *Descriptionis rariorum plantarum, nec non materiae medicae atque œconomiae e terra Surinamensi fragmentum ... Disputarus ... C[hristen] F[riis] Rottböl ... Spactam defendentis venant* (Copenhagen, 1776).
William Aiton, *Hortus Kewensis; or, a catalogue of the plants cultivated in the Royal Botanic Garden at Kew*, 3 vols. (London, 1789).
Henry C. Andrews, *The Botanist's Repository*, vols. 1-10 (London, 1797-1815).
Anon, *Culture of Rhubarb*.
Anon, *The European Magazine and London Review* (London, 1792).
Anon, *The Monthly Review, or Literary Journal, enlarged* (London, 1791).
Anon, *The Monthly Review, or Literary Journal, enlarged* (London, 1792).
Anon, *Natural Shorts Dyeing*.
Anon, *Relative to the manufacture of Sugar*.
Anon, *Report ... Bengal indigo*.
Anon, *Report ... on the state of the Sugar trade*.
Anon, *The Repository. Containing a succinct and clear view of the most considerable transactions ... at home and abroad, of the present times. To which is added a curious collection ... of original letters, speeches, and other papers, ... likewise choice apothegms, ... anecdotes, etc.* (London, 1752).
Anon, *Silk worms and Mulberry*.
Asiatick Researches, volumes 4-7, 10-11 [at least], (Calcutta, 1799-1803, 1809-11).
Benjamin Smith Barton, *Elements of botany; or, outlines of the natural history of vegetables* (London, 1804).
William Beckford [of Jamaica], *A descriptive account of the Island of Jamaica ...* , 2 vols. (London, 1790).
Bernard Forest de Belidor, *Architecture hydraulique; ou, l'art de conduire, d'elever et de menager les eaux pour les differens besoins de la vie*, 2 parts (4 vols.), (Paris, 1737-53).
Bloke and Forrester, *Termini Ichthyologici*. The only relevant title to fit this that has been found is: Marcus Elieser Bloch, *Systema Ichthyologiae iconibus cx illustratum. Post obitum auctoris opus inchoatum absolvit, correxit, interpotavit* (Berlin, 1801).
Botanical Society of Lichfield, *A system of vegetables according to their classes, genera, orders, species with their characteristics and differences ... Translated from the thirteenth edition ... of the Systema vegetabilium of the late Professor Linnaeus ...* , 2 vols. (London, 1782-5).
Francis Buchanan [afterwards Hamilton], *A journey from Madras through the countries of Mysore, Canara, and Malabar*, 3 vols. (London, 1807).
William Urban Buée, *A narrative of the successful manner of cultivating the clove tree in the island of Dominica* (London, 1797).
Georges Louis Leclerc, Comte de Buffon, *Natural history, general and particular ... Translated ... with notes and observations by the translator (W. Smellie)*, 9 vols. (Edinburgh, 1780-5).
E. Burke (ed.), *The Annual Register; or, a view of the history, politicks and literature for the year 1783* (London, 1784).
E. Burke (ed.), *The Annual Register; or, a view of the history, politicks and literature for the year 1784* (London, 1785).

* Letter from William Roxburgh to Sir Joseph Banks, dated 25 November 1811, Mitchell Library, Banks Archive, 810339.
† Anon, *British Museum. General Catalogue of Printed Books*, Photolithographic edition, 263 vols. (London, 1961-6); and Anon, *The National Union Catalog. Pre-1956 Imprints*, 685 vols. London, 1968-79).

E. Burke (ed.), *The Annual Register; or, a view of the history, politicks and literature for the year 1785* (London, 1786).

Joannes Burman, *Thesaurus Zeylanicus, exhibens plantas in insula Zeylana nascentes ... omnia iconibus illustrata ac descripta cura* (Amsterdam, 1737).

Nicolaas Laurens Burman, *Flora Indica; cui accedit series zoophytorum Indicorum, nec non prodromus florae Capensis* (Leiden, 1768).

John Campbell, *A political survey of Britain, etc.*, 2 vols. (London, 1774).

Captain Jonathan Carver, *Travels through the interior parts of North America. To which is added some account of the author by John Coakley Lettsom* (London, 1781).

Antonio José Cavanilles, *Icones et descriptiones plantarum quae aut sponte in Hispania crescunt, aut in hortis hospitantur*, 6 vols. (Madrid, 1791-1801).

Jean Antoine Claude Chaptal, *Elements of chemistry. Translated from the French*, 2nd edition, 3 vols. (London, 1795).

Pierre François Xavier de Charlevoix, *Histoire du Paraguay*, 6 vols. (Paris, 1757).

Francisco Saverio Clavigero, *The history of Mexico ... Translated by ... C. Cullen*, 2 vols. (London, 1787).

Archibald Cochrane, Earl of Dundonald, *The present state of the manufacture of salt explained; and a new mode suggested of refining British salt, so as to render it equal, or superior to the finest foreign salt* (London, 1785).

William Curtis, *Botanical Magazine; or, Flower-Garden displayed*, vol. I onwards (London, 1787).

William Curtis, *Flora Londinensis; or, plates and descriptions of such plants as grow wild in the environs of London*, 3 vols. (London, 1777-87).

Johann Jacob Dillenius, *Historia muscorum: a general history of land and water, &c, mosses and corals* (London, 1763).

John Edwards, *The British herbal, containing one hundred plates of the most beautiful and scarce flowers and useful medicinal plants, which blow in the open air of Great Britain* (London, 1770).

Fitzmorris, *Cultivation of Sugar*.

Johann Reinhold Forster, *Observations made during a voyage round the world, on physical geography, natural history, and ethic philosophy* (London, 1778).

Dr J. C. Fothergill, *The works of J. C. Fothergill*, edited by John Coakley Lettsom, 2 vols. (London, 1783).

A. F. de Fourcroy, *Elements of natural history and of chemistry: being the second edition of the elementary lectures on those sciences ... enlarged and improved by the author ...*, translated by William Nicholson, 4 vols. (London, 1788).

Joseph Gaertner, *De fructibus et seminibus plantarum*, (vol. 3 *C. F. Gaertner ... supplementum carpologia, sui continuatio operis J. Gaertner de fructibus*), 3 vols. (Leipzig, 1788-1807).

Johann Freidrich Gmelin, *Caroli a Linné ... Systema Naturae per regna tria naturae: secundum classes, ordines, genera, species cum characteribus, differentiis, synonymis, locis* (ed. Gmelin), 3 vols. (Leipzig, 1788-93).

Jacques Julien Houtou de Labillardière, *Relation du Voyage à la recherche de la Pérouse, fait par ordre de l'Assemblée Constituante, pendant les année 1791, 1792, et pendant la 1ere et la 2ere année de la République Françoise*, 2 vols. (Paris, 1800); *An Account of a Voyage in search of La Pérouse in the Years 1791, 1792 1793, Performed by Order of the Constituent Assembly. Translated from the French*, 2 vols. (London, 1800).

William Hudson, *Flora Anglia, exhibens plantas per regnum Angliae sponte crescentes, distributas secundum systema sexuale, etc.* (London, 1762).

Nicolaus Josephus Jacquin, *Hortus botanicus Vindobonensis, seu plantarum rariorum, quae in horto botanico Vindobonensi ... coluntur, icones coloratae et sucinctae descriptione.*, 3 vols. (Vienna, 1770-6).

Antoine Laurent de Jussieu, *Genera plantarum secundum ordines naturales disposita* (Paris, 1789).

Andrew Kippis, *The Life of Captain James Cook* (London, 1788).

C. Konig, and J. Sims, *Annals of Botany*, 2 vols. (London, 1805-6).

Jean Baptiste de Lamarck, *Encyclopédie Méthodique. Botanique. Par M. le chevalier de Lamarck (continué par J. L. M. Poiret)*, 8 vols. (Paris, 1783-1808).

La Mathiere, *Analytical Essay ... on fermentation*.

Aylmer Bourke Lambert, *An illustration of the genus cinchona ... Baron Humboldt's account of the cinchona forests of South America; and Laubert's memoir on the different species of Quinona. To which are added several dissertations of Don. H. Ruiz on various plants of South America, etc.* (London, 1821).

Thomas Law, *Letters from Thomas Law, one of the East India Company Collectors, to the Revenue Board relating to the Company's regulations for weavers* (London, 1793).

James Lee, *An Introduction to botany, Containing an explanation of the theory of that science; extracted from the works of Dr Linnaeus*, 4th edition (London, 1788).

Dr John Coakley Lettsom, *The natural history of the tea trade, with observations on the medicinal qualities of the tea and on the effects of tea drinking* (new edition, London, 1799, first published London, 1772).

Dr John Coakley Lettsom, *Hortus Uptonensis; or, a Catalogue of Stove and Green-House Plants in Dr Fothergill's Garden at Upton, at the time of his decease* (London, 1783).

Dr John Coakley Lettsom, *Exposition of the inoculation of the small pox, and cow pock …* , 2nd edition, (London, 1801).

Charles Louis L'Hériter de Brutelle, *Cornus. Specimen botanicum sistens descriptiones et icones specierum corni minus cognitarum* (Paris, 1788).

James Lind, *An essay incidental to Europeans in hot climates … . To which is added an appendix concerning intermittent fevers, etc.* (London, 1768).

Carl Linnaeus, *Genera plantarum corumque characteres naturales secundum numerum, figuram, situm, & proportionem omnium fructificationis partum.*

Carl Linnaeus, *Species plantarum secundum classes, ordines, genera, species, cum characteribus, differentiis, nominibus trivialibus, synomis selectis, et locis natalibus*, 4 vols. (Frankfurt, 1779-80).

Carl Linnaeus, *Flora Zeylanica, sistens plantas Indicas Zeylonae insulae, quae olim 1670-1677 lectae fuere a Paulo Hermanno.*

Carl Linnaeus, *Systema naturae.*

João de Loureiro, *Flora Cochinchinensis; sistens plantas in regno Cochinchina nascentes; quibus accedunt aliae, observatae in sinensi imperio, Africa orientali, Indiaeque locis variis*, 2 vols. (Lisbon, 1790).

John Lunan, *Hortus Jamaicensis; or, a botanical description … of its indigenous plants … as also of the most useful exotics*, 2 vols. (Jamaica, 1814).

Pierre Joseph Macquer, *Art de la teinture en soie. Descriptions des Art et Metiers, faites ou approuvées par Messieur de l'Academie Royale des Sciences* (Paris 1763).

William Marsden, *History of Sumatra, containing an account of the government, laws, … of the native inhabitants, with a description of the natural productions* (London, 1783).

Dr J. Mease, *An Essay on the disease produced by the bite of a mad dog … with a preface and appendix by John Coakley Lettsom* (London, 1793).

John Frederick Miller, *Travels.*

Philip Miller, *The gardeners dictionary, containing the best and newest methods of cultivating and improving the kitchen, fruit and flower garden …*, 8th edition, 2 vols. (London, 1768).

John Hamilton Moore, *A new and complete collection of voyages and travels …* , 2 vols. (London, 1780).

William Nicholson, *The first principles of chemistry* (London, 1790).

Sydney Parkinson, *A Journal of a Voyage to the South Seas, in his Majesty's Ship Endeavour [under the command of Captain James Cook] … embellished with twenty-nine views and designs … . To which is now added, remarks on the preface, by the late John Fothergill … and an appendix, containing an account of the voyages of Commodore Byron, Captain Wallis, Captain Carteret, Monsieur Bourgainville, Captain Cook, and Captain Clerke* (London, 1784).

Thomas Pennant, *Indian Zoology*, 2nd edition (London, 1790).

Philosophical Transactions of the Royal Society.

Leonard Plukenet, *Phytographia, sive stirpium illustriorum … minus cognitarum icones, tabulis aeneis … elaboratae. Pars prior (Almagestum botanicum, etc. – Almagesti mantissa, etc. – Amaltheum botanicum, etc.)*. 7 parts in 4, 2nd edition (London, 1769).

Abraham Rees, *The Cyclopedia; or, An Universal Dictionary of Arts and Sciences … with the Supplement and Modern Improvements* (London, 1778-88).

James Rennell, *Memoir of a map of Hindoostan; or the Mogul empire, etc.* (London, 1788).

Anders Jahan Retzius, *Observationes botanicae, sex fasciculis comprehensae. Quibus accedunt J. G. Koenig Descriptiones monandrarum et epidendrorum in India Orientali factae* (Leipzig, 1779-91).

Henrik Adriaan van Rheede van Draakestein, *Hortus Indicus Malabaricus, continens regni Malabarici apud Indos … omnis generis plantae rariores, Latinis, Malabaricis, Arabicis et Bramanum characteribus nominibusque expressas … Adornatus per H. vn R. van D. [assisted by Matthaeus a S. Joseph] … et J. Casearium … Notis adauxit et commentariis illustravit A. Syen [and J. Commelinus]*, 12 parts (Amsterdam, 1678-1703).

Dr A. Robinson, *Botanist.*

Alexis Marie Rochon, *A Voyage to Madagascar and the East Indies. To which is added, M. Brunel's Memoir on the Chinese Trade … . Translated from the French by J. Trapp* (London, 1793, originally published in French, 1791).

Georg Eberhard Rumpf [also referred to as Georgius Everhardus Rumphius], *Herbarium Amboinense, plutimas conplectens arbores, frutices, herbas, plantas terrestres, & aquaticas, quae in Amboina, at adjacentibus reperiuntur insulis, adcuratissime descriptas juxta earum formas. Quod et insuper exhibet veria insectorum animatumque genera ... Nunc primum in lucem edita, & in latinum sermonen versa, cura et studio Joannis Burmanni*, 6 vols. (Amsterdam, 1750).

Russell, *Account of the Tobasheer.*

Johann Christian Daniel Schreber, *Icones et descriptiones plantarum minus cognitarum*, 4 parts, (Halle, 1765).

Sir Hans Sloane, *A voyage to the Islands Madera, Barbados, Nieve, S. Christopher and Jamaice, with the natural history, ... of the last of those islands*, 2 vols. (London, 1707-25).

W. Smellie (ed.), *Encyclopædia Britannica; or a Dictionary of Arts and Sciences compiled upon a new plan ... with one hundred and sixty copperplates*, 3 vols. (Edinburgh, 1771).

Sir James Edward Smith, *Tracts relating to Natural History* (London, 1798).

James Smollett (ed.), *The Critical Review, extended and improved ...* (London, 1791).

Pierre Sonnerat, *Voyage à la Nouvelle Guinée, dans lequel on trouve la description des lieux, etc.* (Paris, 1776).

Pierre Sonnerat, *Voyage aux Indes Orientales et à la Chine, fait ... depuis 1774 jusqu'en 1781, etc.*, 2 vols. (Paris, 1782).

Tancroix, *Chemistry.*

Carl Peter Thunberg, *De Protea. – De Oxalida. – De Gardenia. – Nova plantarum generar – Novae insectorum species. Acta medicorum Suecicorum, sue Sylloge observatt, et casuum rariorum in variis Medic. partibus, praesertium in historia naturale, praxi medica et chirurgia* (Uppsala, 1783).

Carl Peter Thunberg, *Flora Japonica, sistens plantas insularum Japonicarum, etc.* (Leipzig, 1784).

Pierre Francois Tingry, *The painter and varnisher's guide; or a treatise ... on the art of making and applying varnsihes, etc.* (London, 1804).

Joseph Townsend, *A journey through Spain in ... 1786 and 1787, with particular attention to the agriculture, manufactures and commerce ... of that country, and remarks in passing through a part of France*, 3 vols. (London, 1791).

Transactions of the American Philosophical Society, vol. 5 [at least].

Transactions of the Batavian Society (Rotterdam, Amsterdam and Batavia, 1779).

Transactions of the Linnean Society, vols. 1-5 [at least] (London).

Martin Vahl, *Symbolae Botanicae, sive plantarum, tam earum, quas in itinere, inprimis orientali, collegit Petrus Forskål, quam aliarum, recentius detectarum, exactiores descriptiones, necnon observationes circa quasdam plantae dudum cognitas*, 3 parts (Copenhagen, 1790-4).

Martin Vahl, *Enumeratio platarum vel ab aliis, vel ab ipso observatarum, cum earum differentiis specificis, synonymis ... et descriptionibus*, 2 vols (Copenhagen, 1804).

James Wheeler, *The botanist's and gardener's new dictionary; containing the names, classes, orders, generic characters, and specific distinctions of the several plants cultivated in England, according to the system of Linnaeus ... In which is also comprised, a gardener's calendar ... And to which is prefixed, an introduction to the Linnean system of botany* (London, 1763).

Carl Ludwig Willdenow, *Caroli a Linné Species plantarum ... Editio quarta ... curante C. L. W.*, 5 vols. (Berlin, 1797-1810).

William Withering, *A botanical arrangement of British plants; including the uses of each species, in medicine, diet, rural œconomy and the arts. With an introduction to the study of botany*, 2nd edition, 3 vols. (London, 1787-92).

William Woodville, *Medical botany, containing ... descriptions, with plates, of all the medicinal plants ... comprehended in the catalogues of the materia medica, as published by the Royal College of Physicians of London and Edinburgh. Etc.*, 3 vols. (London, 1790-3).

There are also a number of letters that refer to various additions which are not specified, as well as occasions when Roxburgh referred to 'wants', the latter particularly after he lost everything in the flood of 1787.*

---

* The following letters all refer to additions:
List of papers for Dr Roxburgh, dated 6 December 1789, Natural History Museum, Botany Library, Roxburgh Correspondence.
Commercial Letter from the Court of Directors to Fort St George, dated 25 June 1793, para. 48, in Public Department letter dated 25 June 1793, British Library, OIOC IOR, E/4/879, ff919-22.
Letter from Andrew Ross to William Roxburgh, dated 16 September 1793, Natural History Museum, Botany Library, Roxburgh Correspondence.
Letter from William Roxburgh to Dr Taylor, undated but July 1807, Natural History Museum, Botany Library, Roxburgh Correspondence.
Letter from William Roxburgh to Sir Joseph Banks, dated 25 November 1811, Mitchell Library, Banks Archives, 810339.

WILLIAM ROXBURGH'S FAMILY

The following family trees help to show the relationships between the various members of the family. Key members of the family have been printed in red, those who played essential roles in the life of William Roxburgh are printed in blue, and those who appear more than once on the various family trees are printed in **bold**.

1. WILLIAM ROXBURGH'S PARENTS

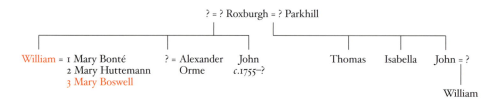

2. WILLIAM ROXBURGH'S FAMILY

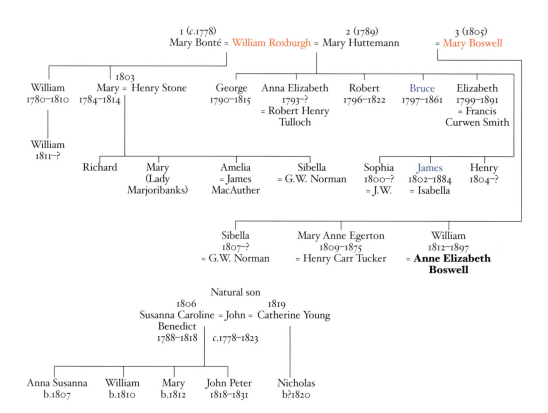

2. MARY BOSWELL'S FAMILY

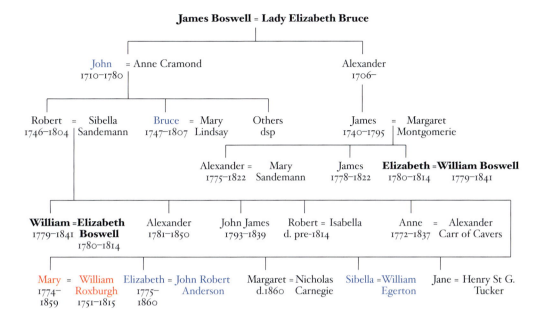

Marriage 1 to Mary Bonté about 1778

Children **William** born about 1780, educated Edinburgh University 1796-97, 1797-98, son (William) born to Sukeowa about May 1811, baptised 20 Feb 1812 at Fort Marlborough (N/7/1/195). Buried 21 Sep 1810 in Padang (N/7/1/223)

Mary born about 1784, married 3 Jan 1803 Henry Stone, Civil Service, in the presence of William & Mary Roxburgh, H. L. Thornhill, Cudbert Thornhill, Maria Macpherson, J. Thornhill (N/1/6/205). Died 30 Jan 1814 and buried in Greyfriars Kirkyard in her 30th year (Brown, 1867)

Marriage 2 to Mary Huttemann 3 Jun 1789 by Rev W. Gericke in Madras (N/2/11/627), buried 25 Nov 1804 (N/1/6/205). Probably sister of George Huttemann (1769-1843), Head of Free School, Calcutta (N/2/11)

Children **George** born 4 Apr 1790, Samulcottah, baptised 1 Sep 1792 at Samulcottah by Lt J. Hewenson, Commandant (N/2/11/75). Cadetship in Bengal Cavalry 12 Jan 1806 aged 15 (L/MIL/9/115/110). Died 6 Dec 1815, Surabaya, Java

Anna Elizabeth born about August 1793, baptised 17 Oct 1793 by Maj. A. Wynch, Commandant (N/2/11/81). Married Robert Henry Tulloch at Serampore, 17 Jul 1811 (N/1/8/439)

Robert born 4 May 1796, baptised privately 13 Oct, received into church 3 Nov 1796 (N/1/4/193). Cadetship (l/MIL/9/125/152-5). Living Park Place, Edinburgh, 1822. Died 1824.

Bruce born 1797, baptised 12 Dec 1797 (N/1/5/11). Cadetship (L/MIL/9/125/445-53). Died 14 Jun 1861, Torquay (DNB)

Elizabeth born 1799, married Francis Curwen Smith, Magistrate of Cawnpore, 6 Apr 1827 (N/1/18/99). Died 1891

Sophia born 1800, married J. W. 6 Nov 1822, St Cuthberts, Edinburgh

James born 25 Feb 1802 and baptised on 9 Aug 1802 in Calcutta (N/1/6/118). Cadetship (L/MIL/9/137/39). Capt 39th Bengal N.I. Assistant Military Auditor General in 1837; Lt Col of Newry, Ireland by 9 May 1857 (L/MIL/9/254/452). Died 11 Jul 1884, Kensington

Henry born 18 Mar 1804, baptised on 14 Aug 1804 in Calcutta (N/1/7/10). Navy

Marriage 3 to Mary Boswell 2 Nov 1805 in Iver, Bucks. Died 18 Jan 1859 in London in her 85th
　　year (Brown, 1867)
　　　　Children　　**Sibella** born 17 May 1807 at sea near the Cape of Good Hope and baptised
　　　　　　　　29 Aug 1807 in Calcutta. Died Ipswich 1896.
　　　　　　　　Mary Anne Egerton born 8 Feb 1809 and baptised 11 Jan 1810 in Calcutta
　　　　　　　　(N/1/8/279). Married Henry Carr Tucker aged 24 of the Civil Service 23 Jul
　　　　　　　　1834. Died 1875
　　　　　　　　William born 15 May 1812 and baptised 3 Sep 1812 at Calcutta (N/1/9/8).
　　　　　　　　Married Anne Elizabeth Boswell 1857. Died 1897

　　　　Natural son　　**John** born about 1777. Married Susanna Caroline Benedict 27 Dec 1806
　　　　　　　　(N/1/7/204) and secondly Catherine Young 24 Mar 1819 (N/1/11/23). Buried
　　　　　　　　18 Sep 1823 aged 46 (N/1/12/593)

Children

William (1780-1810) botanist at Prince of Wales Island, educated Edinburgh University 1796/97,
　　1797/98 (EUML Medical archives), mentioned in letters and sent plants to Calcutta Botanic
　　Garden
　　　　　　　　William (natural son) born about May 1811 (after death of father) to
　　　　　　　　Sukeowa (native woman) and baptised at Fort Marlborough 20 Feb 1812
　　　　　　　　(N/7/1/195)

Mary Roxburgh (1784-1814)
　　Married 3 Jan 1803 Henry Stone, Civil Service (N/1/6/205)
　　　　Children　　**Richard**
　　　　　　　　Mary married Sir ? Marjoribanks
　　　　　　　　Amelia married James MacAuther
　　　　　　　　Sibella married G. W. Norman, son Frederick Henry born about 1827 whose
　　　　　　　　daughter Sibella Charlotte born about 1847 married Henry Bonham Carter
　　　　　　　　whose son Norman born 29 December 1867 married on 9 December 1910
　　　　　　　　Eileen Beatrice Silk (nee Matthew d of Lt Robert M of Indian Medical
　　　　　　　　Service and widow of Albert Edward Silk of Indian Public Works, died
　　　　　　　　1960), killed as 2nd Lt at Battle of Arras on 3 April 1917, having been
　　　　　　　　commissioned into Indian Civil Service, and was grandfather of Peter
　　　　　　　　Bonham Carter (now living in Comrie, Perthshire)

George (1790-1815) Bengal Cavalry

Anna (1792-?) married 17 July 1811 Robert Henry Tulloch at Serampore (N/1/8/439)

Robert (1794-1824)

Bruce (1796-1861) Cadetship in Bengal army (L/MIL/9/125/445-53). Captain by 1832 (*Flora
　　Indica*)

Elizabeth (1799-1891) married 6 April 1827 Francis Curwen Smith, Magistrate of Cawnpore.
　　Died 1891

Sophia (1800-?) married J.W.

James (1802-84) cadetship in Bengal Native Infantry (L/MIL/9/137/39), married Sibella Carnegie,
　　Capt by 1832 (*Flora Indica*), retired as Lt Col to Newry, Ireland by 9 May 1857.
　　　　　　　　James born 21 Aug 1837 and baptised 19 Nov 1837 at Fort William [Calcutta]
　　　　　　　　(N/1/46/212). Madras Cadetship 1857 (L/MIL/9/254/452)

Henry (1804-?)

Sibella (1807-96)

<u>Mary Anne Egerton (1810-75)</u> married 23 Jul 1834 H. C. Tucker of the Civil List
        Mary Ann

<u>William (1812-97)</u> married 1857 Anne Elizabeth Boswell, daughter of Alexander Boswell (1775-1822). Botanist and doctor. Educated at Edinburgh Academy 1827-9 in Mr Ferguson's class (Edinburgh Academy archives)

<u>John (c1777-1823)</u> married Susanna Caroline Benedict 27 Dec 1806, in presence of John Garia and H. Fermie (N/1/7/204). Died 28 Mar 1818 aged 30 (N/1/10/669)
        Anna Susanna born 18 Oct 1807 in Calcutta and baptised 4 Apr 1808 (N/1/8/5)
        William born 11 May 1810 at Calcutta and baptised 27 Aug 1811 (N1/9/110)
        Mary born 31 Aug 1812 in Calcutta and baptised 27 Aug 1813 (N/1/9/110)
        John Peter baptised 20 Sep 1818 at Fort William [Calcutta] (N/1/10/591). Buried 26 Dec 1831 aged 16, Ward of Free School (N/1/31/339)
    Married Catherine Young 24 Mar 1819 in Calcutta (N/1/11/23)
        Nicholas baptised 4 Feb 1821 at Fort William [Calcutta] (N/1/11/597)

GENERA DESCRIBED BY WILLIAM ROXBURGH
COMMEMORATING INDIVIDUALS

The following list is taken from a study of Roxburgh's *Flora Indica* (1832 edition, edited by William Carey, 3 volumes, Serampore). The entry gives Roxburgh's explanation of the creditation to the person concerned. The references at the end of each entry refer to: Roman numeral, the volume, and Arabic numeral, the page.

Andersonia. It was named in memory of the late Dr James Anderson, Physician at Madras. II, 212.

Berria. Named after Dr Andrew Berry of Madras, an eminent Physician and Botanist, to whose abilities and industry, the Botanic garden at Calcutta is much indebted. II, 639.

Boswellia. The genus is so named, in memory of the late Dr John Boswell, Physician in Edinburgh. II, 383.

Buchanania. [Named after Dr Francis Buchanan.]

Careya. [*C. herbaceus*] was found by Dr William Carey, whose name the genus bears. II, 638.

Dalrymplea. In memory of the late Alexander Dalrymple, Esq. Author of the *Oriental Repertory*, etc. I, 555-6.

Doodia. In memory of Mr Samuel Doody, an eminent Botanist, the friend and contemporary of Ray, Plukenet, and Sloane. III, 365.

Flemingia. This genus is assigned to the name of Dr John Fleming, Physician general in Bengal. III, 337.

Hardwickia. [Colonel Thomas Hardwick.] II, 423.

Heynea. [Dr Benjamin Heyne.] II, 389.

Hopea. It is so named in memory of the late Dr John Hope, professor of Botany in Edinburgh. II, 609.

Humea. In honour of the late Lady Amelia Hume ... . Dr Roxburgh takes the liberty of consecrating this genus to the memory of that most amiable lady, by whose death Botany has lost one of its greatest admirers and best benefactors. II, 641.

Hunteria. I consider this to be a well defined, perfectly distinct genus, and have named it after Dr William Hunter of the Bengal Medical Establishment, an eminent Botanist, and author of various papers in the *Asiatic Researches*, of the *History of Pegu*, &c. He was the first to discover of this very plant [*H. Corymbosa*]. I, 696.

Johnia. In honour of the Rev. Dr John of Tranquebar. I, 168.

Jonesia. Sir William Jones, whose name this genus bears. II, 220.

Kydia. Upon the supposition of this forming a new genus, I have ventured to give it the above name, in memory of the late Colonel Robert Kyd, of Bengal, whose attachment to botany and horticulture induce him to retire from the high rank held in the army, to have more leisure to attend to his favourite study, to the advancement of every object which had the good of his fellow-creatures in view, and to the establishment of the Honourable East India Company's Botanic garden at Calcutta, where he was particularly attentive to the introduction of useful plants, and to their being dispersed over every part of the world, for the good of mankind in general. III, 190.

Lettsomia. In honour of John Coakley Lettsom, M.D., F.S.A., author of numerous valuable works, which more than prove him fully entitled to this mark of respect. I, 487.

Millingtonia. Having found it necessary to deprive our countryman the late Sir Thomas Millington of the genus assigned to his memory by the younger Linnaeus, (Suppl. 44 and 201) because, on finding the ripe seed vessel of the only species thereof, I ascertained it to be a true *Bignonia* as I suspected; I have therefore restored that respectable name to the system, under a different dress, by giving it to the two trees which at present constitute this strongly marked family, and which, I am inclined to think, have not, until now, been described. I, 102-3.

Milnea. In honour of Colin Milne, LL.D. author of a *Botanical Dictionary*, *Institutes of Botany*, and other works. I, 637.

Pierardia. This new genus, for so it seems to me, I have named after Francis Pierard, Esq. one of the Honourable East India company's Civil Servants. His abilities as a Botanist, in discovering various new plants, with which he has enriched the Honourable Company's Botanic garden, claims for him this mark of distinction. II, 255.

Roscoea. [William Roscoe.] III, 54.

Rottleria. [John Peter Rottler.] III, 527.

Roydsia. This elegant, strongly marked genus is named in honour of Sir John Royds, one of the puisne Judges of the Supreme Court of Judicature of Bengal, an eminent benefactor to the science. II, 643.

Shorea. In honour of the Right Honourable Lord Teignmouth, late Governor General of Bengal, formerly Sir John Shore. II, 615.

Wrightea. The name which Dr Roxburgh has given to this genus is in honour of William Wright, M.D., F.R.S. and President of the College of Physicians Edinburgh. The plant formerly named after this eminent Physician and Botanist, being found to be a species of *Blakea*. III, 623.

There is one further genus described by Roxburgh, and commemorating an individual, which must be accounted for – a rather ticklish matter, for this is *Roxburghia*.

The genus, and its single species, were first published together as *Roxburghia gloriosoides* in the first volume of the *Plants of the Coast of Coromandel* in 1795. There seems no question but that the genus should be attributed to Roxburgh himself, an apparent act of vanity that must have raised botanical eyebrows at the time, and subsequently. However, there are mitigating factors to be taken into consideration. First the epithet '*gloriosoides*', which initially appears as yet further aggrandisement of the name *Roxburghia*, actually refers to a similarity in the habit of the plant to that of *Gloriosa superba*. Secondly there is the story of the discovery of the plant related in Roxburgh's own *Flora Indica* (1832, vol. 2, p. 236), which is worth quoting: 'This was one of the last plants Dr. König saw. It was brought in when he was on his death bed; he did attempt to examine it, but was unable, the cold hand of death hung over him; he desired I would describe it particularly, for he thought it was uncommonly curious, new, and beautiful. This observation, from a worthy friend, a preceptor, and predecessor, has made me more than unusually minute in describing and drawing it'.

It is was therefore probably König who suggested the name, and Roxburgh no doubt felt torn between observing the dying wishes of his friend, and the risk of appearing vain to his botanical confrères. In *Flora Indica* Roxburgh diverted attention, by attributing the generic name to Banks (who merely supervised the publication of the Coromandel volume) and the specific name to Willdenow (whose publication, however, did not appear till 1799). Subsequent authors have attempted to cover Roxburgh's posthumous blushes: Wallich, in his monumental Catalogue, attributed the name to König, and Joseph Hooker (in *Flora of British India*) to 'Jones in Roxb. Cor. Pl.'. This attribution of the name to Sir William Jones (who died in 1794) is inexplicable, and there is no mention of him in Roxburgh's text; but it has been followed in the standard list of generic names (*Index Nominum Genericorum*), where the authority for the genus is cited as 'W. Jones ex Roxburgh'.

In fact, as Roxburgh himself eventually realised, the plant had been described earlier by Loureiro, as *Stemona tuberosa*, so there is no generic name in current use to commemorate the Father of Indian Botany.

COSTS FOR THE BUILDING AND REPAIR OF THE BOTANIC GARDEN HOUSE AND OTHER BUILDINGS AT SIBPUR, CALCUTTA

	S^{a}R^s
Bricks and Tiles of various kinds	3016.. ..
Surky & Coah	1275.. ..
Sand	450.. ..
Chunam	2924.. ..
Beams & Burgars	2203.. ..
Bricklayers work	2600.. ..
65 Door and Window frames	422.. ..
65 Doors and Windows	1786.. ..
Bamboos, Ropes & Scaffolding	450.. ..
Jars, pots & small articles	250.. ..
Stair & rail for the Veranda	980.. ..
20 Sash Windows	800.. ..
Painters work	130.. ..
S^{a}R^s	17286.. 8..
Sundry small sums due workmen & together with a Cooking room & Bottle cannah, building will amount to about	2713.. 8..
Total S^{a}R^s	20000.. ..
Cash received	15000.. ..
Defficient S^{a}R^s	5000.. ..

Table 1 'Expences incurred for building a House in the Botanical Garden for the Superintendant down to 18th September 1795 including what will be due to the workmen when finished."*

Painters work in the Superintendants House 23 Doors, & 18 Windows, painted green, @ 2Rs each	82	
20 Sash windows, @ 1 Rs each	20	
30 Doors painted with other colours, @ 1 ½ each	45	
250 Beams, (with several thousand feet of Burgers) @ 1 each	250	
stairs, & Verranda rails	52	449
Repairing Doors, Windows, &ca including 21 panes of Glass	165 8	165 8
Bricklayers work, including Sand, Surky, Chunam, Bamboos for Scaffolding, &ca	479	479
Europeans Houses, & two Octagons		
Repairs, including Bricklayers work, Lime, Sand &ca	184	
Painting 54 Doors, & Windows, @ 1 Rs each	54	
Ditto, 110 Beams, (& some hundred feet of Burgers) @ 1 each	110	348
Sicca Rupees	1441 8	

Table 2 'Expences incurred repairing, & Painting the Superintendants House, two Octagons, & the three Houses occupied by the European Nursery Man, Gardener & overseer, in the Hon'ble Companies Botanic Garden, at the close of 1803.'†

* Public Consultations at Fort William, dated 30 November 1795, paragraphs 27-8, letter from William Roxburgh to Colin Shakespear, dated 19 November 1795, British Library, OIOC IOR, P/4/36; Public Department from the Court of Directors to Fort William, dated 5 January 1796, paragraph 75-6, reply to Public Letter, dated 24 December 1794, British Library, OIOC IOR, E/4/644.
† Public Consultations at Fort William, dated 5 January 1804, paragraphs 20-1, letter from William Roxburgh to Thomas Philpot, dated 2 January 1804, British Library, OIOC IOR, P/5/52.

Painters work in the Superintendants House

23	Doors and 18 Windows painted green, @ 2 Rs each	82.. ..
20	Sash windows, @ one Rupee each	20.. ..
30	Doors painted with other colours, @ 1½ each	45.. ..
	Beams (with several thousand feet of Burgers) @1 each	250.. ..
	Stair, and Veranda rails	52.. ..
	Repairing Doors, windows, &c including Glass for the Sash's	110.. 8..
	Bricklayers work, including sand, Surky, Chunam, Bamboos for scaffolding &c	360.. ..
	Europeans Houses, two Octagons, Cenotaph, Bridges and Flood gates, Repairs Including Bricklayers work, Lime, Sand, &ca	234.. ..
	Painting 54 Doors and Windows @ 1 Re each	54.. ..
	Painting 110 Beams, (and some hundred feet of Burgers) @ 1 each	110.. ..
	Sicca Rupees	1317.. 8..

Table 3 'Expences incurred in repairing and painting the Superintendant's House, two octagons, three houses occupied by the European Nureseryman and overseers, Cenotaph, Bridges, and Flood gates in the Honorable Company's Botanic Garden, at the close of 1808, and beginning of 1809." The reference to the Cenotaph is presumably the monument to Kyd by Bacon.

Painters work in the Superintendants House

23	Doors, and 18 Windows painted green, at 2 rupees each	82.. ..
20	Sash Windows at one Rupee each	20.. ..
30	Doors painted with other Colours, at 1½ each	45.. ..
250	Beams (with several thousand feet of Burgahs) at 1 each	250.. ..
	Stairs and verando rails	52.. ..
	Sicca Rupees	449.. ..
	Repairing Doors, Windows &ca including Glass for the Sash's	90.. 8..
	Bricklayers work including Sand, Surky, Chunam, Bamboos for Scaffolding	366.. ..
	European's Houses, two Octagons, Cenotaph, Bridges and Flood-gates, Repairs including Bricklayers work, Lime, Sand, &ca	195.. ..
	Painting 54 Doors, and Windows, @ 1 Rupee each	54.. ..
	Ditto 110 beams, and some hundred feet of Burgers at 1 Rupee each	110.. ..
	Sicca Rupees	1264.. 8..

Table 4 'Expences incurred in repairing and painting the Superintendants Houses, two Octagons, three Houses occupied by the European Nurseryman, and Overseers, Cenotaph, and Flood Gates in the Honourable company's Botanical Garden, at the close of 1811 and beginning of 1812.'[†]

---

* Public Consultations at Fort William, dated 7 July 1809, paragraphs 40-1, letter from William Roxburgh to Henry St G. Tucker, Secretary to the Government Public Department, dated 1 July 1809, British Library, OIOC IOR, P/7/24.

† Public Consultations at Fort William, dated 29 May 1812, paragraphs 37-8, letter from William Roxburgh to C. M. Ricketts, dated 26 May 1812, British Library, OIOC IOR, P/8/4.

PLANT COLLECTING COSTS

To the amount of Sundry Nicknacks, sent as a present with my Letter
Ld Maumet all Kawn, at Nangler   12.. ..
To the amount of presents given to his servants   3.. ..   15.. ..
To the amount expence, for sundry people attending my
       Siccar to Nagler with a letter and a present, for Maumet
       alle Kawn, requesting his assistance to give some of his
       people as a guide and security with my Siccar, for to
       explore the Nangler and Southern Hills, which service
       my Siccar completed in 22 days, the amount the cost
       of this expence was Sicca Rupees   19.. 8..
To sundry people of Maumet Alle Kawn's, that attended my
       Siccar and people upon this excursion, into the Southern
       Mountains as a guide, and for a security, the cost of
       these people was Sicca Rupees   9.. ..   28.. 8..
Total Amount cost, of this excursion into the Southern
       Mountains, Sicca Rupees   43.. 8..

2 plants of the Cossmarre Tree, from this tree a Gum is taken that the Mountaineers
pay the inside of their Turongs, baskets, and sundry utensils, to make them liquid
proof, which Gum, you will see a Muster, of its virtue, by the vessel I send you
Honey in; [later referred to as 'Cossmarre Elastic Rubber']

To the amount expence attending one Pulawah with sundry plants dispatched from
Silhet 9ᵗʰ March 1811
To One Pulawah 1½ Month cost   7..8..
To one Mangay 1½ do do   4.. 8..
To 5 Dandies 1½ do do   19.. 8..   24.. ..
To one Peon 1½ do do   4.. 8..
To Tullah's mullah Tole, for going to Calcutta and returning   3.. ..   34.. 8..
       Sum total Sicca Rupees   226.. ..

Table 1 An Account, in part, what it has cost me [Christopher Smith] from
May 1810 up to the 1st June 1811 for exploring the Southern Mountains, in
search after the Tugger Trees, Petacarre Trees, the Choul Moogree Trees, &ca
&ca and for procuring two or three pieces, of the best Uggur, or Perfume
Wood, taken from the Tugger tree.*

				SaRs	A	P
Dr						
1803						
April	13	to expenses incurred in procuring Nutmeg & Clove plants & seeds at the Moluccas as per Account No. 1 Sp Dollars 1335 @ 213 SaRs per o/o Sp dollars		2843	9	2
		My personal expenses while collecting spice plants and seeds at the Molucca Islands as per Account No. 2 Sp Dollars 955 @ 213 SaRs per o/o Sp Dollars		2034	2	6
1804						
Jan	5	My personal Expenses while at Bencoolen &ca specified in account No. 3 Sp Dollars 2215 @ SaRs213 per o/o Sp Dollars		4717	15	2
			SaRs	9595	10	10
Cr						
1802						
Dec	30	By Cash received from Col Oliver the Commandant at the Moluccas Sp Dollars 1000 exchange 213 SaRs per o/o Sp Dollars		2130		
August	4	Do Do from Walter Ewer Esq the Commissioner at Bencoolen, Sp Dollars 500 @ 213 SaRs per o/o Sp Dollars		1065		

* Attached to a letter from William Roxburgh to C. M. Ricketts, dated 27 June 1811, in Public Consultations at
Fort William, dated 5 July 1811, paragraphs 25-6, British Library, OIOC IOR, P/7/44.

1803
Nov   8   Do Do from Do Sp Dollars 800 @ 213 SaRs per o/o Sp Dollars   1704
          Balance due   4694 10 10
                                                            SaRs   9595 10 10

Para 28
No. 1
Expenses incurred in procuring Nutmeg & Clove Plants, and Seeds at the Molucccas, on
Account of the Hon'ble Company, dated 13 April 1803
1802
Dec   1   Passage from Malacca to Amboyna, & from thence back to
           Bencoolen on the Transit, Captain Lynch, Sp Dollars   300
      31   Cooley and Boat hire since the 20th Inst   21
1803
Jan   31   Eight Labourers employed at Saporooa Island from the 1st
           Instant, in collecting Plants, in carrying Earth to fill Boxes,
           in planting the plants, and sewing Clove Seeds &ca &ca   24
           130 Boxes Bamboos, &ca small expenses for the
           Clove Plants only   160
           Sundry donations to the 20 Chiefs on the Islands Saporooa, and
           Nessaulaut, in Liquor, Clothes, &ca for Clove Plants and Seeds,
           from their respective Districts   300
           Boat & Cooley hire since the 1st Instant, from Amboyna to
           Saporooa, & Nessaulaut, where the Clove Plants & seeds
           were procured, and back to Amboyna with plants & seeds   125
Feb   18   Fifty Gunny Bags, & Cooley hire filling them with Banda
           Earth, to sow the Nutmeg Seeds in at Amboyna   16
           Freight of 26200 Nutmeg Plants & Seeds, and 50 Bags of
           Earth from Banda to Amboyna, as per Account A   100
      28   Cooley and Boat hire since the 1st Inst   31
March 12   30 Boats employed Shipping, & reshipping 240 Boxes
           with Nutmeg & clove plants from the Ship Matilda, to
           the Transit at 5 Dollars per Boat, as per Account B   150
           3 Boats use for carrying spice plants from the Pass to
           Fort Victoria, as per Account C   30
           Wages of a head Malay Gardener while at Amboyna   10
           16 Butts of Water for Spice Plants on board the ship Transit,
           at one Dollar per Butt, as per Account D   16
      31   Cooley and Boat hire since 1st Inst   12
April 12   Provisions for Bengal Molly from 20th December 1802 to
           12 April 1803   15
      Supplies of water for the Spice Plants since the 1st March as per Account E   25
                                                            Sp Dollars   1335

Para 29
No. 2
To my Personal Expenses while on the Public Service collecting Spice plants & Seeds at the
Malucca Islands, dated 13 April 1803

From the 20th to the 31st of December 1802 at Amboyna Spanish Dollars   81
During the Month of January 1803 at Sapowoa & Amboyna   256
Do the Month of February at Banda Pulo – Way and Great Banda   264
Do the Month of March at Amboyna   250
From the 1st to the 13th of April Do   104
                                                            Sp Dollars   955

Para 30
No. 3
To my personal expenses while at Bencoolen & while Travelling into various parts of the
Country by desire of the Commissioner to assist in selecting proper soil & situations for
the Spice Plantations & in assisting to form them & transplant the young Nutmeg & Clove
trees into the Plantations from May to December inclusive 1803, dated 5 January 1804

From the 12th to 31st May 1803   Spanish Dollars   157
June   234
July   222
August 245

September	200
October 213	
November	210
December	234
To Passage from Bencoolen & for the Freight of 12 Chests of Nutmeg	
& various other plants from thence at Calcutta	500
Spanish Dollars	2215

Para 31
Copy of the Receipt granted for the Nutmeg, and Clove Plants. The original was sent to Government, on the 5ᵗʰ September 1803.

Received of William Roxburgh Esq Twenty one thousand four hundred, and eighty three Nutmeg plants, and Six thousand, nine hundred, and ten clove plants; all in excellent order, and in high health; exclusive of those numbers many more are daily rising from seed deposited in the boxes.

Signed by C. Campbell, Assistant Surgeon and Superintendant of Spice Plantations, dated 1 July 1803

Para 32
Account A, dated 11 march 1803

To Freight of 50 Bags of Earth from Banda to Amboyna in the
Ship Clyde   Spanish dollars   100

Account B, dated 12 March 1803

To Thirty Boats use for carrying Spice Plants, 100 Boxes with Earth, having 24,000Nutmeg Seeds, and 140 Boxes with 27,000 Clove Seeds from the Ship Matilda to the Transit, at
5 Sp Dollars per Boat   Spanish Dollars   150

Account C, dated 12 March 1803

To Three Boats use for carrying from the Pass Bagwalla to Fort Victoria 1000Spice Plants at 10 Spanish Dollars
each Boat   Spanish Dollars   30

Account D, dated 11 March 1803

To 16 Butts of Water, for the use of the Spices, Plants on board the ship Transit at 1 Sp Do per But   Spanish Dollars   16

Account E, dated 12 April 1803

To the carriage of 3000 Gallons of Water at 1 dollar per 120 Gallons for Watering the Hon'ble Company's Spice Plants   Spanish Dollars   25

Table 2 Copy of the accounts of William Roxburgh junior while collecting nutmeg and cloves in the Moluccas during the period December 1802 to December 1803.*

1. Origins and Education 1751-1776

1. These excerpts are taken from the letter from William Roxburgh to Sir Joseph Banks, dated 28 November 1814, Sutro Library, MS HLC950504/011.03814.
2. Letter from William Roxburgh to Dr Edward Smith, dated 7 February 1807, Linnean Society, *Smith Letters*, vol. 25, f65.
3. See Appendix 5, List of Correspondents, taken from people who supplied him with plants while at Calcutta Botanic Garden and published in his *Hortus Bengalensis*, for details of those with whom he corresponded while in India. As many of his letters have not survived, including many with these individuals, this must be a very incomplete list.
4. Obituary in *The Annual Biography and Obituary of 1816*, 1-15 (London, 1816).
5. Anon, 'The Glenfield Ramblers', *Kilmarnock Standard*, 4 July, 1931. The Kennedy and Cassillis families were owners of Culzean at this time.
6. Anon, *Annual Biography and Obituary*, p. 3.
7. Scottish Record Office, CC8/8/141, pp. 452-63.
8. Edinburgh University Library, Special Collections, Matriculation Records.
9. Letter from Sir Joseph Banks to William Roxburgh, 7 January 1799, British Library, Add MSS 33980 f170.
10. General Register Office for Scotland, Kimarnock Parish Register of Births and Baptisms, entry for 10 November 1751.
11. A search of the General Register Office for Scotland website Origins was used for the following names for the period 1745 to 1780: Parkhill, John and Thomas; Roxburgh, Adam, John and William. This was backed up by a study of the Parish Registers for the parishes surrounding Kilmarnock, for which no suitable matches could be found. With the patchiness of contemporary records, this does not, however, prove that William Roxburgh's birth was being 'hidden' because of possible illegitimacy.
12. T. C. Smout, *A History of the Scottish People, 1560-1830* (London, 1969), p. 88.
13. Anon, *The Annual Biography and Obituary*, p. 3.
14. See, for example, his letters dated 1 August 1790 to Dr Patrick Russell, Sutro Library, Banks MSS, EI 1:46, and 8 July, 1791 to Sir Joseph Banks, Sutro Library, Banks MSS, EI 1:44.
15. Edinburgh University Library, Special Collections, Register of Matriculations, 1762-1785. Roxburgh's name has been written in pencil for the following year, but as will be explained below, this attendance is unlikely. Alexander Monro *secundus* (1733-1817) had studied in London, Leiden, Paris and Berlin, and succeeded his father as Professor of Anatomy in 1754 until his retirement in 1808. As a result of his popularity, the number of students attending his lectures increased from under 200 in 1760 to over 400 by 1800.
16. See, for example: his various obituaries; in two letters to Sir Joseph Banks, dated 18 December 1784 (British Library, Add MSS 33977 ff272-5) and 16 September 1785 (British Library, Add MSS 33978 ff31-2), Roxburgh sent Hope plants; W. Roxburgh, *Flora Indica*, 3 vols. (Serampore, 1832), vol. 2, p. 609; Ray Desmond, *The European Discovery of the Indian Flora* (Oxford, 1992), p. 47 states 'at Edinburgh University where he had the benefit of being initiated in the rudiments of botany and Linnaean taxonomy by Professor John Hope.' This is also mentioned in his entry in the DNB. Professor John Hope (1725-86) had studied at Edinburgh under Alexander Monro *primus* and then succeeded Charles Alston as Professor of Botany and Materia Medica in 1761 by which time he had established a thriving medical practice which he retained until his death.
17. Dr John Boswell (1707-80) had studied at Leiden under Boerhaave and was the last British student to be promoted by him. He was Treasurer of the Royal College of Physicians of Edinburgh for two periods (1748-56 and 1758-63) and was its President during the time Roxburgh was staying with him (1770-2). Boswell's Court on the Royal Mile of Edinburgh was named after him.
18. A. C. Chitnis, *The Scottish Enlightenment. A Social History* (London, 1976), p. 91.
19. A. J. Youngson, *The Making of Classical Edinburgh, 1750-1840* (Edinburgh, 1966), p. 81.
20. P. B. Wood, *The Aberdeen Enlightenment: the arts curriculum in the eighteenth century* (Aberdeen, 1993), p. xiii.
21. N. T. Phillipson, 'Commerce and Culture: Edinburgh, Edinburgh University, and the Scottish Enlightenment', in T. Bender (ed.), *The University and the City, from Medieval Origins to the Present* (Oxford and New York, 1988), p. 101.
22. S. Shapin, 'The audience for science in eighteenth century Edinburgh', *History of Science*, 12 (1974), p. 99.
23. Phillipson, 'Commerce and Culture', p. 107.
24. Herman Boerhaave (1668-1738) had occupied five professorial chairs at Leiden University from the late 17th century, when the founding professors of the rejuvenated Edinburgh Medical School had been his students. His teachings stressed the importance of reading the works of the earlier clinicians but at the same time questioning their findings in the light of new evidence.
25. W. S. Craig, *History of the Royal College of Physicians of Edinburgh* (London, 1976), p. 958.
26. Sir Walter Scott, *Guy Mannering; or, The Astrologer* (Edinburgh, 1815, Nelson's New Century Library edition, London, nd), pp. 296-8.
27. Shapin, 'The audience for science', p. 97.
28. Warren McDougall, 'Charles Elliot's Medical Publications and the International Book Trade', p. 216, in Withers and Wood, pp. 215-54.
29. W. J. Stuart, *The History of the Aesculapian Club* (Edinburgh, 1949), p. 15.
30. Charles D. Waterston, 'Late Enlightenment Science and Generalism: the Case of Sir George Mackenzie of Coul, 1780-1848', in Withers and Wood, pp. 301-26.
31. A. G. Morton, *John Hope. 1725-1786. Scottish Botanist* (Edinburgh, 1986), passim.
32. See for example, John Hope, 'Lectures in Botany 1780', Lecture 28, Royal Botanic Garden Edinburgh Library.
33. John Hope, 'Lectures in Botany. Taken by Francis Buchanan in 1780', 1780. MSS in Royal Botanic Garden Edinburgh Library.
34. John Hope, 'Lectures in Botany', 1777-8. MSS in Royal Botanic Garden Edinburgh Library, Lecture 29.
35. Anon [I. B. Balfour], 'Eighteenth century records of British plants', *Notes from the Royal Botanic Garden Edinburgh*, 4 (1907), 123-92.
36. Morton, *John Hope*, p. 22.
37. John Hope, 'Lectures in Botany', 1777-8, Royal Botanic Garden Edinburgh Library.
38. Ibid., Lecture 1.
39. Ibid.
40. Ibid, Lecture 12, 1783.
41. See the above study for the detailed reasoning behind this. It is reinforced by such studies as I. G. Stewart and Joan P. S. Ferguson, 'Rhododendrons, doctors and India, 1780-1860', *Proceedings of the Royal College of Physicians Edinburgh*, 26 (1996), 282-94.
42. John Hope Class Lists, Edinburgh University Library, Special Collections; and the numbers of doctors in the Indian Medical Service are taken from D. G. Crawford, *Roll of the Indian Medical Service 1615-1930* (London, Calcutta and Simla, 1930), passim. The reasons for so many students not graduating from Edinburgh Medical

School were at least twofold. First, it was more expensive to get an MD from Edinburgh than from many other Medical Schools, and secondly it was not necessary to have graduated to become a Surgeon's Mate on one of the East India Company ships: a license from the London College of Surgeons or patronage were the usual routes, and Roxburgh appears to have followed the second (see below for further details).

43. John Hope, 'Lectures in Botany', undated. F125. Royal Botanic Garden Edinburgh Library.

44. A full list of Hope's Lecture notes appears in the Bibliography.

45. John Hope, 'Lectures in Botany', 1777-8, Lecture 2.

46. The Library of the Royal Botanic Garden Edinburgh, contains the notes taken down from John Hope's lectures by various students for the years 1777-8 and 1780.

47. Henry J. Noltie, *Indian Botanical Drawings 1793-1868, from the Royal Botanic Garden Edinburgh* (Edinburgh, 1999), p. 13.

48. Hope, 'Lectures in Botany', 1777-8, Lecture 6.

49. Hope, 'Lectures in Botany', 1780, Lecture 34.

50. William Roxburgh, *Hortus Bengalensis, or a catalogue of the plants growing in the Honourable East India Company's Garden at Calcutta*, edited by William Carey (Serampore, India, 1813-14), passim.

51. Hope, 'Lectures in Botany', 1777-8, Lecture 2.

52. Ibid. The term 'missionary' in this context refers to people sent on botanical missions, rather than the later proselytising meaning.

53. Hope, 'Lectures in Botany', 1780, Lecture 17.

54. Letter from William Roxburgh to Sir Joseph Banks dated 18 December 1784. British Library, Add MSS 33977 ff272-5.

55. Letter from William Roxburgh to Sir Joseph Banks dated 16 September 1785. British Library, Add MSS 33978 ff31-2.

56. The numbers of students are taken from John Hope, MSS Class Lists, Edinburgh University Special Collections. The numbers of doctors in the Indian Medical Service are taken from Crawford.

57. Hope, 'Lectures in Botany', 1777-8, Lecture 2.

58. Hope. MSS Class Lists.

59. Morton, John Hope, p. 45.

60. Hope, 'Lectures in Botany', 1780, Lecture 29.

61. Hope, 'Lectures in Botany', 1780, Lecture 27. In all cases, the spelling used by contemporary authors has been used.

62. R. H. Grove, *Green Imperialism. Colonial expansion, tropical island Edens and the origins of environmentalism, 1600-1800* (Cambridge, 1995), p. 13.

63. Ibid. p. 339.

64. P. L. Farber, 'Aspiring naturalists and their Frustrations: The case of William Swainson (1789-1855)', in A. Wheeler and J. H. Price (eds), *From Linnaeus to Darwin: Commentaries on the History of Biology and Geology* (London, 1985), p. 52.

65. Hugo Arnot, *The History of Edinburgh, Edinburgh and London*, 1779, p. 416.

66. Lady Smith (ed.), *Memoirs and Correspondence of the late Sir James Edward Smith, M.D.* (London, 1832), vol. I, pp. 20-3. Letter dated 2 November 1781.

67. W. Huggins, *Sketches in India, Treating on Subjects Connected with the Government; Civil and Military Establishments; Characters of the Europeans, and Customs of the Native Inhabitants* (London, 1824), pp. 65-66.

68. Quoted in Edward Smith, *The Life of Sir Joseph Banks, President of the Royal Society with some notices of his friends and contemporaries* (London, 1911), p. 139.

69. In a letter to Sir Joseph Banks of 17 August 1792 (British Library, Add MSS 33979 ff171-3), Roxburgh, referring to himself, wrote 'that man at Dinner at Sir John Pringles nearly twenty years ago'.

70. William Roxburgh, 'A meteorological diary, Ec. Kept at Fort St George in the East Indies. Communicated by Sir John Pringle, Bart. P.R.S. read January 29, 1777, *Philosophical Transactions of the Royal Society*, 68 (1778), 180-95; letter from William Roxburgh to Dr James Edward Smith, dated 4 August 1792, Linnean Society, *Smith Letters*, volume 8, f182; and letter from William Roxburgh to Dr Taylor, dated 16 February 1807, British

Library, OIOC IOR EUR/D809.

71. The *Houghton*'s log for the journey to and from India, lasting from November 1772 until June 1774 is in British Library, OIOC IOR, L/MAR/B/438. All the details of the voyage are taken from this source. As stated above, this would preclude Roxburgh from having attended Edinburgh University for the academic year of 1772/73, which suggests that the pencilled entry in the Register of Matriculations for that year was erroneously entered at some time afterwards.

72. H. B. Carter, *Sir Joseph Banks, 1743-1820* (London, 1988), p. 64.

73. William Spavens, *Memoirs of a Seafaring Life. The Narrative of William Spavens* (first published 1796, reprinted London, 2000), pp. 133-4. William Spavens pointed out that this Trinidad was not the island in the West Indies but 'in the South Atlantic, off the coast of Brazil.'

74. William Hickey, *Memoirs of William Hickey* (1749-1809), 4 vols. (London, 1913-25), edited by Alfred Spencer, vol. 1, pp. 164-7.

75. Spavens, *Memoirs of a Seafaring Life*, pp. 204-5.

76. The log for the *Queen* for the journey from October 1774 until her return to Deptford two years later is in British Library, OIOC IOR, L/MAR/B/356E. The information that follows is taken from this.

77. Anon, *The Annual Biography and Obituary of 1816*, p. 2.

78. William Roxburgh, 'A meteorological diary', 1778; and 'A continuation of a meteorological diary, kept at Fort St George, on the coast of Coromandel. Communicated by Joseph Banks. Read January 20, 1780', *Philosophical Transactions of the Royal Society of London*, 70 (1780), 246-71.

2. Coromandel Coast Years 1776-1793

1. Crawford, *Roll of the Indian Medical Service*, entry B285. In the British Library, OIOC IOR Service Army List L/MIL/11/72 both the entries for who Roxburgh was 'Nominated by' and 'Admitted to the service from' are blank, confirming Crawford's comment.

2. Lt Col W. J. Wilson, *History of the Madras Army* compiled by Lieutenant Colonel W. J. Wilson, retired list, Madras Army, 5 vols. (Madras, 1882-9), vol. 1, p. 345.

3. A. M. Davies, *Warren Hastings, Maker of British India* (London, 1935), pp. 65-6.

4. Ibid., p. 85.

5. D. G. Crawford, *A History of the Indian Medical Service, 1600-1913*, 2 vols. (London, Calcutta and Simla, 1914), vol. 2, p. 142.

6. H. D. Love, *Vestiges of Old Madras 1640-1800. Traced from the East India Company's Records Preserved at Fort St George and the India Office, and from Other Sources*, 3 vols. (London, 1913), vol. 3, p. 178.

7. See, for instance, B. M. Johri and M. A. Rau, 'Plant sciences in India: Yesterday, today and tomorrow', p. 4, and Rau, M. A., 'Plant exploration in India and floras', p. 18, both in B. M. Johri (ed.), *Botany in India. History and Progress*, 2 vols. (Lebanon, New Hampshire, U.S.A., 1994). Benjamin Heyne died at Madras in 1819 having succeeded Roxburgh as Company Botanist on the Carnatic in 1793; Gerhard König had travelled to Iceland in 1765, and from Madras to Siam and Malacca during 1778-9; John Peter Rottler (1749-1836) arrived in Ceylon in 1788 and collected plant material on the Coromandel 1795-6, in James Britten and George S. Boulger, *A Biographical Index of Deceased British and Irish Botanists*, 2nd edition revised by A. B. Rendle (London, 1931), passim.

8. Carter, *Sir Joseph Banks*, pp. 132, 266.

9. Letter from William Roxburgh to Sir Joseph Banks, 8 March 1779, British Library, Add MSS 33977 ff93-5.

10. Letter from William Roxburgh to Sir Joseph Banks, 24 November 1782, British Library, Add MSS 33977 ff181-2.

11. Joannes Burman, *Thesaurus Zeylanicus, exhibens plantas in insula Zeylana nascentes ... omnia iconibus illustrata ac descripta cura* (Amsterdam, 1737). For further information about Roxburgh's library, see Appendix 6 which lists the book, periodicals and some of the manuscript copies which he collected after 1785.

12. Letter from William Roxburgh to Sir Joseph Banks, 18 December 1784, British Library, Add MSS 33977 ff272-5.
13. Alexander Dalrymple, *Oriental Repertory*, 2 vols. (London, 1793-7), vol. 2, p. 70; and letter from William Roxburgh to Sir Joseph Banks, 8 July 1791, Sutro Library, Banks MSS, EI 1:44.
14. Letter from Patrick Russell to Sir Joseph Banks, 11 June 1787, British Library, Add MSS 33978 ff129-30. A fuller description, by William Parsons and Roxburgh appeared in Dalrymple's *Oriental Repertory*, vol. 1, pp. 87-97.
15. Public Consultations at Fort St George, 18 June 1787, British Library, OIOC IOR, P/240/68, ff1325-8. A Candy was an Indian measure of weight averaging about 500 pounds.
16. Love, *Vestiges of Old Madras*, vol. 1, p. 29.
17. Alexander Dalrymple, *Oriental Repertory*, vol. 2, passim.
18. Crawford, *A History of the Indian Medical Service*, vol. 2, p. 262.
19. For instance see letter from C. John to William Roxburgh, 18 March 1790, British Library, OIOC, MSS/EUR/D809, 'I received with Lindams Vessel your most agreeable Letters, Drawings, Plants, Descriptions, Linnen, Chints, etc, for my Household which filled Mr Rottler, Mrs John, Your little Jack and my children & even our Society with Pleasure & Gratitude.'
20. J. L. Cranmer-Byng (editor), *An Embassy to China, Being the Journal Kept by Lord Macartney during his Embassy to the Emperor Chi'en-lung 1793-1794* (London, 1962), passim.
21. Letter from Patrick Russell to Sir Joseph Banks, 11 June 1787, British Library, Add MSS 33979, ff129-30.
22. Letter from Andrew Ross to William Roxburgh, 11 December 1789, Natural History Museum, Botany Library, Roxburgh Correspondence.
23. Letter from William Roxburgh to N. Kindersley, 12 January 1790, Linnean Society, *Smith Letters*, vol. 25, f47.
24. There is some confusion as to which, for in a letter from Patrick Russell to Sir Joseph Banks (9 July 1785, British Museum, Add MSS 33978, ff19-23), Russell stated: 'In Bengal, a proposal was made to him [König] to undertake an expedition to Thibet, which on several accounts he declined for the present'. However Banks, writing to Olof Swartz (10 October 1787, Natural History Museum, Botany Library, Banks Collection Correspondence, JSB 931020/005.21787), wrote 'instead of being near to embarking for Europe when he died he was preparing for a journey towards Boutan. He had been with Hastings the Governor General on the subject who had increased his appointment & allowed him to charge his travelling expenses.' Without checking the latter correspondence, it has not been possible to determine which proposed destination was correct, but both could be as travellers often went through Bhutan to Tibet.
25. Letter from Sir Joseph Banks to Olof Swartz, 10 October 1787, Natural History Museum, Banks Collection Correspondence JSB 931020/005.21787.
26. Letter from Patrick Russell to Sir Joseph Banks, 9 July 1785, British Library, Add MSS 33978 ff19-23. The confirmation of König's papers arriving at Soho Square is given in letter from Sir Joseph Banks to Dr Edward Smith, 5 August 1786, Linnean Society, *Smith Letters*, vol. 1, ff77-8: 'König's MSS which he left me by will & his Specimens are arrived.'
27. Public Consultations at Fort St George, 4 November 1785, British Library, OIOC IOR, P/240/61, ff1102-3.
28. Letter from Patrick Russell to Sir Joseph Banks, 21 December 1788, British Library, Add MSS 33978, f215.
29. Public Department at Fort St George to the Court of Directors, 6 April 1789, British Library, OIOC IOR, E/4/319. Also Court of Directors to Public Department at Fort St George, dated 19 May 1790, British Library, OIOC IOR, E/4/876 f412.
30. Letter from Sir Joseph Banks to Dr James Edward Smith, 15 August 1787. Smith, Memoirs and Correspondence of the late Sir James Edward Smith, vol. 1, p. 269.
31. Sir George King, 'The early history of Indian botany', *Journal of Botany*, 37 (1899), 454-63.
32. Letter from the Rev. C. John to William Roxburgh, 18 August 1789, British Library, OIOC IOR, EUR/D809.
33. Dalrymple, *Oriental Repertory*, vol. 2, p. 471.
34. Letter from William Roxburgh to Sir Joseph Banks, 25 April 1795, British Library, Add MSS 33980 ff9-10.
35. P. J. Anderson (ed.), *Marischal College, Officers, Graduates and Alumni* (Aberdeen, 1897), p. 132.
36. Minutes of meetings held on 2 and 13 February 1790, Royal College of Physicians Edinburgh.
37. Anon, 'Appendix to the History of the Society', *Transactions of the Royal Society of Edinburgh*, 3 (1791), 24. The catalogue of the herbarium collection is a hand-written manuscript, made by William Wright and held in the RBGE Library.
38. Letter from Andrew Ross to William Roxburgh, 19 December 1788, Natural History Museum, Roxburgh Correspondence; and letter from Patrick Russell to Sir Joseph Banks, 21 December 1788, British Library, Add MSS 33978 f215.
39. Letter from William Roxburgh to Patrick Russell, 1 August 1790, Sutro Library, EI 1:50.
40. Extract from General Letter from Fort St George to the Court of Directors, 6 April 1789, Sutro Library, EI 1:56, and British Library, OIOC IOR, E/4/319.
41. Letter from the Court of Directors to Fort St George, dated 16 February 1787, Sutro Library, EI 1:56.
42. Letter from Sir Joseph Banks to the Court of Directors, 22 February 1787, British Library, OIOC IOR, E/1/80, f169.
43. Letter from Sir Joseph Banks to the Court of Directors, 25 November 1788, Sutro Library, EI 1:56.
44. Dr Patrick Russell, *Account of Indian Serpents, Collected on the Coast of Coromandel* (London, 1796-1807).
45. Desmond, *The European Discovery of the Indian Flora*, p. 47. This symbiotic relationship between the two men is highlighted in a letter from William Roxburgh to Dr Patrick Russell, dated 20 January 1790, Sutro Library, EI 1:52, when Roxburgh mentioned work on both plants and snakes, including the light touch that 'no snakes have ventured abroad since the orders of the Government were issued for their [snake pills] being administered for their bite, nor have I heard of any mad dogs.'
46. Letter from the Rev. C. John to William Roxburgh, dated 24 November 1789, Natural History Museum, Botany Library, Roxburgh Correspondence; and letter from William Roxburgh to Patrick Russell, 1 August 1790, Sutro Library, EI 1:50.
47. Letter from Dr Patrick Russell to the Hon. John Holland, 23 February 1789, Sutro Library, EI 1:56.
48. Letter from Dr Patrick Russell to Sir Joseph Banks, 21 December 1788, British Museum, Add MSS 33978 f215.
49. Ibid.
50. Letter from Andrew Ross to William Roxburgh, 12 May 1791, Natural History Museum, Botany Library, Roxburgh Correspondence.
51. Ibid., 13 May 1791.
52. Ibid., 20 May 1791.
53. Ibid., 29 March 1792 and 17 July 1792.
54. Letter from Rev. John to William Roxburgh, 26 November 1792, British Library, OIOC IOR, EUR/D809.
55. Letter from Andrew Ross to William Roxburgh, 16 June 1793, Natural History Museum, Botany Library, Roxburgh Correspondence.
56. Ibid., 6 March 1793.
57. T. H. Beaglehole, *Thomas Munro and the Development of the Administration Policy in Madras, 1792-1870. The Origins of the Munro System* (Cambridge, 1996), passim. This book is a study of the settlement and Munro's suggestions for it, which had far-reaching effects on the Permanent Settlement of 1793.
58. Letter from Andrew Ross to William Roxburgh, 24 April 1793, Natural History Museum, Botany Library, Roxburgh Correspondence.
59. Ibid., 17 February 1793.
60. Letter from William Roxburgh to David Haliburton, 8 May 1793, British Library, OIOC IOR, EUR/D809. David Haliburton was the Acting President to the Board of Revenue at Madras.
61. William Roxburgh, 'Process for making Indigo on the Coast near Ingeram', in J. F. Charpentier de Cossigny de Palma, *Memoir, Containing an Abridged Treatise on the Cultivation and Manufacture of Indigo* (Calcutta, 1789), p. 157.

62. Dalrymple, *Oriental Repertory*, vol. 2, pp. 56-57.
63. Letter from Andrew Ross to William Roxburgh, 28 March 1793, Natural History Museum, Botany Library, Roxburgh Correspondence.
64. Dalrymple, *Oriental Repertory*, vol. 2, p. 235.
65. Letters from Andrew Ross to William Roxburgh, 5 January 1793 and Major Beatson to Andrew Ross, 12 January 1792, Natural History Museum, Botany Library, Roxburgh Correspondence; Revenue Department at Fort St George, 28 January 1793, paragraphs 5-6, British Library, OIOC IOR, E/4/322; and Dalrymple, 'Watering the Circars', *Oriental Repertory*, vol. 2, pp. 33-84
66. Letter from Andrew Ross to William Roxburgh, 24 March 1793, Natural History Museum, Botany Library, Roxburgh Correspondence.
67. Ibid., 8 May 1793.
68. Ibid., 24 April 1793.
69. Ibid., 25 April 1793.
70. Ibid., 5 January 1793.
71. Ibid., 24 March 1793.
72. Letter from William Roxburgh to David Haliburton, 9 May 1793, Natural History Museum, Roxburgh MSS, and printed from a copy at Calcutta Botanic Garden in M. P. Nayar and A. R. Das, 'Glimpse of William Roxburgh through unpublished letters and his interest in Indian economic botany', *Journal of Economic and Taxonomic Botany*, 5 (1984), 1159-67.
73. Ibid., 5 June 1793.
74. Letter from Andrew Ross to David Haliburton, 16 and 20 June 1793, Natural History Museum, Botany Library, Roxburgh Correspondence.
75. Ibid., 21 June 1793.
76. Letters from Andrew Ross to William Roxburgh, 21 and 24 June 1793, Natural History Museum, Botany Library, Roxburgh Correspondence.
77. Ibid., 28 June 1793.
78. Ibid., 29 June 1793; letter from Andrew Ross to William Roxburgh, 5 July 1793, to which is attached the Minutes of the Revenue Board, 21 June 1793, Natural History Museum, Botany Library, Roxburgh Correspondence.
79. Letter from William Roxburgh to Andrew Ross, 8 July 1793, Natural History Museum, Botany Library, Roxburgh Correspondence.
80. Ibid., 8 and 10 July 1793.
81. Ibid., 11 and 14 July 1793.
82. Ibid., 15 and 16 July 1793.
83. Ibid., 25 and 30 July 1793.
84. Ibid., 9 August 1793.
85. Ibid., 16 September 1793.
86. Ibid., 15 August 1793.
87. Ibid., 21 and 22 August 1793.
88. Ibid., 26 August 1793.
89. Ibid., 2 September 1793.
90. Revenue Department at Fort St George to the Court of Directors, 20 September 1793, paragraphs 20-1, British Library, OIOC IOR, E/4/323.
91. Letter from Andrew Ross to William Roxburgh, 11 October 1793, Natural History Museum, Botany Library, Roxburgh Correspondence.
92. Letter from William Roxburgh to C. Shakespear, Sub-Secretary at Fort William, 1 January, 1794, in Public Consultations at Fort William, 3 January, paragraph 4, British Library, OIOC IOR P/4/26, ff17-22.
93. Letter from Andrew Ross to William Roxburgh, 3 March 1794, Natural History Museum, Botany Library, Roxburgh Correspondence. The words in square brackets are to indicate those parts which are lost in holes in the manuscript.
94. Letter from Andrew Ross to William Roxburgh, 10 March 1794, Natural History Museum, Botany Library, Roxburgh Correspondence.
95. Translation of a letter from the Gentoos [Telugu] to William Roxburgh, 2 March 1794, included with a letter from Andrew Ross to William Roxburgh, 29 March 1794, Natural History Museum, Botany Library, Roxburgh Correspondence.
96. Letter from Andrew Ross to William Roxburgh, 30 March 1794, Natural History Museum, Botany Library, Roxburgh Correspondence.
97. Public Consultations at Fort St George to the Court of Directors, 25 July 1794, paragraphs 26-27, British Library, OIOC IOR, E/4/323.
98. Letter from the Court of Directors to the Revenue Department at Fort St George, 4 October 1797, paragraphs 17-19, British Library, OIOC IOR, E/4/883, ff704-10.
99. Letter from the Revenue Department at Fort St George to the Court of Directors, 15 October 1798, paragraphs 8 and 128-33, British Library, OIOC IOR, E/4/326, and letter from the Court of Directors to the Revenue Department at Fort St George, 31 August 1801, paragraph 5, British Library, OIOC IOR, E/4/888.
100. C. A. Bayly, *Indian Society and the Making of the British Empire* (Cambridge, 1988), p. 44.

3. Superintendent of the Calcutta Botanic Garden 1793-1813

1. King, *The early history of Indian botany*.
2. Letter from Robert Kyd to the Bengal Government, 1 June 1786, Natural History Museum, Botany Library, Dawson Turner Correspondence, vol. VII, ff57-67.
3. Letter from Andrew Ross to William Roxburgh, 16 June 1793, Natural History Museum, Botany Library, Roxburgh Correspondence.
4. Ibid., 21 June 1793.
5. Public Department letter from the Governor General at Calcutta to the Governor in Council at Fort St George, 5 July 1793, paragraph 6, British Library, OIOC IOR, P/4/21.
6. Public Department letter from the Court of Directors to Fort William, 3 July 1795, paragraph 17, British Library, OIOC IOR, E/4/643.
7. Letters from Andrew Ross to William Roxburgh, 24 June and 5 July 1793, Natural History Museum, Botany Library, Roxburgh Correspondence.
8. Letter from the Rev. C. S. John to William Roxburgh, 25 July 1793, Natural History Museum, Botany Library Roxburgh MSS.
9. Letter from Andrew Ross to William Roxburgh, 26 July 1793, Natural History Museum, Botany Library, Roxburgh Correspondence; the reference to the iron mines is in British Library, OIOC, MSS EUR/D809.
10. Letter from Andrew Ross to William Roxburgh, 17 August 1793, Natural History Museum, Botany Library, Roxburgh Correspondence.
11. Ibid., 21 August 1793.
12. Ibid., 16 September 1793; and she was baptised on 17 October 1793, British Library, OIOC IOR, N/2/11/75.
13. Public Consultations letter from William Roxburgh to Edward Hay, Secretary to the Government at Fort William, 16 November, paragraph 9, British Library, OIOC IOR, P/4/24.
14. Letter from William Roxburgh to Sir Joseph Banks, 1 December 1793, British Library Add MSS 33979 f224.
15. Public Consultations, 16 December 1793, letter from William Roxburgh to Edward Hay, 12 December 1793, paragraph 26, British Library, OIOC IOR, P/4/24.
16. Public Consultations, 3 January 1794, letter from William Roxburgh to Edward Hay, 1 January 1794, paragraph 4, British Library, OIOC IOR, P/4/26 ff17-22.
17. R. E. Holttum, 'The historical significance of Botanic Gardens in S. E. Asia', *Taxon*, 19 (1970), 707-14.
18. Letter from William Roxburgh to N. Kindersley, 14 January 1794, Linnean Society, *Smith Letters*, vol. 25, f45.
19. Public Consultations, 18 July 1794, letter from William Roxburgh to Edward Hay, 10 June 1794, paragraph 12, British Library, OIOC IOR, P/4/29 ff697-701.
20. Public Consultations at Fort William, 4 August 1794, paragraph 14, letter from William Roxburgh to Colin Shakespear, 30 July 1794, British Library, OIOC IOR, P/4/30, ff70-3.
21. Public Consultations at Fort William, 22 August 1794, paragraph 17, letter from William Roxburgh to Colin Shakespear, 11 August 1794, British Library, OIOC IOR, P/4/30, ff258-71.
22. Public Consultations at Fort William, 29 September

1794, paragraph 8, letter from William Roxburgh to Colin Shakespear, 22 September 1794, British Library, OIOC IOR, P/4/30, ff570-1; and Public Consultations at Fort William, 4 May 1795, paragraph 9, letter from William Roxburgh to Colin Shakespear, 4 March 1795, British Library, OIOC IOR, P/4/35, ff13-18.

23. Public Consultations at Fort William, 30 November 1795, paragraphs 27-8, letter from William Roxburgh to Colin Shakespear, 19 November 1795, British Library, OIOC IOR, P/4/36; Public Department from the Court of Directors to Fort William, 5 January 1796, paragraph 75-6, reply to Public Letter, 24 December 1794, British Library, OIOC IOR, E/4/644.

24. Public Consultations at Fort William, 13 October 1803, paragraph 21, letter from William Roxburgh to Thomas Philpot, 7 October 1803, British Library, OIOC IOR, P/5/48, for request; and Public Consultations at Fort William, 20 October 1803, paragraph 26, letter from William Roxburgh to John Lumsden, 18 October 1803, British Library, OIOC IOR, P/5/48.

25. Public Consultations at Fort William, 5 January 1804, paragraphs 20-1, letter from William Roxburgh to Thomas Philpot, 2 January 1804, British Library, OIOC IOR, P/5/52.

26. Public Consultations at Fort William, 9 October 1806, paragraph 42, letter from William Roxburgh junior to Thomas Brown, 6 October 1806, British Library, OIOC IOR, P/6/34, giving costs in 1803; and Public Consultations at Fort William, 22 January 1807, paragraph 45, letter from William Roxburgh junior to Thomas Brown, 13 January 1807, British Library, OIOC IOR, P/6/39, giving costs for 1807.

27. Public Consultations at Fort William, 7 July 1809, paragraphs 40-1, letter from William Roxburgh to Henry St G. Tucker, Secretary to the Government Public Department, 1 July 1809, British Library, OIOC IOR, P/7/24.

28. Public Consultations at Fort William, 29 May 1812, paragraphs 37-8, letter from William Roxburgh to C. M. Ricketts, 26 May 1812, British Library, OIOC IOR, P/8/4.

29. Letter from William Roxburgh to Colin Shakespear, 30 December 1793, in the Public Consultations at Fort William, dated 3rd January 1794, paragraph 3, British Library, OIOC IOR, P/4/26, ff11-16.

30. Ray Desmond, *Dictionary of British and Irish Botanists and Horticulturalists including Plant Collectors and Botanical Artists* (London, 1977); and Britten and Boulger, *A Biographical Index of Deceased British and Irish Botanists*, p. 279; and Carter, *Sir Joseph Banks*, pp. 276-80.

31. Letter from the Court of Directors, Public Department, 19 February 1794, paragraph 21, British Library, OIOC IOR E/4/641.

32. Letter from William Roxburgh to Colin Shakespear, 26 January 1795, in the Public Consultations at Fort William, 30 January 1795, paragraph 4, British Library, OIOC IOR, P/4/33, ff229-32.

33. Letter from Captain E. H. Bond of the ship *Royal Admiral* to Sir John Shore, 23 April 1795, in the Public Consultations at Fort William, 24 April 1795, paragraph 2, British Library, OIOC IOR, P/4/34, ff519-21.

34. Letter from William Roxburgh to Colin Shakespear, 4 March 1795, in the Public Consultations at Fort William, 4 May 1795, paragraph 9, British Library, OIOC IOR, P/4/35, ff13-18.

35. Letter from Christopher Smith to Edward Hay, 4 June 1795, in the Public Consultations at Fort William, 8 June 1795, paragraph 10, British Library, OIOC IOR, P/4/35, ff161-2.

36. Letter from William Roxburgh to Sir Joseph Banks, 19 December 1795, British Library, Add MSS 33980, ff41-2.

37. Letter from William Roxburgh to Colin Shakespear, 26 December 1795, in the Public Consultations at fort William, 28 December 1795, paragraph 14, British Library, OIOC IOR, P/4/37, ff401-5.

38. Letter from Messr Campbell & Clark to G. H. Barlow, 19 January 1798, in the Public Consultations at Fort William, 26 January, paragraph 55, British Library, OIOC IOR, P/4/57, ff565-7.

39. Letter from Christopher Smith to Sir Joseph Banks, 23 June 1799, Mitchell Library, Banks Archives, CY 36800/279-94.

40. Letter from William Roxburgh to John Stracey, 18 January 1798, in the Public Consultations at Fort William, 2 February 1798, paragraph 9, British Library, OIOC IOR, P/4/358, ff13-18.

41. Letter from James Fleming to Duncan Campbell, 14 July 1798, in the Public Consultations, at Fort William, 16 July 1798, paragraph 22, British Library, OIOC IOR, P/4/61, ff235-9.

42. Letter from J. Hall, Civil Pay Master, to Duncan Campbell, 13 August 1798, in the Public Consultations at Fort William, 16 August, paragraph 26, British Library, OIOC IOR, P/4/61, ff788-90.

43. Letter from James Fleming to Duncan Campbell, 11 October 1798, in the Public Consultations, at Fort William, 15 October 1798, paragraph 18, British Library, OIOC IOR, P/4/62, ff751-61.

44. Letter from James Fleming to Duncan Campbell, 16 November 1798, in the Public Consultations, at Fort William, 16 November 1798, paragraph 25, British Library, OIOC IOR, P/4/63, ff365-6.

45. Letter from Christopher Smith to Sir Joseph Banks, 23 June 1799, Mitchell Library, Banks Archives, CY 36800/279-94.

46. Letter from William Roxburgh to Sir Joseph Banks, 1 July 1799, British Library, Add MSS 33980 ff189-90.

47. Letter from William Roxburgh to C. R. Crommelin, 2 May 1800, in the Public Consultations at Fort William, dated 22nd May 1800, paragraph 46, British Library, OIOC IOR, P/5/12.

48. Letter from William Roxburgh to C. R. Crommelin, 20 February 1802, in the Public Consultations at Fort William, 4 March 1802, paragraph 49, British Library, OIOC IOR, P/5/32.

49. Letter from William Roxburgh to Thomas Brown, 6 June 1804, in the Public Consultations at Fort William, 5 July 1804, paragraph 42, British Library, OIOC IOR, P/5/55, ff3556-88.

50. Maggie Campbell-Culver, *The Origin of Plants. The People and Plants that have shaped Britain's Garden History since the Year 1000* (London, 2001), p. 198.

51. Letter from William Roxburgh to C. R. Crommelin, 10 August 1800, in the Public Consultations at Fort William, 21 August 1800, paragraph 37, British Library, OIOC IOR, P/5/14; and 29 June 1801, in the Public Consultations at Fort William, 30 July 1801, paragraph 77, British Library, OIOC IOR, P/5/23.

52. Letter from Thomas Douglas to C. R. Crommelin, 17 October 1801, in the Public Consultations at Fort William, 13 November 1800, paragraph 52, British Library, OIOC IOR, P/5/26.

53. Letter from Thomas Douglas to Maj. Malcolm, 3 January 1803, in the Public Consultations at Fort William, 28 January 1803, paragraphs 49-50, British Library, OIOC IOR, P/5/41.

54. Letter from Thomas Douglas to Thomas Brown, 14 May 1805, in the Public Consultations at Fort William, 27 June 1805, paragraphs 41-3, British Library, OIOC IOR, P/6/14, ff6635-42.

55. Letter from Thomas Douglas to Thomas Brown, 23 July 1805, in the Public Consultations at Fort William, 29 August 1805, paragraphs 70-1, British Library, OIOC IOR, P/6/16.

56. Letter from Thomas Douglas to J. E. Elliot, Private Secretary to the Governor General, 5 September 1807, in the Public Consultations at Fort William, 15 January 1808, paragraph 53, British Library, OIOC IOR, P/7/3.

57. Letter from Thomas Douglas to William Roxburgh, 22 November 1808, in the Public Consultations at Fort William, 25 November 1808, paragraphs 39-40, British Library, OIOC IOR, P/7/13.

58. Letter from William Roxburgh to Thomas Brown, 29 November 1808, in the Public Consultations at Fort William, 2 December 1808, paragraph 29, British Library, OIOC IOR, P/7/14.

59. Draft instructions from Sir Joseph Banks to Peter Good, 5 June 1794, Mitchell Library, Banks Archives 800374-5.

60. Public Department letter at Fort William to the Court of Directors, 8 March 1795, paragraph 7, British Library, OIOC IOR, E/4/55; and letter from William Roxburgh to Colin Shakespear, 8 April 1795, in Public Consultations at Fort William, 10 April 1795, British Library, OIOC IOR, P/4/34, ff432-3.

61. Letter from William Roxburgh to Sir Joseph Banks, 25 April 1795, British Library Add MSS 33980, ff9-10; and letter from William Roxburgh to Edward Hay, 13 April 1795, in Public Consultations at Fort William, 20 April 1795, British Library, OIOC IOR, P/4/34, ff507-13.

62. Agreement on the duties of the exploratory journey of the Investigator, 29 April 1801, British Library, Add MSS 32439 ff31-2.

63. Letter from William Roxburgh to Duncan Campbell, 25 July 1797, in Public Consultations at Fort William, 11 August 1775, paragraph 12, British Library, OIOC IOR, P/4/52, ff92-5.

64. Reply to Public Letter from Calcutta to the Court of Directors, 12 January 1794, paragraphs 61-5, in letter from the Public Department of the Court of Directors, 3 July 1795, paragraph 82, British Library, OIOC IOR, E/4/643.

65. Letter from William Roxburgh to C. R. Crommelin, 2 May 1800, in the Public Consultations at Fort William, 22 May 1800, paragraph 46, British Library, OIOC IOR, P/5/12.

66. Letter from William Roxburgh junior to Thomas Brown, 5 September 1805, in the Public Consultations at Fort William, 5 September 1805, paragraph 74, British Library, OIOC IOR, P/6/17.

67. Letter from William Roxburgh junior to Thomas Brown, 5 June 1806, in the Public Consultations at Fort William, 12 June 1806, paragraph 52, British Library, OIOC IOR, P/6/27, ff5611-13.

68. Letter from Edward Thornton, magistrate at Zillah, to George Dowdeswell, Secretary, 25 November 1806, in the Public Consultations at Fort William, 4 December 1806, paragraph 55, British Library, OIOC IOR, P/6/36.

69. Letter from William Roxburgh to Thomas Brown, 9 November 1807, in the Public Consultations at Fort William, 13 November 1807, paragraph 41, British Library, OIOC IOR, P/6/52.

70. Letter from William Roxburgh junior to Thomas Brown, 29 July 1806, in the Public Consultations at Fort William, 31 July 1806, paragraph 46, British Library, OIOC IOR, P/6/30, ff8274-6.

71. Letter from William Roxburgh junior to Thomas Brown, 2 April 1806, in the Public Consultations at Fort William, 9 April 1806, paragraph 34, British Library, OIOC IOR, P/6/42.

72. Letter from William Roxburgh to Sir John Shore, 10 June 1794, in the Public Consultations at Fort William, 18 July 1794, paragraph 12, British Library, OIOC IOR, P/4/29, ff697-701.

73. For the information about the octagon at Culzean, I am grateful to Derek Alexander, archaeologist for the West of Scotland, with the National Trust for Scotland; and the Bauer watercolour appears as Plate XV in Walter Lack, *A Garden for Eternity, the Codex Liechtenstein*, translated by Martin Walters (Berne, 2000).

74. Letter from William Roxburgh to Colin Shakespear, 22 September 1794, in the Public Consultations at Fort William, 29 September 1794, paragraph 8, British Library, OIOC IOR, P/4/30, ff570-1.

75. Letter from William Roxburgh to Colin Shakespear, 4 October 1794, in the Public Consultations at Fort William, 6 October 1794, paragraph 7, British Library, OIOC IOR, P/4/30, ff613-15.

76. Letter from G. R. Foley, Civil Paymaster, to Sir John Shore, 20 December 1793, in the Public Consultations at Fort William, 20 December 1793, paragraph 18, British Library, OIOC IOR, P/4/24.

77. Letter from William Roxburgh to Colin Shakespear, 2 February 1796, in the Public Consultations at Fort William, 15 February 1796, paragraph 29, British Library, OIOC IOR, P/4/39, ff745-51.

78. Letter from William Roxburgh to Colin Shakespear, 1 June 1796, in the Public Consultations at Fort William, 6 June 1796, paragraph 28, British Library, OIOC IOR, P/4/42, ff578-9.

79. Letter from William Roxburgh to Colin Shakespear, September 1796, in the Public Consultations at Fort William, 16 September 1796, paragraph 58, British Library, OIOC IOR, P/4/44, ff119-21.

80. These figures are taken from the following documents: June to October 1793, from letter from G. R. Foley to Sir John Shore, 20 December 1793, from the Public Consultations at Fort William, 20 December 1793, paragraph 18, British Library, OIOC IOR, P/4/24; March to May 1794, from letter from G. R. Foley to Sir John Shore, 9 July 1794, from the Public Consultations at Fort William, 18 July 1794, paragraph 11, British Library, OIOC IOR, P/4/29; October 1794 to January 1795, from letter from J. Hall, Civil Paymaster to Sir John Shore, dated 11 February 1795, from the Public Consultations at Fort William, 13 February 1795, paragraph 17, British Library, OIOC IOR, P/4/33; and February and March 1795, from letter from J. Hall to Sir John Shore, 19 May 1795, from the Public Consultations at Fort William, 22 May 1795, paragraph 4, British Library, OIOC IOR, P/4/35, ff91-3.

81. Valentia, *Voyages and Travels in India*, entry for 12 February 1803.

82. Maria Graham, *Journal of a Residence in India* (Edinburgh, 1812), entry for 30 November 1810.

83. Viscount Valentia, *Voyages and Travels to India*, entry for 23 October 1802.

84. Letter from William Roxburgh to G. H. Barlow, 13 November 1799, in the Public Consultations at Fort William, 19 November 1799, paragraph 23, British Library, OIOC IOR, P/5/7, ff729-30.

85. Crawford, *A History of the Indian Medical Service*, vol. 2, p. 262.

86. Crawford, *Roll of the Indian Medical Service*, p. 265 (entry number Madras 205).

87. Letter from William Roxburgh to Edward Hay, 1 January 1794, in the Public Consultations at Fort William, 3 January 1794, paragraph 4, British Library, OIOC IOR, P/4/26, ff17-22.

88. Letter from Andrew Ross to William Roxburgh, 11 December 1789, Natural History Museum, Botany Library, Roxburgh Correspondence.

89. Ibid., 3 February 1790 and 7 February 1790.

90. Ibid., 5 January 1793.

91. Ibid., 28 March 1793.

92. Ibid., 24 April and 8 May 1793.

93. Ibid., 26 July 1793.

94. Letter from William Roxburgh to Edward Hay, 12 January 1795, in Public Consultations at Fort William, 16 January 1795, paragraph 34, British Library, OIOC IOR, P/4/33, ff117-19.

95. Letter from William Roxburgh to Colin Shakespear, 26 January 1795, in the Public Consultations at Fort William, 30 January 1795, paragraph 4, British Library, OIOC IOR, P/4/33, ff229-32.

96. Letter from William Roxburgh to Colin Shakespear, 6 February 1795, in the Public Consultations at Fort William, 13 February 1795, paragraph 16, British Library, OIOC IOR, P/4/33, ff323-4.

97. The request for Roxburgh to be paid his furlough salary is documented in the Military Department letter at Fort St George, 8 March 1805, paragraph 311, British Library OIOC IOR, E/4/332; and its approval is in the Minutes of the Court Book, 8 January 1806, British Library, OIOC IOR, B/142, f992. The medical hierarchy was dependent solely on seniority by time spent in India with certain qualifications such as periods on active service, starting as Assistant Surgeon, then Surgeon, Head Surgeon, Member of the Medical Board and finally Head of the Medical Board.

98. Military Department at Fort St George, 21 October 1807, paragraphs 236, 765-6, British Library, OIOC IOR, E/4/336.

99. Letter from William Roxburgh to the Court, 30 June 1814, referred to in the Minutes of the Court Book, 6 July 1814, British Library, OIOC IOR, B/159, f297; and letter from William Roxburgh to the Court, 27 August 1814, referred to in the Minutes of the Court Book, 2 September 1814,

British Library, OIOC IOR, B/159, f492.

100. Letter from Alexander Boswell to the Court, 27 April 1815, referred to in the Minutes of the Court Book, 9 May 1815, British Library, OIOC IOR, B/161, f104; Letter from Alexander Boswell to the Court, 10 July 1815, referred to in the Minutes of the Court Book, 14 July 1815, British Library, OIOC IOR, B/161, f317; and Minutes of the Court Book, 20 October 1815, British Library, OIOC IOR, B/162, f618.

101. Letter from Francis Buchanan to Dr James Edward Smith, 4 November 1806, Linnean Society, *Smith Letters*, vol. 2, f218.

102. Letter from William Roxburgh to [Mr Aiton], 20 July 1805, Royal Botanic Garden Kew Library, Record Book, 1793-1809, ff239-40.

103. Letter from Francis Buchanan to Dr James Edward Smith, 22 November 1806, Linnean Society, *Smith Letters*, vol. 2, f219.

104. Letter from William Roxburgh to Sir Joseph Banks, 13 July 1797, British Library, Add MSS 33980, ff101-3.

105. Letter from Sir Joseph Banks to William Roxburgh, 9 August 1798, British Library, Add MSS 33980, ff159-60. There is an irony in this statement, for Banks published a number of Buchanan's descriptions and drawings in the *Philosophical Transactions of the Royal Society*, under his own name, thereby effectively 'stealing' them from Buchanan.

106. Letter from Francis Buchanan to Dr J. E. Smith, 15 November 1797, Linnean Society, *Smith Letters*, vol. 2, ff202-5.

107. Charles Allen, *The Buddha and the Sahibs. The Men who Discovered India's Lost Religion* (London, 2002), p. 19.

108. Public Letter from Fort William, 31 August 1796, paragraph 53, in P. C. Gupta (ed.), *Fort William – India House Correspondence, 1796-1800* (Delhi, 1959), p. 262.

109. Edinburgh University Library, Special Collections, Alumni and Staff Catalogue.

110. Allen, *The Buddha and the Sahibs*, p. 14.

111. Letter from Francis Buchanan to William Roxburgh, 21 May 1795, Natural History Museum, Botany Library, Buchanan MSS.

112. Ibid., 11 April and 10 May 1797.

113. Letter from William Roxburgh to Sir Joseph Banks, 26 November 1802, British Library, Add MSS 33981, ff667-70.

114. Letter from William Roxburgh to Dr James Edward Smith, 11 December 1806, Linnean Society, *Smith Letters*, vol. 25, ff63-4.

115. Letter from William Roxburgh to Aylmer Bourke Lambert, 2 October 1814, Linnean Society, *Lambert Papers*, ff166-166a.

116. Letter from Dr J. Fleming to Francis Buchanan, 30 November 1811, Scottish Record Office, GD 161/19/4.

117. Britten and Boulger, *A Biographical Index*, p. 50.

118. Letter from Nathaniel Wallich to Francis Buchanan, 25 November 1814, Scottish Record Office, GD 161/19/4.

119. Noltie, *Indian Botanical Drawings*, p. 21.

120. Desmond, *The European Discovery of the Indian Flora*, p. 81.

121. Letter from William Roxburgh to Nathaniel Wallich, 4 October 1809, Edinburgh University Library, Special Collections, La. II.643/52.3. This was probably the Blistering fly that Burt sent to Banks in about 1810, when Roxburgh had told the former that it was a species of Trianthema: the insect was described by Fleming in his second edition of the *Catalogue of Indian Medicines* as *Meloe trianthemae*. Letter from Adam Burt to Sir Joseph Banks, 4 November 1810, Natural History Museum, Botany Library, BC Correspondence, ff112-13.

122. Desmond, *The European Discovery of the Indian Flora*, passim.

4. Family

1. Holmes, Richard, *Sahib, The British Soldier in India* (London, 2005) has a chapter on the changing attitudes from the late 18th century for the following century (pp. 436-58).

2. William Dalrymple, *White Mughals. Love and Betrayal in Eighteenth Century India* (London, 2002), p. 50.

3. Siân Rees, *The Floating Brothel. The Extraordinary True Story of an Eighteenth Century Ship and its Cargo of Female Convicts* (London, 2001), p. 41.

4. James Britten, '"John" Roxburgh', *Journal of Botany*, 56 (1918), 202-3.

5. Sir David Prain, '"John" Roxburgh', *Journal of Botany*, 57 (1919), 28-34.

6. The registration of his burial, on 18 September 1823, states that he was 46 years old, British Library, OIOC IOR, N/1/12/593.

7. Letter from William Roxburgh to Dr James Edward Smith, 7 June 1799, Edinburgh University Library, Special Collections, La. II.643/52.1.

8. Letter from the Rev. C. John to William Roxburgh, 15 October 1789, British Library OIOC, MSS EUR/D809.

9. Ibid., 18 August 1789.

10. Ibid., 29 September 1789.

11. Ibid., 29 January 1790.

12. Ibid., 20 June 1791.

13. Ibid., 28 October 1789.

14. Ibid., 2 April 1793.

15. Letter from the Rev. C. John to William Roxburgh, 25 July 1793, Natural History Museum, Botany Library, MSS Rox (Manuscript correspondence from John, Hardwicke, Smith, Etc.).

16. Letter from John Roxburgh to William Roxburgh, 24 October 1789, Natural History Museum, Botany Library, Roxburgh MSS.

17. Letter from William Roxburgh to Dr James Edward Smith, 7 June 1799, Edinburgh University Library, Special Collection, La. II.643/52.1.

18. Instructions for Mr John Roxburgh, at the Cape of Good Hope, 1799, British Library, OIOC IOR, E/1/140 f301.

19. Letter from John Roxburgh to Sir Joseph Banks, 6 January 1801, British Library, Add MSS 33980 f261.

20. Letter from Aylmer Lambert to Dr James Edward Smith, 6 May 1801, Linnean Society, *Smith Letters*, vol. 6 f38.

21. Letter from John Roxburgh to Aylmer Lambert, 6 January 1802, British Library, Add MSS 28545 f180.

22. Letter from Viscount Valentia to Sir Joseph Banks, 13 December 1802, Natural History Museum, Botany Library, Dawson Turner Collection, vol. 13, pp. 324-8.

23. Letter from Aylmer Lambert to Dr J. E. Smith, 6 May 1801, Linnean Society, *Smith Letters*, vol. 6, f38.

24. Letter from William Roxburgh to Aylmer Lambert, 12 February 1812, Linnean Society, Richard Pulteney Correspondence.

25. Letter from William Roxburgh to John Roxburgh, 19 October 1804, British Library, OIOC IOR, E/1/140, f302.

26. Memorial from John Roxburgh to Joseph Dart, 29 December 1919, British Library, OIOC IOR, E/1/140, f300.

27. Public Consultations at Fort William, 5 January 1810, paragraph 72, British Library, OIOC IOR, P/7/29.

28. Letter from Dr J. Hare to C. M. Ricketts, 11 January 1816, British Library, OIOC IOR, E/1/140, f304.

29. Desmond, *The European Discovery of the Indian Flora*, p. 82.

30. Prain, ' "John" Roxburgh', p. 33.

31. British Library, OIOC IOR, N/1/7/204 for their marriage, N/1/10/669 for Susanna's death, and N/1/10/591 for John's baptism.

32. British Library, OIOC IOR, N/1/31/339.

33. William Huggins, *Sketches in India, Treating on Subjects Connected with the Government; Civil and Military Establishments; Character of the Europeans, and Customs of the Native Inhabitants* (London, 1824), p. 77.

34. Bayly, *Indian Society and the Making of the British Empire*, p. 70.

35. King, 'A brief memoir of William Roxburgh', 1-9.

36. Brown, *The Epitaphs and Monumental Inscriptions*, pp. 252-4.

37. Letter from Sir Joseph Banks to William Roxburgh, 7 January 1799, British Library, Add MSS 33980 f170.

38. Letter from William Roxburgh to Sir Joseph Banks, 24 April 1798, British Library, Add MSS 33980 ff137-41.

39. Letter from William Roxburgh to Dr James Edward Smith, 23 April 1798, Linnean Society, *Smith Letters*, vol. 25, f53.

40. Letter from Sir Joseph Banks to William Roxburgh, 29 May 1796, British Library, Add MSS 33980 ff65-6.

41. Letter from William Roxburgh to Dr James Edward Smith, 7 June 1799, Edinburgh University Library, Special Collections La. II.643/52.1.

42. Letter from William Roxburgh to Sir Joseph Banks, 1 July 1799, British Library, Add MSS 33980 ff189-90.

43. Edinburgh University Library, Special Collections, Alumni and Staff catalogue.

44. Record of their arrival back in Calcutta is in Public Consultations at Fort William, 19 November 1799, British Library, OIOC IOR P/57, No. 23, ff729-30; and young William's idleness in letter from William Roxburgh to Sir Joseph Banks, 17 December 1799, British Library, Add MSS 33980 ff210-12.

45. Letter from the Court of Directors to the Governor General at Fort William, 18 April 1800, in Gupta, *Fort William*, p. 139.

46. Letter from Francis Buchanan to William Roxburgh, 12 November 1800, Natural History Museum, Botany Library, Buchanan MSS.

47. Letter from William Roxburgh to Sir Joseph Banks, 18 July 1802, British Library, Add MSS 33981, f42.

48. Letter from William Roxburgh to Sir Joseph Banks, 26 November 1802, British Library, Add MSS 33981, ff67-70.

49. Public Consultations at Fort William, 31 March 1803, paragraphs 47-9, British Library, OIOC IOR, P/5/43.

50. Public Consultations at Fort William, 14 April 1803, paragraph 37, British Library, OIOC IOR, P/5/44. If William was paid 250 rupees per month, this appears as 25 Spanish dollars per month, giving an equivalent of £1 2s 6d per Spanish dollar, and the total cost for these nine months was therefore about £1,671 15s.

51. Public Consultations at Fort William, 5 July 1804, paragraph 42, British Library OIOC IOR, P/5/55, ff3556-88.

52. See, for instance, the letter from William Roxburgh junior to his father, 6 June 1804, in Public Consultations at Fort William, 5 July 1804, paragraph 44, British Library OIOC IOR, P/5/55, ff3592-3600.

53. Roxburgh, *Hortus Bengalensis*, passim.

54. Extract of Public Letter from the Court of Directors to Bengal, 17 August 1803, British Library, OIOC IOR, Board's Collection, F/4/345.

55. Military Department at Fort St George, 8 March 1805, paragraph 311, British Library, OIOC IOR, E/4/332.

56. Public Consultations at Fort William, 27 June 1805, paragraph 43, British Library OIOC IOR, P/6/14, ff6635-42; to Public consultations at Fort William, 9 April 1807, paragraphs 34-5, British Library, OIOC IOR, P/6/42; and the hand-over from son to father in Public Consultations at Fort William, 7 August 1807, paragraph 88, British Library, OIOC IOR, P/6/49.

57. Letter from William Roxburgh to Sir Joseph Banks, 26 April 1808, Natural History Museum, Botany Library, Roxburgh Correspondence.

58. Ibid., 14 January 1809, ff99-100.

59. John Mathison, and Alexander Way Mason, *The East India Register and Directory for 1807; ... from official returns received at the East India House* (London, 1809).

60. British Library, OIOC IOR, N/7/1/223.

61. British Library, OIOC IOR, N/7/1/195.

62. Public Consultations at Fort William, 18 March 1814, paragraph 42, British Library, OIOC IOR, P/8/30.

63. British Library, OIOC IOR, N/1/6/205.

64. British Library, OIOC IOR, N/2/11.

65. For his baptism, see British Library, OIOC IOR, N/2/11/75, and according to Sir George King's 'A Brief Memoir of William Roxburgh', was killed by lightning in Java in December 1815.

66. Letter from William Roxburgh to Aylmer B. Lambert, 12 February 1801, British Library, Add MSS 28545, ff169-71.

67. Letter from Andrew Ross to William Roxburgh, 14 March 1793, Natural History Museum, Botany Library, Roxburgh Correspondence.

68. Ibid., 16 March 1793.

69. Letter from Francis Buchanan to William Roxburgh, 30 November 1797, Natural History Museum, Botany Library, Buchanan Correspondence.

70. Stephen Daniels, *Humphry Repton, Landscape Gardening and the Geography of Georgian England* (New Haven, Conn., and London, 1999), pp. 38-9.

71. Mary Roxburgh's burial on 25 November 1804 is recorded in British Library, OIOC IOR, N/1/6/205; and after getting permission to proceed to Europe on sick leave, confirmed by the Military Department at Fort St George (Roxburgh remained on this Establishment all his years in India) on 8 March, paragraph 311, British Library, OIOC IOR, E/4/332, he wrote a letter on board the ship *Holstein* on 31 January, recorded in the Public Consultations at Fort William, dated 9 May 1805, paragraph 58, British Library, OIOC IOR, P/6/11 ff4451-2.

72. Roxburgh's arrival is recorded in the Court Book Minutes, 17 July 1805, f354; and his residences during that month in a letter from William Roxburgh to [William Aiton], 20 July 1805, Royal Botanic Garden Kew, Archives, Record Book, 1793-1809, ff239-40.

73. Letter from Francis Buchanan to William Roxburgh, 16 October 1798, Natural History Museum, Botany Library, Buchanan Correspondence.

74. Public Consultations at Fort William, 23 April 1801, paragraph 86, British Library, OIOC IOR, P/5/21.

75. Royal College of Physicians of Edinburgh, Boswell Family Collection.

76. Public Consultations at Fort William, 25 November 1802, paragraph 70, British Library OIOC IOR P/5/39.

77. This took place on 2 November, 1805, at Iver, Buckinghamshire, Peter Roxburgh, personal communication.

78. Letter from Aylmer B. Lambert to Dr James E. Smith, 28 October 1805, Linnean Society, *Smith Letters*, vol. 6, ff90-1.

79. Letter from William Roxburgh to William Ramsay, 8 September 1806, British Library, OIOC IOR, E/1/114 f63.

80. Permission for Roxburgh, Mary and Anne with the female native servant, called Elizabeth Christen, was granted in the Court Book Minutes, 28 January 1807, British Library OIOC IOR, B/144, ff1171, 1175; refusal of the man servant was given in the Court Book Minutes, 10 February 1807, British Library, OIOC IOR, B/144, ff1206-7. The control over entry to India is indicated in the Public Letter at Fort St George, 31 July 1787, paragraph 13, British Library, OIOC IOR, E/4/873 ff616-17, 'By several Acts of Parliament, ... the Power vested in our Government abroad for sending to England Persons resident in India without our License, are clearly explained – but if instead of putting in force those Powers, our Governments in India, not only connive at the residence of unlicensed persons, but even proceed so far as to give them lucrative appointments; every Regulation adopted by the Legislature or by us to obviate this grievance, will be in vain, and we may suffer the mortification of having valuable Offices filled up by Persons entire Strangers to us, and of whose abilities and integrity, we have no Assurance.'

81. Letter from William Roxburgh to Dr C. Taylor, 28 July 1807, Natural History Museum, Botany Library, Roxburgh Correspondence (Letters from T. Hardwicke ...).

82. Letter from William Roxburgh to Aylmer B. Lambert, 20 September 1807, British Library, Add MSS 28545 ff176-7.

83. Information on youngest William's education at the Edinburgh Academy comes from the Edinburgh Academy Register 1824-1914, kindly supplied by Elizabeth Mackay, Librarian; and information regarding his time otherwise, from Alumni and Staff of Edinburgh University, Edinburgh University Library, Special Collections Department. There is a brief outline of his life in Frederick Boase, *Modern English Biography containing many thousand concise memoirs of persons who have died between the years 1851-1900 with an index of the most interesting matter*, vol. 6, 2nd impression (London, 1965), p. 507.

84. Personal communication from David White, descendant of Roxburgh and Rouge Croix Pursuivant at the College of Arms, London. The arms were: Or a Chevron Azure

between in chief two leaves of 'Roxburghia' and in base on a mount Vert a Palm Tree all proper a Sun in splendour Or.

5. After Calcutta 1813-1815

1. Letter from William Roxburgh to C. M. Ricketts, 6 January 1813, in the Public Consultations at Fort William, 8 January 1813, paragraph 31, British Library, OIOC IOR, P/8/13.
2. Letter from William Roxburgh to C. M. Ricketts, 2 February 1813, in the Public Consultations at Fort William, 11 February 1813, paragraph 128, British Library, OIOC IOR, P/8/14.
3. British Library, OIOC IOR, L/MIL/9/115/110.
4. British Library, OIOC IOR, N/1/8/439.
5. British Library, OIOC IOR, L/MIL/9/125/152-5, and L/MIL/9/125/445-53 respectively, with Bruce's transfer in Minutes of Court Book, 27 July 1814, B/159, ff350-1.
6. Letter from William Roxburgh to Sir Joseph Banks, 3 March 1813, Sutro Library, Banks Archives, HLC 950719/009.03813.
7. Letter from William Roxburgh to Sir Joseph Banks, 17 June 1813, Mitchell Library, Banks Archives, 810346.
8. Letter from William Roxburgh to Dr C. Taylor, 31 August 1813, *Transactions of the Society for the Encouragement of Arts, Manufactures, and Commerce*, 32 (1814), 207.
9. Minutes of Court Book, 9 September 1814, British Library, OIOC IOR B/159, f515.
10. Letter from James H. Brooke, Secretary at St Helena, to George Dowdeswell, 20 November 1813, in the Public Consultations at Fort William, 17 June 1814, paragraph 11, British Library, OIOC IOR P/8/35.
11. Minutes of Court Book, 18 May 1814, British Library, OIOC IOR B/159, f115.
12. Letter from Dr James Edward Smith to William Roscoe, 5 July 1814, Linnean Society, *Smith Letters*, vol. 17, f185.
13. Letter from William Roxburgh to Dr James Edward Smith, 15 August 1814, Linnean Society, *Smith Letters*, vol. 25, f67.
14. Minutes of Court Book, 2 September 1814, British Library, OIOC IOR, B/159, f492.
15. Letter from William Roxburgh to Aylmer B. Lambert, 2 October 1814, Linnean Society, Lambert Papers, ff166-166a.
16. Letter from William Roxburgh to Sir Joseph Banks, 28 November 1814, Sutro Library, Banks MSS HLC950504/011.03814.
17. Letter from William Roxburgh to Robert Brown, 30 December 1814, British Library Add MSS 32440 ff65-6.
18. Anon, *The Post Office Annual Directory from Whitsunday 1814 to Whitsunday 1815* (Edinburgh, 1814).
19. Letter from Dr William Wright to Robert Brown, 25 April 1815, British Library, Add MSS 32440, ff77-8.
20. Will of William Roxburgh, recorded 9 July 1815, Scottish Record Office, CC8/8/141.
21. G. J. Bryant, 'Scots in India in the eighteenth century', *Scottish Historical Review*, 65 (1985), 22-41.
22. Allen, *The Buddha and the Sahibs*, p. 123.
23. Richard Wilson and Alan Mackley, *Creating Paradise. The Building of the English Country House 1660-1880* (London, 2000), p. 290.
24. James Forbes, *Oriental Memoirs: selected and abridged from a Series of Familiar Letters written during Seventeen Years Residence in India: including Observations on parts of Africa and South America, and a Narrative of Occurrences in Four Indian Voyages*, 4 vols. (London, 1813), vol. 4, p. 211.
25. Letter from Sir William Jones to Sir Joseph Banks, 18 October 1791, in G. Cannon (ed.), *The Letters of Sir William Jones*, 2 vols. (Oxford, 1970), vol. 2, p. 892.
26. Letter from William Roxburgh to Patrick Russell, 20 January 1790, Sutro Library, EI 1:52.
27. Letter from William Roxburgh to Sir Joseph Banks, 25 April 1795, British Library, Add MSS 33980 ff9-10.
28. Letter from William Roxburgh to Colin Shakespear, 6 February 1796, in the Public Consultations at Fort William, 2 February 1796, paragraph 29, British Library, OIOC IOR, P/4/39, ff45-51.
29. Letter from William Roxburgh to Sir Joseph Banks, 16 February 1796, Sutro Library, BG 1:76.
30. Letter from William Roxburgh to Dr J. E. Smith, 7 June 1799, Edinburgh University Library, Special Collections, La. II.643/52.1.
31. Letter from William Roxburgh to Aylmer B. Lambert, 21 February 1801, British Library, Add MSS 28545, ff173-4.
32. Letter from William Roxburgh to Dr James Edward Smith, 23 April 1798, Linnean Society, *Smith Letters*, vol. 25, f53.
33. Letter from William Roxburgh to Sir Joseph Banks, 1 July 1799, British Library, Add MSS 33980 ff189-90.
34. Anon, *Asiatic Journal and Monthly Register* (London, 1815), pp. 28-30.
35. Anon, *The Annual Biography and Obituary of 1816*, p. 10.
36. W. Roxburgh, 'A Botanical and Economical Account of Bassia Butyracea, or East India Butter Tree', *Asiatick Researches*, 8 (1805), 477-85.
37. Ann Lindsay Mitchell and Syd House, *David Douglas, Explorer and Botanist* (London, 1999), p. 70.
38. Letter from William Roxburgh to Dr James Edward Smith, 25 January 1806, Linnean Society, *Smith Letters*, vol. 25, ff59-60.
39. Ibid., 4 December 1806, vol. 25, f62.
40. Letter from William Roxburgh to Sir Joseph Banks, 5 September 1809, Mitchell Library, Banks Archives, 20 55.
41. Letter from William Roxburgh to Nathaniel Wallich, 4 October 1809, Edinburgh University Library, Special Collections, La.II.643/52.3.
42. Letter from William Roxburgh to Sir Joseph Banks, 3 March 1813, Sutro Library, BG 1:88.
43. Ibid., 17 and 19 June 1813, 20.59 and 20.60.
44. Letter from William Roxburgh to William G. Mackenzie, 28 November 1811, National Library of Scotland, MSS 365, ff79-80.

6. The Botanist

1. Frans A. Stafleu, *Linnaeus and the Linnaeans. The Spreading of their Ideas in Systematic Botany, 1735-1789* (Utrecht, Netherlands, 1971), pp. 201-2.
2. Ibid. pp. 217 and 227.
3. Smith, *The Life of Sir Joseph Banks*, pp. 115-16.
4. George Forster, *A Journey from Bengal to England, through the Northern Part of India, Kashmire, Afghanistan, and Persia, and into Russia, by the Caspian Sea*, 2 vols. (London, 1798), vol. 1, p. 10.
5. Forbes, *Oriental Memoirs*, vol. 1, p. 80.
6. Valentia, *Voyages and Travels to India*, vol. 1, p. 70.
7. Letter from Patrick Russell to Sir Joseph Banks, 1 October 1789, British Library, Add MSS 33978 f263.
8. Letter from William Roxburgh to Sir Joseph Banks, 3 May 1811, Mitchell Library, Banks Archives, 810337.
9. Letter from William Roxburgh junior to Thomas Brown, 22 July 1805, in Public Consultations, 25 July 1805, paragraph 72, British Library, OIOC IOR, P/6/15; and letter from Francis Buchanan to Henry Darell, 9 July 1799, in Public Consultations, 11 July 1799, paragraph 8, British Library, OIOC IOR, P/5/5, ff186-8.
10. Letter from William Roxburgh to Sir Joseph Banks, 15 December 1796, Sutro Library, Banks Manuscripts BG 1:84.
11. Kenneth Lemmon, *The Golden Age of Plant Hunters* (London, 1968), p. 74.
12. Letter from William Roxburgh to Dr Taylor, 16 February 1807, British Library, OIOC, EUR/D809.
13. Letter from William Roxburgh to Sir Joseph Banks, 17 August 1792, British Library, Add MSS 33979 ff171-3.
14. Letter from William Roxburgh to W. G. Mackenzie, 22 June 1811, National Library of Scotland, MSS 6366, ff60-2.
15. These are shown in Desmond, *The European Discovery of the Indian Flora*, pp. 312-13, and Ray Desmond, *Kew. The History of the Royal Botanic Gardens* (London, 1995), p. 119.
16. Lemmon, *The Golden Age of Plant Hunters*, p. 76. John Ellis, *Directions for Bringing over Seeds and Plants* (London, 1770).

17. 'Directions for taking care of growing plants at sea', British Library, OIOC IOR, F/4/134, ff59-132; printed in *Transactions of the Society for the Encouragement of Arts, Manufactures, and Commerce*, 27 (1809), 237-8.

18. A series of letters, from a number of people including William Roxburgh, William Salisbury (London), Nicolay (Vyberg), Mr A. Vanneck (Sierra Leone), and Sir George Young and Mr J. Sutherland (both from St Vincent), published in *Transactions of the Society for the Encouragement of Arts, Manufactures, and Commerce*, 30 (1812), 193-203.

19. Letter from Sir Joseph Banks to William Roxburgh, 9 August 1798, British Library, Add MSS 33980 ff159-60.

20. Letter from William Roxburgh to Thomas Brown, 11 April 1808, in the Public Consultations at Fort William, 22 April 1808, paragraph 21, British Library, OIOC IOR, P/7/7.

21. Letters from Francis Buchanan to William Roxburgh, Natural History Museum, Botany Library, Buchanan MSS.

22. Letter from Thomas Brown to William Roxburgh, 6 May 1808, in the Public Consultations at Fort William, 6 May 1808, paragraph 81, British Library, OIOC IOR, P/7/8.

23. Letter from William Roxburgh to Thomas Brown, 18 January 1809, in the Public Consultations at Fort William, 27 January 1809, paragraph 36, British Library, OIOC IOR, P/7/16.

24. For the years 1793 to 1805, the figures have been taken from the Kew archives: Record Book, 1793-1809, for 1808 from Record Book 1804-26, for seeds from 1805 to 1807 from Inwards Book, 1805-09, and for 1811 to 1814 from Inwards Book, 1809-14. See also Carter, *Sir Joseph Banks*, Appendix VII, p. 556.

25. References to plants to be forwarded to Edinburgh occur, for example, in the letters 15 December 1796, 12 February 1801 and 8 June 1810; and plants for Henry Dundas in the letter 16 December 1791.

26. Letter from William Moorcroft to William Roxburgh, 29 November 1812, Sutro Library, Banks MSS, BO 1:69.

27. Letter from William Roxburgh to Colin Shakespear, 15 January 1794, in the Public Consultations at Fort William, 20 January 1794, paragraph 19, British Library, OIOC IOR, P/4/26, ff348-50.

28. Letter from Sir Joseph Banks to William Roxburgh, 29 May 1796, British Library, Add MSS 33980, ff65-6.

29. Letter from Aylmer B. Lambert to Dr James Edward Smith, 2 March 1801, Linnean Society, *Smith Letters*, vol. 6, ff40-1.

30. Ibid., 29 September 1801, vol. 6, f46.

31. Letter from William Roxburgh to C. R. Crommelin, Secretary to the Government, 2 May 1800, in Public Consultations at Fort William, 22 May 1800, paragraph 46, British Library, OIOC IOR, P/5/12.

32. Deb, 'The Indian Botanic Garden', 263-8.

33. Letters from Aylmer Lambert to Dr James Edward Smith, 2 March and 29 September 1801, Linnean Society, *Smith Letters*, vol. 6, pp. 40-1 and 46.

34. Anonymous review of J. E. Smith's *Plantarum Icones*, 1st fasciculus, from the *Monthly Review*, August 1789, pp. 112-13, copied in British Library, OIOC IOR, EUR/D809.

35. Mildred Archer, *Natural History Drawings in the India Office Library* (London, 1962), p. 5.

36. Ray Desmond, *Wonders of Creation. Natural History Drawings in the British Library* (London, 1986), p. 120.

37. Ray Desmond, *The India Office Museum, 1801-1879* (London, 1982), p. 51.

38. Richard Mabey, *The Flowering of Kew* (London, 1988), p. 92.

39. Holttum, 'The Historical Significance of Botanic Gardens in S. E. Asia', 707-14; and Alain C. White and Boyd L. Sloane (eds.), *The Stapelieae*, 2 edition, 3 vols. (Pasedena, California, 1937), vol. 1, p. 172.

40. See, for example, letter from the Rev. C. John to William Roxburgh, 24 November 1789, British Library, OIOC, MSS EUR/D/809.

41. Letter from Andrew Ross to William Roxburgh, 7 February 1790, Natural History Museum, Botany Library, Roxburgh Correspondence; and letter from William Roxburgh to Patrick Russell, 1 August 1790, Sutro Library, Banks Manuscripts, EI 1:50.

42. List of drawings and descriptions that went by the *Houghton* in September 1790, Sutro Library, Banks Manuscripts, EI 1:45; and letter from William Roxburgh to Sir Joseph Banks, 30 December 1790, British Library, Add MSS 33979, ff64-5.

43. Carter, *Joseph Banks*, pp. 267-8, 465-6, 483-4; J. R. Sealy, 'The Roxburgh Flora Indica Drawings at Kew', *Kew Bulletin*, 2 (297-348) and 3 (349-399) (1956) which lists the drawings held at Kew and Calcutta; and Ray Desmond, 'William Roxburgh's Plants of the Coast of Coromandel', *Hortulus Aliquando*, 2 (1977), 22-41, which gives a full study of the publication, although the year given for *Fasciculus 2* is a printing error.

44. See, for example, Archer, *Natural History Drawings in the India Office Library*; Blunt and Stearn, *The Art of Botanical Illustration*; Desmond, *Wonders of Creation*; Mabey, *The Flowering of Kew*; and Noltie, *Indian Botanical Drawings*.

45. Letter from Andrew Ross to William Roxburgh, 7 August 1792, Natural History Museum, Botany Library, Roxburgh Correspondence.

46. Ibid., 5 June 1793.

47. Letter from William Roxburgh to Sir Joseph Banks, 17 August 1792, British Library, Add MSS 33979 ff171-3.

48. Letter from William Roxburgh to Dr James Edward Smith, 27 December 1794, Linnean Society, *Smith Letters*, volume 8, f192.

49. Letter from William Roxburgh to Sir Joseph Banks, 19 December 1795, British Library, Add MSS 33980 ff41-2.

50. Letter from Sir Joseph Banks to William Roxburgh, 29 May 1796, British Library, Add MS 33980 ff65-6.

51. William Curtis, *Flora Londinensis* (London, 1775-98), which was a model for Roxburgh's *Plants of the Coast of Coromandel*; William Curtis, *Botanical Magazine*, vol. 1, 1787; and James E. Smith, *English Botany* (London, 1790-1814). A full list of the contents of Roxburgh's library is given in Appendix 6.

52. Letter from William Roxburgh to Sir Joseph Banks, 30 December 1794, British Library, Add MSS 33979 ff292-3.

53. Public Consultation at Fort William, 11 October 1814, paragraph 40, British Library, OIOC IOR, P/8/40.

54. Bengal Public Letter, 21 July 1818, British Library, OIOC IOR, F/4/588 ff218-25; and F/4/714 ff1-28.

55. Extract from a letter from Henry Colebrooke to William Egerton, 3 April 1818, Bengal Public Letter, 21 July 1818, British Library, OIOC IOR, Board's Collection F/4/588, ff218-25.

56. J. Robert Sealy, 'The Roxburgh Flora Indica drawings at Kew'; and M. Sanjappa, K.Thothathri and A. R. Das, 'Roxburgh's Flora Indica drawings at Calcutta', *Bulletin of the Botanical Survey of India*, 33 (1991), 1-232.

57. Letter from William Carey to Lord Minto, 21 October 1812, in the Public Consultations at Fort William, 23 October 1812, paragraph 56, British Library, OIOC IOR, P/8/9.

58. Letter from William Carey to C. M. Ricketts, 18 June 1814, in the Public Consultations at Fort William, 24 June 1814, paragraph 35, British Library, OIOC IOR, P/8/36. The Catalogue was published as William Roxburgh, *Hortus Bengalensis; or, a catalogue of the plants growing in the Honourable East India Company's Garden at Calcutta*, edited by William Carey (Serampore, India, 1813-14).

59. C. B. Robinson, 'Roxburgh's Hortus Bengalensis', *Philippine Journal of Science*, C. Botany, 7 (1912), 411-19.

60. D. J. Mabberley, 'Francis Hamilton's commentaries with particular reference to the Meliaceae', *Taxon*, 26 (1977), 523-40.

61. Carrie Karegeannes, 'Roxburgh: chronicles of Indian begonias', *The Begonian*, 46 (1979), 261-80.

62. Leonard Huxley, *Life and Letters of Sir Joseph Dalton Hooker, O.M., G.C.S.I. Based on materials collected and arranged by Lady Hooker*, 2 vols. (London, 1918), vol. 1, p. 473.

63. William Roxburgh, 'Descriptions of several of the Monandrous Plants of India, belonging to the natural order called Scitamineae by Linnaeus, Cannae by Jussieu, and Drimyrhizae by Ventenat', *Asiatick Researches*, 11 (1810), 318-62.

64. Letter from William Roscoe to Dr James Edward Smith, 27 June 1814, Linnean Society, *Smith Letters*, vol. 17, ff183-4.

65. Letter from Dr James Edward Smith to William Roscoe, 5 July 1814, Linnean Society, *Smith Letters*, vol. 17, f185.

66. Letter from William Roxburgh to Dr James Edward Smith, 15 August 1814, Linnean Society, *Smith Letters*, vol. 25, ff67-8.

67. William Roscoe, 'Remarks on Dr Roxburgh's Description of the Monandrous Plants of India; in a Letter to the President', *Transactions of the Linnean Society*, 11 (1815), 270-82; William Roscoe, *Monandrian Plants of the Order Scitamineae, chiefly drawn from living specimens in the Botanic Garden at Liverpool. Arranged according to the System of Linnaeus, with Descriptions and Observations*, 2 vols. made up of 15 fasciculi (Liverpool, 1824-9).

68. Letter from William Roxburgh to Sir Joseph Banks, 17 June 1813, Mitchell Library, Banks MSS 810346.

69. William Roxburgh, 'Cryptogamous plants', Calcutta *Journal of Natural History*, 4 (1844), 463-520.

70. Desmond, *The Discovery of the Indian Flora*, p. 69.

71. Letter from William Roxburgh to Dr James Edward Smith, 12 January 1815, Linnean Society, *Smith Letters*, vol. 25, f69.

72. Elmer Drew Merrill, 'The fugitive names of "Hamoa" and "Romoa" in Roxburgh's *Flora Indica*, errors for Honimoa – Sapura', *Taxon*, 1 (1952), 124-5.

73. D. J. Mabberley, 'Francis Hamilton's commentaries with particular reference to Meliaceae'.

74. Thomas Thomson, 'Notes on the Herbarium of the Calcutta Botanic Garden, with especial reference to the completion of the Flora Indica', *Journal of Botany and Kew Miscellany*, 9 (1857), 10-14 and 33-41.

75. Elmer Drew Merrill, 'The Botany of Cook's voyages', *Chronica Botanica*, 2 (1936), 206.

76. Robert Wight, 'Illustrations of Indian Botany, principally of the southern parts of the Peninsula', *Botanical Miscellany*, 2 (1831), 90-7.

77. I am grateful to Leander Wolstenholme of the Botany Department at Liverpool Museum, for his help and patience looking through a sample of this material.

78. A. Lasègne, *Musée Botanique de M. Benjamin Delessert* (Paris, 1845), pp. 73 and 145; and H. S. Miller, 'The herbarium of Aylmer Bourke Lambert. Notes on its acquisition, dispersal and present whereabouts', *Taxon*, 19 (1970), 489-553 states that 'from 2000 to 2250 plants, with a large Cabinet containing them was bought by Rich for £36'. The quotation from Lasègne reads 'the herbarium of Doctor Roxburgh, now in the possession of M. Delessert, is made up of a large collection that he made in the Indian continent, on the Coast of Coromandel and at Banda, Amboyna and other islands in the East Indian archipelago, and some plants that he received from the Cape of Good Hope.'

79. C. V. Morton, 'William Roxburgh's Fern Types', *Contributions from the United States National Herbarium*, 38 (1974), 383-96. See also Merrill, 'The Botany of Cook's voyages', 161-383, especially pp. 171-2.

80. Letter from William Roxburgh to Sir Joseph Banks, 19 June 1813, Mitchell Library, Banks Archives 20.60 810350.

81. A. Beatson, *Tracts Relative to the Island of St Helena*; written during a residence of five years (London, 1816), contains 'Description by Dr Roxburgh, of the most remarkable of the plants indigenous to St Helena', pp. 295-326.

82. Q. C. B. Cronk, *The Endemic Flora of St Helena* (Oswestry, Shropshire, 2000), p. 37.

83. Grove, *Green Imperialism*, p. 355.

84. Desmond, Ray, *Sir Joseph Dalton Hooker, Traveller and Plant Collector* (London, 1999), p. 32.

85. 'Florida Stae Helenae', Natural History Museum, Botany Library, Robert Brown MSS, B66 (iv); and letter from Dr William Wright to Robert Brown, 25 April 1815, British Library, Add MSS 32440 ff77-8.

86. Letter from William Roxburgh to C. M. Ricketts, 11 January 1813, in the Public Consultations at Fort William, 15 January 1813, paragraph 34, British Library, OIOC IOR, P/8/13.

87. Grove, *Green Imperialism*, pp. 355-565.

88. Ibid., p. 357.

89. Letter from William Roxburgh to Nathaniel Wallich, undated, Edinburgh University Library, Special Collections, La.II.643/52.2.

7. Wider Scientific Interests

1. Letter from William Roxburgh to the President and Members of the American Philosophical Society, 26 December 1793, American Philosophical Society, Manuscripts Library.

2. Minutes of the American Philosophical Society, 30 May 1794, and see Appendix 5 for Hamilton's generosity over the years.

3. Letter from William Roxburgh to Dr J. R. Coxe, 14 February 1803, American Philosophical Society, Manuscripts Library.

4. Letter from Caspar Wistar to John Vaughan, 26 April 1803, American Philosophical Society, Manuscripts Library.

5. Letters from William Roxburgh to John Vaughan, 29 December 1803 and 26 January 1804, American Philosophical Society, Manuscripts Library.

6. Personal communication, Dr Alison Lewis, American Philosophical Society, Manuscripts Library.

7. Letter from William Roxburgh to Dr James Edward Smith, 14 January 1794, Linnean Society, *Smith Letters*, vol. 8, f186.

8. Letter from the Rev. C. John to William Roxburgh, 15 October 1789, British Library, OIOC, MSS EUR/D809.

9. Letter from William Roxburgh to Patrick Russell, 20 January 1790, Sutro Library, EI 1:52.

10. Letter from William Roxburgh to Sir Joseph Banks, 28 October 1797, British Library, Add MSS 33980 ff117-18.

11. Letters from William Roxburgh to Dr James Edward Smith, 8 November 1797, Linnean Society, *Smith Letters*, vol. 25, f52; William Roxburgh to Sir Joseph Banks, 24 April 1798, British Library, Add MSS 33980 ff137-41; and Sir Joseph Banks to William Roxburgh, 9 August 1798, British Library, Add MSS 33980 ff159-60.

12. William Roxburgh, 'An Account of a new Species of Delphinus, an Inhabitant of the Ganges', *Asiatick Researches*, 7 (1801), 170-4. The original is now in the Royal College of Surgeons with a plaster cast in the British Museum. There have been problems over who described it first as a description was also published by Heinrich Julius Lebeck in 1801. The latter was a friend of Roxburgh's and was indeed staying with him in 1801. G. Pillerie credits Roxburgh with the description in G. Pillerie, 'William Roxburgh (1751-1815), Heinrich Julius Lebeck (†1801) and the Discovery of the Ganges Dolphin (Platanista gangetica Roxburgh 1801)', *Investigations on Cetaceae*, 9 (1978), 11-21.

13. H. T. Colebrooke, 'Description of a Species of Ox, named Gayal', *Asiatick Researches*, 8 (1805), 511-24.

14. William Roxburgh, 'Remarks made about the year 1790', unpublished, National Library of Scotland, Dundas Papers, Acc 7664.

15. William Roxburgh, 'A Description of the Plant Butea', *Asiatick Researches*, 3 (1790), 469-74.

16. Vincent C. Harlow, *Founding the Second British Empire, 1763-1793, Vol. II. New Continents and Changing Values* (London, 1964), p. 784.

17. Garland Cannon, *The Life and Mind of Oriental Jones. Sir William Jones, the Father of Modern Linguistics* (Cambridge, 1990), p. 229.

18. Lord Macartney, *An Embassy to China*, entry for 7 January 1794.

19. Ibid., p. 207.

20. *Asiatick Researches*, 6 (1799), 589 (Rules).

21. Sir William Jones, 'Introduction', *Asiatick Researches*, 1 (1788), iii.

22. Ibid., xiii.

23. Sir William Jones, 'The Second Anniversary Discourse', *Asiatick Researches*, 1 (1788), 407-8.

24. Bayly, *Empire and Information*, p. 253.

25. Cannon, *The Life and Mind of Oriental Jones*, p. 253.

26. John Bentley, 'Remarks on the Principal Æras and Dates of the Ancient Hindus', *Asiatick Researches*, 5 (1799), 315-43.

27. Sir William Jones, 'On the Gods of Greece, Italy, and India, written in 1784 and since revised', *Asiatick Researches*, 1 (1788), 221-75.

28. Blagdon, *A Brief History of Ancient and Modern India*, p. vi.

29. Cannon, *The Life and Mind of Oriental Jones*, p. 347.
30. Sir William Jones, 'The Design of a Treatise on the Plants of India', *Asiatick Researches*, 2 (1790), 345-52.
31. Ibid., p. 348.
32. From the *Monthly Review* for July 1789, British Library, OIOC IOR, EUR/D809.
33. Cannon, *The Life and Mind of Oriental Jones*, p. 277.
34. Ibid., pp. 332-3. The last phrase is from a letter from Jones to Banks, 18 October 1791.
35. Richard Gombrich, 'Introduction' in Alexander Murray (ed.), *Sir William Jones 1746-1794. A commemoration* (Oxford, 1998), p. 3.
36. William Roxburgh, 'Botanical Observations on the Spikenard of the Ancients, intended as a Supplement to the late Sir William Jones's papers on that Plant', *Asiatick Researches*, 4 (1795), 433-6.
37. William Roxburgh, 'A Description of the Jonesia', *Asiatick Researches*, 4 (1795), 355-7.
38. Allen, *The Buddha and the Sahibs*, p. 53.
39. *Asiatick Researches*, 5 (1797).
40. William Roxburgh's will, Scottish Record Office, CC8/8/141.
41. Grove, *Green Imperialism*, passim.
42. Carter, *Sir Joseph Banks*, p. 173.
43. Letter from William Roxburgh to Aylmer B. Lambert, 21 February 1801, British Library, Add MSS 28545 ff173-4.
44. Grove, *Green Imperialism*, p. 265.
45. William Roxburgh, 'Remarks on the Land Winds and their Causes', *Transactions of the Medical Society of London*, 1 (1810), 189-211.
46. Much of the background information for this has been taken from Allen, *The Buddha and the Sahibs*, pp. 117-23; the letters between Roxburgh and Mackenzie in the National Library of Scotland are in their MSS 6365 and 6366.
47. Love, *Vestiges of old Madras 1640-1800*, vol. 1, passim, and vol. 3, p. 229 'for two years before [Lord] Macartney's arrival [at Fort George on the Swallow in April 1781], Madras had suffered from scarcity, which was due mainly to scanty rainfall.'
48. Mention of his recordings appear in his article 'Remarks on the land winds and their causes', when he states that 'we experienced for almost a fortnight in the year 1799 in the Northern Circars, when the thermometer at eight o'clock in the night stood at 108°, and at noon at 112°.' As late as November 1814, he wrote from Edinburgh to Joseph Banks that 'the Thermr has been already as low as 20°', letter from William Roxburgh to Sir Joseph Banks, 28 November 1814, Sutro Library, Banks MS HLC950504/011.03814.
49. Letter from William Roxburgh to Sir Joseph Banks, 18 December 1784, British Library, Add MSS 33977 ff272-5.
50. Dalrymple, *Oriental Repertory*, vol. 2, p. 70; and letter from William Roxburgh to Sir Joseph Banks, 8 July 1791, Sutro Library, Banks MSS, EI 1:44.
51. Letter from William Parsons to Andrew Ross, 7 June 1787, in Dalrymple, *Oriental Repertory*, vol. 1, p. 88.
52. Letter from William Roxburgh to Andrew Ross, 25 August 1788, in Dalrymple, *Oriental Repertory*, vol. 1, p. 95.
53. Letter from Edward Otto Ives to Edward Hay, 4 January 1794, in the Public Consultations at Fort William, 6 January 1794, paragraph 3, British Library, OIOC IOR, P/4/26, ff194-6.
54. Letter from William Roxburgh to Colin Shakespear, 11 August 1794, in the Public Consultations at Fort William, 22 August 1794, paragraph 17, British Library, OIOC IOR, P/4/30, ff258-61.
55. Letter from William Roxburgh to Colin Shakespear, 22 September 1794, in the Public Consultations at Fort William, 29 September 1794, paragraph 8, British Library, OIOC IOR, P/4/30, ff570-1.
56. Letter from William Roxburgh to Edward Hay, 13 January 1795, in the Public Consultations at Fort William, 16 January 1795, paragraph 35, British Library, OIOC IOR, P/4/33, ff119-20.
57. Letter from William Roxburgh to Colin Shakespear, 26 January 1795, in the Public Consultations at Fort William, 30 January 1795, paragraph 4, British Library, OIOC IOR, P/4/33, ff229-32.
58. Letter from William Roxburgh to Colin Shakespear, 14 April 1796, in the Public Consultations at Fort William, 18 April 1796, paragraph 28, British Library, OIOC IOR, P/4/41.
59. This is referred to in a number of documents, for instance, 'Extract of a general letter from England in the Public Department 9 May 1797, paragraph 13, Subject Dr Heyne's Iron Works, in a collection of Roxburgh Manuscripts which make up British Library, OIOC MSS/EUR/D809.
60. Letter from Andrew Ross to William Roxburgh, 29 September 1793, Natural History Museum, Botany Library, Roxburgh Correspondence.
61. Note, undated in Natural History Museum, Botany Library, Roxburgh MSS, Correspondence from the Rev. C. John, Hardwicke, Smith, etc.
62. Letter from William Roxburgh to Dr Charles Taylor, 18 April 1808, Natural History Museum, Botany Library, Roxburgh Correspondence.
63. 'Description of two Subular Petrified Shells from the Island of Battas on the S. W. Coast of Sumatra, where they were first discovered, after repeated shocks of an Earthquake, & supposed to have been thrown from a chasm in the earth, caused by the earthquakes' by Dr Roxburgh, British Library, Add MSS 33980, ff143-5.
64. James Hutton, *A Theory of the Earth* (Edinburgh, 1785).
65. Sir William Jones, 'Discourse the Ninth. On the Origin and Families of Nations', *Asiatick Researches*, 3 (1792), 479-92.

8. Background

1. See, for example, McCracken, *Gardens of the Empire*, p. 13.
2. Letter from William Roxburgh to Patrick Russell, 1 August 1790, Sutro Library, Banks MSS, EI 1:50.
3. McCracken, *Gardens of the Empire*, p. 11.
4. Grove, *Green Imperialism*, passim.
5. Extract from Revenue Letter from Bengal, 5 September 1809, paragraph 59, British Library, OIOC IOR, F/4/99 2028, f78.
6. Crawford, *Roll of the Indian Medical Service*, p. 265.
7. For instance, in a letter of 24 November 1782, he described the doses he gave and the form in which given, of the roots of *Clitoria ternatea*, stating the native practice and that given by Burman in his *Thesaurus Zeylanica* (letter from William Roxburgh to Sir Joseph Banks, 24 November 1782, British Library Add MSS 33977 ff181-2); and two years later he wrote 'In this Country some of our Surgeons have been obliged to use the Bark of the Melia Azadirachta as a substitute for Peruvian Bark & found it succeed wounderfully well.' Letter from William Roxburgh to Sir Joseph Banks, 18 December 1784, British Library, Add MSS 33977 ff272-5.
8. Letter from William Roxburgh to Sir Joseph Banks, 24 November 1782, British Library, Add MSS 33977, ff181-2, in which he also gives the preparation and doses for the use of this drug.
9. Pharmacopeia, British Library, OIOC IOR EUR E120, and mentioned in Bayly, *Empire and Information*, p. 272. This pharmacopeia consists of 184 folios of very rough, locally made paper that also contains purely botanical information as well as the full range of materia medica, plus more unusual remedies such as the use of meat and faeces of various animals from tigers, crocodiles and elephants to dogs, bears and hyenas.
10. The success with this is shown by 'Your Bark Swietenia febrifuga gets here the highest Reputation by Dr Klein & Dr Folly & is applied in the hospital here.' British Library, OIOC, MSS EUR/D809.
11. For his work on coffee, it was first mentioned in the Revenue Department letter from Fort St George to the Court of Directors, 14 October 1786, paragraph 28, British Library, OIOC IOR, E/4/316; a fuller reference to Roxburgh cultivating it appears in the Madras Despatches from the Court of Directors, 20 April 1789 (British Library, OIOC IOR, E/4/875 ff29-30): 'in our letter to this Commercial Department of the 31 July 1787, we approved of the measures you had taken to assist Dr

Roxburgh in the cultivation of Coffee and Pepper Plants in the Rajahmundry Circar.' The interest in Caducay galls went right back to a letter from Roxburgh to Sir Joseph Banks, 24 November 1782 (British Library, Add MSS 33977, ff181-2); indigo is first mentioned by Andrew Ross, letter from Andrew Ross to William Roxburgh, 2 June 1788, Natural History Museum, Botany Library, Roxburgh Correspondence; and lac is first mentioned in a letter from William Roxburgh to Patrick Russell, 20 January 1790, Sutro Library, Banks MSS EI 1:52. 'At the request of Mr William Roxburgh, We transmit herewith a Copy of a Letter, which he has addressed to us on the subject of the Bread Fruit Tree,' Public Letter from Fort St George to the Court of Directors, 27 July 1787, paragraph 22, British Library OIOC IOR, E/4/319. Amongst a list of papers for Roxburgh from Ross, 6 December 1789, is 'Remarks on the Culture and Manufacture of Silk', Natural History Museum, Botany Library, Roxburgh Correspondence. 'I am glad that you have seed applications for more of the Teak Seeds, for the Governor General' in letter from Andrew Ross to William Roxburgh, 11 December 1789, Natural History Museum, Botany Library, Roxburgh Correspondence; but see p. 130 for a more detailed study of Roxburgh's work on teak. See, for example, the letter from William Roxburgh to N. Kindersley, 12 January 1790, Linnean Society, *Smith Letters*, volume 25, f47. 'I am endeavoring to stimulate our Government to set about introducing the culture of the sago Palm,' letter from William Roxburgh to Patrick Russell, 20 January 1790, Sutro Library, Banks MSS, EI 1:52. 'For several years past I have turn'd many of my leisure hours to the cultivation & improvement of Sugar' reproduced in *Journal of Economic and Taxonomic Botany*, 5 (1984), 1160, but see Chapter 12 for a fuller study of Roxburgh's work on sugar.

12. Letter from William Roxburgh to Patrick Russell, 1 August 1790, Sutro Library, Banks MSS, EI 1:50.

13. Public Letter at Madras to the Court of Directors, 27 July 1789, paragraph 22, British Library, OIOC IOR, E/4/319.

14. Letter from Andrew Ross to William Roxburgh, 11 December 1789, Natural History Museum, Botany Library, Roxburgh Correspondence; and letters from the Rev. C. John to William Roxburgh, 18 March 1790, 26 November 1792 and 2 April 1793, British Library, OIOC, EUR/D809.

15. Letter from William Roxburgh to Sir Joseph Banks, 8 July 1791, Sutro Library, Banks MS, EI 1:44.

16. Letter from William Roxburgh to Sir Joseph Banks, 17 August 1792, British Library, Add MSS 33979 ff171-3.

17. Carter, *Sir Joseph Banks*, p. 257.

18. Ibid., pp. 315-17.

19. Letter from William Roxburgh to Colin Shakespear, 24 October 1795, in the Public Consultations, 26 October 1795, paragraph 14, British Library, OIOC IOR, P/4/36.

20. Letter from William Roxburgh to Colin Shakespear, 14 April 1796, in the Public Consultations, 18 April 1796, paragraph 28, British Library, OIOC IOR, P/4/41, ff249-54.

21. Letter from William Roxburgh to Sir John Shore, 3 May 1797, in the Public Consultations, 3 May 1797, paragraph 77, British Library, OIOC IOR, P/4/50 ff225-8.

22. Ibid., paragraphs 79-83.

23. Letter from Sir John Sinclair to Sir John Shore, 27 March 1797, in the Public Consultations, 27 February 1798, paragraph 12, British Library OIOC IOR, P/4/58, ff422-4.

24. Letter from William Roxburgh to H. V. Darell, 1 December 1799, in the Public Consultations at Fort William, 5 December 1799, paragraph 11, British Library, OIOC IOR, P/5/8, ff123-37.

25. Letter from William Roxburgh to Thomas Philpot, 26 April 1803, in the Public Consultations at Fort William, 28 April 1803, paragraphs 28-9, British Library, OIOC IOR, P/5/44.

26. Letter from William Roxburgh to Colin Shakespear, 29 February 1796, in Public Consultations at Fort William, 14 March 1796, paragraph 47, British Library, OIOC IOR, P/4/40.

27. Letter from William Roxburgh to Henry Darell, 26 November 1799, included in 'Cultivation of the Teak Tree', British Library, OIOC IOR, F/4/99, 2028.

28. Letter from William Roxburgh to Colin Shakespear, 30 December 1793, British Library, OIOC IOR, P/4/26, ff11-16.

29. Letter from Thomas Parr to Colin Shakespear, 19 January, 1794, British Library, OIOC IOR, P/4/26, 24 January 1794.

30. Letter from William Roxburgh to William Duncan, 23 November 1797, in Public Consultations at Fort William, 27 November 1797, paragraph 48, British Library, OIOC IOR, P/4/55, ff133-7.

31. Letter from William Roxburgh junior to Thomas Brown, 5 September 1805, in Public Consultations at Fort William, 5 September 1805, paragraph 74, British Library, OIOC IOR, P/6/17.

32. Letter from William Roxburgh to Thomas Philpot, 5 September 1803, in Public Consultations at Fort William, 22 September 1803, paragraph 13, British Library, OIOC IOR, P/5/47.

33. Letter from William Roxburgh to Colin Shakespear, 29 February 1796, in Public Consultations at Fort William, 14 March 1796, paragraph 47, British Library, OIOC IOR, P/4/40.

34. Huggins, *Sketches in India*, pp. 75-6.

35. Letter from Roxburgh to Henry J. Darell, Acting Secretary to Government, 26 Nov 1799, included in 'Cultivation of the Teak Tree', British Library, OIOC IOR, F/4/99, 2028.

36. Revenue Letter from Calcutta to the Court of Directors, 5 September 1809, paragraph 61, included in 'Cultivation of the Teak Tree', British Library, OIOC IOR, F/4/99, 2028.

37. Letter from William Roxburgh to C. M. Ricketts, 4 November 1811, in Board's Collection, Report on Teak Plantations at Rampu Boalia, British Library, OIOC IOR, F/4/427.

38. Letter from George Ballard to E. Barnett, Acting Collector at Bauleah, 21 August 1812, in Public Consultations at Fort William, 23 October 1812, paragraph 6, British Library, OIOC IOR, P/8/9.

39. William Roxburgh, 'Some Account of the Teak Tree of the East Indies', *Nicholson's Journal*, 33 (1812), 348-54. This article which appears to have been sent first to Dr Taylor, Secretary of the Royal Society of Arts but does not seem to have been published in its *Transactions*, went into the growth, description of how it was to be planted and cared for, and ending with its economics.

40. William Roxburgh, 'Notes to illustrate some few of the Trees in the Annexed Table of Measurements', *Transactions of the Society for the Encouragement of Arts, Manufactures and Commerce*, 32 (1814), 207-10.

41. Letter from Andrew Ross to William Roxburgh, 2 October 1793, Natural History Museum, Botany Library, Roxburgh Correspondence.

42. H. Santapau, 'The story of tea', *Bulletin of the Botanical Survey of India*, 8 (1966), 103-7.

43. Cranmer-Byng, *An Embassy to China*, entry for Sunday, 17 November 1793, p. 182. From the timing of this reference, it is obvious that Macartney was unaware that Kyd had died earlier that year.

44. Public Letter at Fort William to the Court of Directors, 5 February 1795, paragraph 72, British Library, OIOC IOR, E/4/55.

45. Public Letter from the Court of Directors to the Governor General at Fort William, 5 January 1796, in Gupta, *Fort William – India House Correspondence*.

46. Letter from William Roxburgh to Colin Shakespear, 15 March 1796, in Public Consultations at Fort William 21 March 1796, paragraph 21, British Library, OIOC IOR, P/4/40.

47. Letter from William Roxburgh to D. Seton, 7 April 1797, in Public Consultations at Fort William, 3 May 1797, paragraph 85, British Library, OIOC IOR, P/4/50.

48. Letter from William Roxburgh to Sir Joseph Banks, 24 April 1798, British Library, Add MSS 33980 ff137-41.

49. J. Forbes Royle, 'Report on the Progress of the Culture of the China Tea Plant in the Himalayas, from 1835 to 1847', *Journal of the Royal Asiatic Society of Great Britain and Ireland*, 12 (1850), 125-52.

9. Dyes

1. Derek Hudson and Kenneth W. Luckhurst, *The Royal Society of Arts, 1754-1954* (London, 1954), p. 8.
2. Ibid., p. 89.
3. British Library, OIOC IOR EUR/D809.
4. Letter from William Roxburgh to Sir Joseph Banks, 8 March 1779, British Library, Add MSS 33977, ff93-5.
5. Letter from the Rev. C. John to William Roxburgh, 15 October 1789, British Library, OIOC MSS/EUR/D809.
6. From the *English Review* for Jan 1790, Article IV on Dyeing, British Library, OIOC MSS/EUR/D809.
7. Carter, *Sir Joseph Banks*, p. 274.
8. Ibid.
9. Letter from Patrick Russell to Sir Archibald Campbell, 8 March 1787, in the Public Consultations at Fort St George, 9 March 1787, British Library, OIOC IOR P/240/66, ff463-4.
10. Desmond, *European Discovery of the Indian Flora*, p. 209.
11. Ibid., p. 210.
12. Extracts of papers relating to the East India trade, compiled April 1788, British Library, OIOC IOR, EUR/D809.
13. Copies of letters sent by Mr Ross in May 1792, letter from Sir George Young to possibly Dr Anderson, 8 December 1791, British Library, OIOC IOR, EUR/D809.
14. Letter from William Roxburgh to Edward Hay, 26 June 1795, in the Public Consultations at Fort William, 29 June 1795, paragraph 8, British Library, OIOC IOR P/4/35, ff283-94.
15. Desmond, *The European Discovery of the Indian Flora*, p. 212.
16. Letter from Edward Hay to Captain Neilson, 29 June 1795, in the Public Consultations at Fort William, 29 June 1795, paragraph 9, British Library, OIOC IOR P/4/35, ff295-7.
17. Public Letter from Fort William to the Court of Directors, 28 August 1795, paragraph 34, British Library, OIOC IOR E/4/56.
18. Desmond, *The European Discovery of the Indian Flora*, p. 211, quoting a letter from Smith to Banks, 17 September 1795, Natural History Museum, Botany Library, Banks letters.
19. Letter from William Roxburgh to Colin Shakespear, 13 July 1795, in the Public Consultations at Fort William, 20 July 1795, paragraph 7, British Library, OIOC IOR P/4/35, ff488-90.
20. Letter from Dr James Dinwiddie to the Governor General, 22 October 1795, in the Public Consultations at Fort William, 26 October 1795, paragraph 15, British Library, OIOC IOR P/4/36.
21. Letter from William Roxburgh to Colin Shakespear, 2 February 1796, in the Public Consultations at Fort William, 15 February 1796, paragraph 29, British Library, OIOC IOR P/4/39, ff745-51.
22. Letter from William Roxburgh to Sir Joseph Banks, 16 February 1796, Sutro Library, BG 1:76.
23. Letter from William Roxburgh to Colin Shakespear, 29 February 1796, in the Public Consultations at Fort William, 14 March 1796, paragraph 47, British Library, OIOC IOR P/4/40.
24. Letter from William Roxburgh to Patrick Russell, 20 January 1790, Sutro Library, EI 1:52. The lac insect in this case refers to the insect from which the red dye Red Lake is derived, rather than the lac insect which produces varnish.
25. Letter from William Roxburgh to Sir Joseph Banks, 30 December 1790, British Library, Add MSS 33979, ff64-5.
26. Description of Lacc, by William Roxburgh, 15 September 1790, British Library, Add MSS 33979, ff195-6.
27. Letter from William Roxburgh to Sir Joseph Banks, 17 August 1792, British Library, Add MSS 33979, ff171-3.
28. William Roxburgh, 'On the Lácshà, or Lac, Insect', *Asiatick Researches*, 2 (1790), 361-4; and William Roxburgh, 'Chermes Lacca', *Philosophical Transactions of the Royal Society*, 81 (1791), 228-35.
29. William Roxburgh, 'A description of the plant Butea', *Asiatick Researches*, 4 (1792), 469-74.
30. Letter from Andrew Stephens, published in *Transactions of the Society for the Encouragement of Arts, Manufactures, and Commerce*, 19 (1801), 352-63.
31. Anon, *The Annual Biography and Obituary*, p. 6.
32. Revenue Department at Fort St George to the Court of Directors, 20 February 1798, paragraphs 62-4, British Library, OIOC IOR E/4/324.
33. Court of Directors to the Revenue Department at Fort St George, 3 September 1800, paragraph 20, British Library, OIOC IOR E/4/887, ff285-8.
34. Copies of letters sent by Mr Ross in May 1792, letter from Sir George Young, 8 December 1791, British Library, OIOC IOR, EUR/D809.
35. Letter from David Scott to Herbert Harris, 16 May 1796, in C. H. Phillips, *The Correspondence of David Scott, Director and Chairman of the East India Company, relating to Indian Affairs 1787-1805*, 2 vols. (London, 1951).
36. 'Introduction of Cochineal to India', 12 January 1820, British Library, OIOC IOR O/6/8, ff551-60, and O/6/10, ff677-92.
37. Letters from David Scott to William Fairlie, 14 July 1797, and to Herbert Harris, 5 August 1797, in Phillips, *The Correspondence of David Scott*.
38. 'Process for making Indigo on the Coast near Ingeram', in J. F. Charpentier de Cossigny de Palma, *Memoir, containing an Abridged Treatise on the Cultivation and Manufacture of Indigo* (Calcutta, 1789), p. 157.
39. Letter from Andrew Ross to William Roxburgh, 2 June 1788, Natural History Museum, Botany Library, Roxburgh Correspondence.
40. Letter from Andrew Ross to William Roxburgh, 19 December 1788, Natural History Museum, Botany Library, Roxburgh Correspondence.
41. 'List of papers for Dr Roxburgh', 6 December 1789, Natural History Museum, Botany Library, Roxburgh Correspondence.
42. Letter from the Rev. C. John to William Roxburgh, 24 November 1789, British Library, OIOC IOR, MSS EUR/D809.
43. This was placed in a new genus named by Robert Brown after William Wright, another pupil of Hope's who worked in the West Indies.
44. William Roxburgh, 'A Botanical Description, and Drawing of a new Species of Nerium (Rose-Bay) with the Process for extracting, from It's Leaves, a very beautiful Indigo', in Dalrymple, *Oriental Repertory*, vol. 1, pp. 39-44.
45. Letter from Andrew Ross to William Roxburgh, 11 December 1789, Natural History Museum, Botany Library, Roxburgh Correspondence.
46. Letter from William Roxburgh to Patrick Russell, 20 January 1790, Sutro Library, EI 1:52.
47. Ibid., 1 August 1790, EI 1:50.
48. Letter from Andrew Ross to William Roxburgh, 12 January 1791, Natural History Museum, Botany Library, Roxburgh Correspondence.
49. Ibid., 11 May 1791.
50. Ibid., 14 May 1791.
51. Ibid., 6 October 1791.
52. Ibid., 26 July 1793.
53. Letter from William Roxburgh to David Haliburton, 8 May 1793, Natural History Museum, Botany Library, Roxburgh Correspondence.
54. Public Letter at Fort St George, 28 January 1793, paragraph 50, British Library, OIOC IOR, E/4/322; Public Letter at Fort St George, 5 February 1793, paragraph 5, British Library, OIOC IOR, E/4/322; and letter from William Roxburgh to Dr James Edward Smith, 20 March 1793, Linnean Society, *Smith Letters*, vol. 8, f184.
55. Letter from Andrew Ross to William Roxburgh, 30 July 1793, Natural History Museum, Botany Library, Roxburgh Correspondence.
56. Letter from William Roxburgh to Sir Joseph Banks, 1 December 1793, British Library, Add MSS 33979, f224.
57. Letter from William Roxburgh to Sir Joseph Banks, 25 April 1795, British Library, Add MSS 33980, ff9-10.
58. From the *Calcutta Gazette*, 11 October 1792, British Library, OIOC IOR, EUR/D809.
59. Report from the Court of Directors 30 May 1792 relative to the favourable state of the Market for Bengal Indigo. Additional supplement to the *Calcutta Gazette*, Thursday 15 December 1792, British Library, OIOC IOR, EUR/D809.

60. Letter from William Roxburgh to Sir Joseph Banks, 11 November 1782, British Library, Add MSS 33977, ff181-2.
61. Letter from William Roxburgh to Dr C. Taylor, 16 June 1804, *Transactions of the Society for the Encouragement of Arts, Manufactures, and Commerce*, 23 (1805), 407-14.

10. Hemp and Fibrous Plants

1. Mentioned in Grove, *Green Imperialism*, p. 331n (Banks Letters).
2. Letter from William Roxburgh to Robert Wissett, 27 February 1801, published in *Transactions of the Society for the Encouragement of Arts, Manufactures, and Commerce*, 22 (1804), 363-8.
3. William Roxburgh, 'Observations on the Culture, Properties, and comparative Strength of Hemp, and other vegetable fibres, the Growth of the East Indies', *Transactions of the Society for the Encouragement of Arts, Manufactures, and Commerce*, 24 (1806), 143-53.
4. Letter from Dr James Edward Smith to William Roscoe, 5 July 1814, Linnean Society, *Smith Letters*, vol. 17, f185.
5. Letter from William Roxburgh to Robert Wissett, 27 February 1801, published in *Transactions of the Royal Society for the Encouragement of Arts, Manufactures, and Commerce*, 22 (1804), 363-8.
6. Letter from William Roxburgh to Robert Wissett, 24 December 1799, published in *Transactions of the Royal Society for the Encouragement of Arts, Manufactures, and Commerce*, 22 (1804), 363-8.
7. Hudson and Luckhurst, *The Royal Society of Arts, 1754-1954* (London, 1954), pp. 90-1 and 158.
8. Letter from William Roxburgh to Sir Joseph Banks, 13 July 1797. British Library, Add MSS 33980 ff101-3.
9. Public Letter from Court of Directors in London to the Governor General at Fort William, 27 July 1797, in Gupta, *Fort William – India House Correspondence*, 1796-1800.
10. Letter from William Roxburgh to Sir Joseph Banks, 24 April 1798. British Library, Add MSS 33980 ff137-41.
11. William Roxburgh, 'Observations on the Culture, Properties, and comparative Strength of Hemp, Sun, Jute, and other Vegetable fibres, the Growth of India', *Transactions of the Royal Society for the Encouragement of Arts, Manufactures, and Commerce* (1804), 372-96.
12. Ibid. p. 377.
13. William Roxburgh, 'Observations on the Culture, Properties, and comparative Strength of Hemp, and other vegetable fibres, the Growth of the East Indies', *Transactions of the Royal Society for the Encouragement of Arts, Manufactures, and Commerce* (1806), 143-53.
14. Roxburgh, *Flora Indica*, vol. 2, p. 581.
15. Ibid, vol. 3, pp.259-63.
16. Letter from William Roxburgh to Sir Joseph Banks, 17 December 1799. British Library, Add MSS 33980 ff210-12.
17. Letter from William Roxburgh to Robert Wissett, 24 December 1799, published in *Transactions of the Royal Society for the Encouragement of Arts, Manufactures, and Commerce*, 22 (1804), 363-4.
18. 'In the Council Chamber the 19 December 1800'. Sutro Library, Banks Papers HP 1:47.
19. Letter from William Roxburgh to Aylmer Lambert, 1 March 1801. British Library, Add MSS 28545 f175.
20. Letter from William Roxburgh to Sir Joseph Banks, 27 September 1801. British Library, Add MSS 33980 ff318-20.
21. 'Contract For the Culture of Hemp in Bengal'. Mitchell Library, Sir Joseph Banks Electronic Archives, 20: 83.
22. Letter from William Roxburgh to Sir Joseph Banks, 26 November 1802. British Library, Add MSS 33981 ff67-70.
23. Letter from Charles Favargill, Simon Benstead, Joseph Seymour, William Hogarth and Mark Everson to Sir Joseph Banks, 8 June 1803. Mitchell Library, Banks Archives, 70: 30.
24. Letter from William Moorcroft to William Roxburgh, 29 November 1812. Sutro Library, Banks Papers, BO 1:69.
25. Letters from Joseph Cotton to Dr Charles Taylor et al., published in *Transactions of the Society for the Encouragement of Arts, Commerce, and Manufactures*, 33 (1815), 182-95.

26. Hudson and Luckhurst, *The Royal Society of Arts*, p. 158. Its modern name is *Boehmeria nivea* var. *tenacissima*, showing again, the problems of nomenclature, which in this case has assigned Roxburgh's species to a variety.
27. Desmond, *The European Discovery of the Indian Flora*, p. 205.

11. Pepper and Spices

1. Desmond, *The European Discovery of the Indian Flora*, p. 58.
2. Om Prakesh, *European commercial enterprise in pre-colonial India* (Cambridge, 1998), Table 4.1.
3. Letters from William Roxburgh to Sir Joseph Banks, 8 July and 30 August 1791. Sutro Library, EI 1:44 and 1:47.
4. Letter from William Roxburgh to Colin Shakespear, 20 September 1795, in Public Consultations, 28 September 1795, paragraph 4. British Library, OIOC IOR, P/4/36.
5. Letter from William Roxburgh to C. R. Crommelin, 20 February 1802, in Public Consultations, 4 March 1802, paragraph 49. British Library, OIOC IOR, P/5/332.
6. Letter from William Roxburgh to Sir Joseph Banks, 14 January 1794. British Library, Add MSS 33979 f240.
7. Letter from William Roxburgh to Sir Joseph Banks, 25 April 1795. British Library, Add MSS 33780 ff9-10.
8. Letter from William Roxburgh to C. R. Crommelin, 20 February 1802, in Public Consultations, 4 March 1802, paragraph 49. British Library, OIOC IOR, P/5/32.
9. Letter from William Roxburgh to Thomas Brown, 6 June 1804, in Public Consultations, 5 July 1804, paragraph 42. British Library, OIOC IOR, P/5/55, ff3556-88.
10. Letter from William Roxburgh to Sir Joseph Banks, 18 December 1784. British Library, Add MSS 33977 ff272-5.
11. Extracts of letter from William Roxburgh to Andrew Ross 25 April 1786. British Library, IOR H/209, ff199-202 covers five letters on the subject, from April to June 1786.
12. Public Letter from the Court of Directors to the President and Council at Fort St George, 31 July 1787. British Library, OIOC IOR, E/4/873 ff183-4.
13. Letter from Andrew Ross to William Roxburgh, 2 June 1788. BM(NH), Rox. Corr.
14. Ibid., 19 December 1788.
15. Madras Despatches, 20 April 1789. British Library, OIOC IOR, E/4/875 ff729-80.
16. This forms part of a letter from William Roxburgh to Sir Joseph Banks, 1 November 1789. Sutro Library, Banks Papers, EI 1:49.
17. Letter from Andrew Ross to William Roxburgh, 11 December 1789, Natural History Museum, Botany Library, Roxburgh Correspondence. Ross and Roxburgh found similar problems with this Council when they were trying to rent a Zemindary in 1793.
18. Letter from Andrew Scott to William Roxburgh, 16 December 1789. Sutro Library, Banks Papers, EI 1:51.
19. Letter from William Roxburgh to Patrick Russell, 20 January 1790. Sutro Library, Banks Papers, EI 1:52.
20. Letter from William Roxburgh to Sir Joseph Banks, 30 December 1790. British Library, Add MSS 33979 ff64-65.
21. Letter from William Roxburgh to Patrick Russell, 20 January 1790. Sutro Library, Banks Papers, EI 1:54; and Roxburgh, *Flora Indica*, vol. 1, pp. 151-4.
22. Letter from William Roxburgh to Sir Joseph Banks, 28 April 1791. Sutro Library, Banks Papers, EI 1:46.
23. Letter from William Roxburgh to Sir Joseph Banks, 25 February 1792, British Library, OIOC IOR, K/148 ff249-52.
24. Letter from William Roxburgh to Sir Joseph Banks, 17 August 1792. British Library, Add MSS 33979 ff171-3.
25. Letter from Andrew Ross to William Roxburgh, 17 August 1793, Natural History Museum, Botany Library, Roxburgh Correspondence.
26. Quoted in Ray Desmond, *The European Discovery of the Indian Flora*, Oxford, 1992, p. 203.
27. Dalrymple, *Oriental Repertory*, vol. 1, pp. 1-38, 451-66.
28. Sheila Lambert (ed), *House of Commons Sessional Papers of the Eighteenth Century. George III. Volume 74, 1789-1790* (covers 1786-87 to 89-90). *Volume 84, 1791. Volume 85, 1792. Volume*

91, 1793-1796. Volume 106, 1796-1797. Volume 126, 1799-1800 (covers 1797-98 to 1799-1800) (Wilmington, Delaware, 1975).

29. Letters from Andrew Ross to William Roxburgh, 26 February and 16 September 1793. Natural History Museum, Botany Library, Roxburgh Correspondence.

30. Letter from William Roxburgh to Sir Joseph Banks, 28 December 1794. British Library, Additional MSS 33979, ff290-1.

31. Letter from Sir Joseph Banks to William Roxburgh, 29 May 1796. British Library, Additional MSS 33980, ff65-6.

32. Letter from William Roxburgh to D. Seton, 24 April 1797, in Public Consultations 3 May 1797, paragraph 76, British Library, OIOC IOR P/4/50 ff 222-56.

33. Letter from William Roxburgh to Sir John Shore, 15 May 1797, in Public Consultations 22 May 1797, paragraph 16. British Library, OIOC IOR P/4/50.

34. Public Consultations, 22 May 1797, paragraphs 18-20. British Library, OIOC IOR, P/4/50.

35. Taken from Botany Deputation of Mr William Roxburgh. British Library, OIOC IOR, F/4/142, ff5-24.

36. Letter from William Roxburgh to Duncan Campbell, 25 July 1797, in Public Consultations, 11 August 1797, paragraph 12. British Library, OIOC IOR, P/4/52, ff92-5.

37. Letter from William Roxburgh to Duncan Campbell, 23 November 1797, in Public Consultations, 27 November 1797, paragraph 48. British Library, OIOC IOR, P/4/55, ff133-7.

38. Letter from William Roxburgh to Duncan Campbell, 18 December 1797, in Public Consultations, 16 January 1798, paragraph 25. British Library, OIOC IOR, P/4/57, ff328-9.

39. Letter from William Roxburgh to John Stacey, 18 January 1798, in Public Consultations, 2 February 1798, paragraphs 9-10. British Library, OIOC IOR, P/4/58, ff13-28.

40. Letter from Dr J. Fleming to Duncan Campbell, 14 July 1798, in Public Consultations, 16 July 1798, paragraphs 22-3. British Library, OIOC IOR, P/4/61, ff235-9.

41. Letter from Dr J. Fleming to Duncan Campbell, 11 October 1798, in Public Consultations, 15 October 1798, paragraph 18. British Library, OIOC IOR, P/4/62, ff751-61.

42. Letter from Francis Buchanan to Henry Darell, 1 July 1798, in Public Consultations, 9 July 1798, paragraph 28. British Library, OIOC IOR, P/5/5, ff152-3.

43. Letter from William Roxburgh to Henry Darell, 1 December 1799, in Public Consultations, 5 December 1799, paragraph 11. British Library, OIOC IOR, P/5/8, ff123-37.

44. Letter from William Roxburgh to Thomas Philpot, 29 July 1802, State of the Spice Plantations on Prince of Wales Island. British Library, OIOC IOR, F/4/142, ff25-8.

45. Letter from William Roxburgh to C. R. Commelin, 20 February 1802, Cultivation of Spices – State of the Plants &ca brought from Amboyna. British Library, OIOC IOR, F/4/134, ff133-44.

46. Public Consultations, 31 March 1803, British Library, OIOC IOR, P/5/43, paragraphs 48-9.

47. Letter from William Roxburgh to John Lumsden, 3 March 1804, in Public Consultations, 8 March 1804, paragraph 40. British Library, OIOC IOR, P/5/53, ff 1394-402.

48. Letter from William Roxburgh to Thomas Brown, 6 June 1804, in Public Consultations, 5 July 1804, paragraph 42. British Library, OIOC IOR, P/5/55, ff3556-88.

49. Sale of the Company's Spice Plantations at Prince of Wales Island. British Library, OIOC IOR, O/6/15, f285.

50. Letter from William Roxburgh junior to Thomas Brown, 5 June 1806, in Public Consultations, 12 June 1806, paragraph 52. British Library, OIOC IOR, P/6/27, ff5611-13.

12. Sugar

1. From the Diary or Woodfall's Register, 31 January 1792, British Library, OIOC IOR, EUR/D809.

2. From the *Gentleman's Magazine* for March 1792, British Library, OIOC IOR, EUR/D809.

3. Adam Smith, *An Inquiry into the Nature and Causes of th Welath of Nations* (1776, London, 1884 ed), Book I, chapter X, p. 50.

4. Commercial Letter from Fort St George to the Court of Directors, 15 January 1790, paragraph 51, British Library, OIOC IOR, E/4/319.

5. Letter from Andrew Ross to William Roxburgh, 20 May 1791, Natural History Museum, Botany Library, Roxburgh Correspondence.

6. William Roxburgh, 'An Account of the Hindoo method of Cultivating the Sugar Cane & Manufacturing the Sugar & Jagary in the Rajahmundry Circar ...', 20 June 1792, Sutro Library, SU 1:7; in Dalrymple, *Oriental Repertory*, vol. 2, pp. 497-514; and reprinted in Tilloch's *Philosophical Magazine*, 21 (1805), 264-75; confirmation of despatch to the Court of Directors, in Revenue Department at Fort St George, 2 May 1793, paragraphs 40-3, British Library, OIOC IOR E/4/323. More information also appeared in *Oriental Repertory*, vol. 2, pp. 54-5.

7. Review of *Voyage à Madagascar* by Abbé Rochon, British Library, OIOC IOR, EUR/D809.

8. British Library, OIOC IOR, EUR/D809.

9. Letter from Andrew Ross to William Roxburgh, 17 February 1793, Natural History Museum, Botany Library, Roxburgh Correspondence.

10. Ibid., 7 August 1792.

11. Ibid., 27 August 1792.

12. Ibid., 20 October 1792.

13. Ibid., 14 March 1793.

14. Ibid., 10 March 1794.

15. Ibid., 29 September 1793.

16. Cultivation of Sugar and Indigo, British Library, OIOC IOR, O/210(4), ff141-84.

17. Letter from the Court of Directors, 3 July 1795, paragraphs 17-19, British Library, OIOC IOR, E/4/881, ff674-91.

18. Commercial Department Letter at Fort St George, 8 June 1796, paragraphs 107-9, British Library, OIOC IOR, E/4/882, ff562-8.

19. Commercial Department Letter at Fort St George, 14 February 1795, paragraphs 84-5, British Library, OIOC IOR, E/4/324.

20. Letter from William Roxburgh to Colin Shakespear, 15 March 1796, in Public Consultations at Fort William, 21 March 1796, paragraph 21, British Library, OIOC IOR, P/4/40.

21. Letter from William Roxburgh to Sir John Shore, 22 November 1796, in the Public Consultations at Fort William, 2 December 1796, paragraph 14, British Library, OIOC IOR, P/4/45, ff371-5.

22. Letter from William Roxburgh to H. Macleod, 22 December 1796, in the Public Consultations at Fort William, 26 December 1796, paragraph 53, British Library, OIOC IOR, P/4/46, ff670-3.

23. Letter from William Roxburgh to D. Seton, 7 April 1797, in the Public Consultations at Fort William, 3 May 1797, paragraph 85, British Library, OIOC IOR, P/4/50, ff251-3.

24. Letter from William Roxburgh to Duncan Campbell, 23 November 1797, in the Public Consultations at Fort William, 27 November 1797, paragraph 48, British Library, OIOC IOR, P/4/55, ff133-7.

25. Letter from William Roxburgh to H. V. Darell, 1 December 1799, in the Public Consultations at Fort William, 5 December 1799, paragraph 11, British Library, OIOC IOR, P/5/8, ff123-37.

26. Letter from William Roxburgh to C. R. Crommelin, 18 February 1802, in the Public Consultations at Fort William, 4 March 1802, paragraph 50, British Library, OIOC IOR, P/5/32. A full description of it and its cultivation appeared in his *Flora Indica*, vol. 1, pp. 239-43, and of *S. officinarum* in vol. 1, pp. 237-9.

27. Letter from William Roxburgh to Thomas Brown, 6 June 1803, in the Public Consultations at Fort William, 5 July 1804, paragraph 42, British Library, OIOC IOR, P/5/55, ff3556-88.

13. Watering the Circars

1. Love, *Vestiges of Old Madras*, vol. 1, p. 29.

2. Philip Mason, *The Men who Ruled India* (London, 1985), p. 220.

3. Love, *Vestiges of Old Madras*, vol. 3, p. 229.
4. Ibid., p. 236.
5. Question 9 in letter from William Roxburgh to Andrew Ross, 14 February 1793, in Dalrymple, *Oriental Repertory*, vol. 2, p. 73.
6. Public Letter from the Court of Directors to Fort St George, 19 March 1793, paragraph 4, in reply to Public Letter from Fort St George, 22 October 1791, in British Library, OIOC IOR, E/4/879, ff124-46.
7. Dalrymple, *Oriental Repertory*, vol. 1, pp. 91-2.
8. Ibid, p. 34.
9. Letter from Andrew Ross to David Haliburton, not dated but by its position in the archives probably 20 June 1793, Natural History Museum, Botany Library, Roxburgh Correspondence.
10. Love, *Vestiges of Old Madras*, p. 409.
11. Public Letter from Fort St George to the Court of Directors, 16 January 1792, paragraph 56, British Library, OIOC IOR, E/4/321.
12. Letter from William Roxburgh to Sir Joseph Banks, 30 August 1791, British Library, OIOC IOR, EUR/K148, ff243-7.
13. Thomas Newte, *Prospects and Observations; on a Tour in England and Scotland: Natural, Œconomical, and Literary* (London, 1791), p. 101.
14. Thomas Newte, *A Tour in England and Scotland in 1785* (London, 1788), pp. 269-70.
15. Letter from Andrew Ross to David Haliburton, not dated but by its position in the archives probably 20 June 1793, Natural History Museum, Botany Library, Roxburgh Correspondence.
16. Letter from Andrew Ross to David Haliburton, 15 July 1793, Natural History Museum, Botany Library, Roxburgh Correspondence.
17. Letter from William Roxburgh to Sir Joseph Banks, 30 August 1791, Sutro Library, EI 1:47; and British Library, OIOC IOR, MSS/EUR/K148, ff243-7.
18. Letter from William Roxburgh to Sir Joseph Banks, 16 December 1791, British Library, Add MSS 33979 ff116-17.
19. Letter from William Roxburgh to Henry Darell, 26 November 1799, in Cultivation of the Teak Tree, ff4-11, British Library, OIOC IOR, F/4/99 2028.
20. Letter from William Roxburgh, 23 January 1793, in Cultivation of the Teak Tree, ff27-55, British Library, OIOC IOR, F/4/99 2028.
21. This tree will soon be common in Bengal as it thrives in the Botanic Garden most luxuriant, and begins to bear seed, which, with additional supplies from the Eastern Island, will soon produce plants in abundance; of the real Sago, there are only a very few plants in the Botanic Garden.
22. Letter from William Roxburgh, 23 January 1793, in Cultivation of the Teak Tree, ff27-55, British Library, OIOC IOR, F/4/99 2028.
23. Letter from the Rev. C. John to William Roxburgh, 2 April 1793, British Library, OIOC IOR, MSS/EUR/D809.
24. Public Department Letter from Fort St George, 2 May 1793, British Library, OIOC IOR, E/4/323.
25. Public Letter from the Court of Directors to Fort William, in reply to a letter from Fort William 29 January 1793, 15 April 1795, paragraph 123, British Library, OIOC IOR, E/4/642.
26. Letter from William Roxburh to H. V. Darell, 20 November 1799, in Public Consultations at Fort William, 5 December 1799, paragraph 8, British Library, OIOC IOR, P/5/8, ff81-6.
27. Letter from William Roxburgh to C. R. Crommelin, 20 February 1802, in Public Consultation at Fort William, 4 March 1802, paragraph 49, British Library, OIOC IOR, P/5/32.
28. Letter from William Roxburgh to Thomas Philpot, 22 March 1803, in Public Consultation at Fort William, 31 March 1803, paragraph 46, British Library, OIOC IOR, P/5/43.
29. Letter from William Roxburgh to Thomas Philpot, 9 April 1803, in Public Consultation at Fort William, 14 April 1803, paragraph 36, British Library, OIOC IOR, P/5/44.
30. Letter from William Roxburgh junior to Thomas Brown, 5 September 1805, in Public Consultation at Fort William, 5 September 1805, paragraph 74, British Library, OIOC IOR, P/6/17.
31. Letter from William Roxburgh junior to Thomas Brown, 5 June 1806, in Public Consultation at Fort William, 12 June 1806, paragraph 52, British Library, OIOC IOR, P/6/27, ff5611-13.
32. Public Consultations at Fort William, 18 August 1803, paragraph 11, British Library, OIOC IOR, P/5/46.
33. Letter from William Roxburgh to Thomas Philpot, 5 September 1803, in Public Consultation at Fort William, 22 September 1803, paragraph 13, British Library, OIOC IOR, P/5/47.
34. Letter from William Roxburgh to Thomas Brown, 6 June 1804, in Public Consultation at Fort William, 5 July 1804, paragraph 42, British Library, OIOC IOR, P/5/55, ff3556-88.
35. British Library, OIOC IOR, EUR/D809; Newte, *Prospects and Observations*.
36. Newte, *Prospects and Observations*, p. 432.
37. Ibid, pp. 254-5.
38. British Library, OIOC IOR, EUR/D809.
39. Public Letter from the Court of Directors to Fort St George, 16 May 1792, paragraph 10, British Library, OIOC IOR, E/4/878.
40. Letter from William Roxburgh to Col. Kyd, 17 October 1792, in Dalrymple, *Oriental Repertory*, vol. 1, p. 36.
41. Letter from William Roxburgh to Andrew Ross, 20 January 1793, in Dalrymple, *Oriental Repertory*, vol. 1, p. 45.
42. Letter from Andrew Ross to Colonel Alexander Ross, 5 December 1792, Natural History Museum, Botany Library, Roxburgh Correspondence.
43. Dalrymple, *Oriental Repertory*, vol. 1, p. 50.
44. Letter from William Roxburgh to Andrew Ross, 21 January 1793, in Dalrymple, *Oriental Repertory*, vol. 1, pp. 51-2.
45. Ibid., 22 March 1793, pp. 77-8.
46. Dalrymple, *Oriental Repertory*, vol. 1, p. 65.
47. Letter from William Roxburgh to Andrew Ross, 31 January 1793, in Dalrymple, *Oriental Repertory*, vol. 1, p. 53.
48. Ibid., 13 April, p. 78.
49. Letter from Andrew Ross to William Roxburgh, 5 January 1793, Natural History Museum, Botany Library, Roxburgh Correspondence.
50. Letter from Major A. Beatson to Andrew Ross, 12 January 1793, Natural History Museum, Botany Library, Roxburgh Correspondence.
51. Letter from Andrew Ross to William Roxburgh, 15 January 1793, Natural History Museum, Botany Library, Roxburgh Correspondence.
52. Ibid., 14 March 1793, referring to a letter from Roxburgh of 6 January that is missing.
53. Ibid., 10 February 1793.
54. Ibid., 17 February 1793.
55. Ibid., 18 February 1793.
56. Ibid., 21 March 1793.
57. Ibid., 24 March 1793.
58. Ibid., 5 May and copied in letter of 8 May 1793.
59. Ibid.
60. Sir John William Kaye, *The Administration of the East India Company; a History of Indian Progress* (London, 1853), p. 305.

14. Roxburgh at the Cape of Good Hope

1. Harlow, *The Founding of the Second British Empire*, vol. 1, pp. 107-8.
2. 1792, Masson & c., Mitchell Library, Banks Archives, 810538.
3. Campbell-Culver, *The Origin of Plants*, p. 189.
4. John P. Rourke, *The Proteas of Southern Africa* (Cape Town, Johannesburg and London, 1980), p. 10.
5. Letter from William Roxburgh to Sir Joseph Banks, 10 December 1796, Sutro Library Archives, BG 1:83.
6. Toby Musgrave, Chris Gardner and Will Musgrave, *The Plant Hunters. Two Hundred Years of Adventure and Discovery around the World* (London, 1998), p. 50.

7. Letter from Lt Owen to Sir John Shore, 6 February 1797, in the Public Consultations at Fort William, 3 May 1797, paragraph 87, British Library, OIOC IOR, P/4/50, ff255-6.

8. Letter from William Roxburgh to Sir Joseph Banks, 25 April 1795, British Library, Add MSS 33980, ff9-10.

9. Letter from William Roxburgh to Sir Joseph Banks, 19 December 1795, British Library, Add MSS 33980, ff41-2.

10. Letter from William Roxburgh to Colin Shakespear, 2 February 1796, in the Public Consultations at Fort William, 15 February 1796, paragraph 29, British Library, OIOC IOR, P/4/39, ff745-8.

11. Letter from William Roxburgh to Hugh Macleod, 5 December 1796, in the Public Consultations at Fort William, 16 December 1796, paragraph 31, British Library, OIOC IOR, P/4/46, ff45-54.

12. Letter from William Roxburgh to Dr James Edward Smith, 16 December 1796, Linnean Society, *Smith Letters*, vol. 8, f196.

13. Letter from William Roxburgh to Hugh Macleod, 11 January 1797, in the Public Consultations at Fort William, 16 January 1797, paragraph 42, British Library, OIOC IOR, P/4/48, f249.

14. Letter from William Roxburgh to Duncan Campbell, 23 November 1797, in the Public Consultations at Fort William, 27 November 1797, paragraph 48, British Library, OIOC IOR, P/4/55, ff133-7.

15. Letter from William Roxburgh to Duncan Campbell, 8 January 1798, in the Public Consultations at Fort William, 16 January 1798, paragraph 26, British Library, OIOC IOR, P/4/57, ff329-32.

16. Letter from William Roxburgh to Dr James Edward Smith, 23 April 1798, Linnean Society, *Smith Letters*, vol. 25, f53.

17. Letter from William Roxburgh to Sir Joseph Banks, 24 April 1798, British Library, Add MSS 33980, ff137-41.

18. Letter from Sir Joseph Banks to William Roxburgh, 7 January 1799, British Library Add MSS 33980 f170.

19. Letter from William Roxburgh to Sir Joseph Banks, 30 March 1799, British Library, Add MSS 33980, f184.

20. Letter from William Roxburgh to Dr James Edward Smith, 7 June 1799, Edinburgh University Library, Special Collections, La. II.643/52.1.

21. Letter from William Roxburgh to Sir Joseph Banks, 1 July 1799, British Library, Add MSS 33980, f189-90.

22. Letter from William Roxburgh to H. V. Darell, 1 December 1799, in the Public Consultations, 5 December 1799, paragraph 11, British Library, OIOC IOR, P/5/8, ff123-37.

23. Letter from William Roxburgh to G. H. Barlow, 16 October 1799, in the Public Consultations at Fort William, 5 November 1799, paragraph 12, British Library, OIOC IOR, P/5/7, ff541-2.

24. Letter from William Roxburgh to C. R. Crommelin, 20 August 1800, in the Public Consultations at Fort William, 21 August 1800, paragraph 26, British Library, OIOC IOR, P/5/14.

25. E. Charles Nelson and John P. Rourke, 'James Niven (1776-1827), a Scottish botanical collector at the Cape of Good Hope: his *Hortus Siccus* at the National Botanic Gardens, Glasnevin, Dublin (DBN), and the Royal Botanic Gardens, Kew (K)', *Kew Bulletin*, 48 (1993), 663-82.

26. Instructions from William Roxburgh to John Roxburgh

at the Cape of Good Hope, 1799, British Library, OIOC IOR, E/1/140 f301.

27. Richard Anthony Salisbury, 'Species of Erica', *Transactions of the Linnean Society*, 6 (1802), 316-88; and Robert Brown, 'On the Proteaceae of Jussieu', *Transactions of the Linnean Society*, 10 (1811), 15-226.

28. Personal communications from Dr E. G. H. Oliver, Compton Herbarium, National Botanical Institute, Kirstenbosch. I am most grateful to Dr Oliver for the information that he has supplied for this section.

29. Letter from William Roxburgh to Aylmer Bourke Lambert, 12 February 1812, Linnean Society, Richard Pulteney Correspondence.

30. Anon, 'William Roxburgh, M.D.', *Asiatic Journal and Monthly Register for British India and its Dependencies*, 1 (January-July 1816), 28-30.

31. Valentia, *Voyages and Travels to India*, entry for 23 October 1802.

15. The Founding Father of Indian Botany

1. Anon, *Asiatic Journal and Monthly Register*, 1 (1816), 28-30.

2. Anon, *The Gentleman's Magazine*, 85 (1815), 476.

3. Thomson, 'Notes on the Herbarium of the Calcutta Botanic Garden', 10-14 and 33-41; Huxley, *Life and Letters of Sir Joseph Dalton Hooker*, vol. 2, p. 283; and Cronk, *The Endemic Flora of St Helena*, p. 37.

4. Roscoe, 'Remarks on Dr Roxburgh's Description of the Monandrous Plants of India', pp.270-82.

5. Nayar and Das, 'Glimpse of William Roxburgh through his unpublished letters', pp.1159-67.

6. One of the first times it appeared was in the Memorial of John Roxburgh to the Court of Directors, 31 March 1819, British Library, OIOC IOR, E/1/140 f311, and has been repeated frequently since, for example in Anon, *Icones Roxburghianae or Drawings of Indian Plants* (Calcutta, 1964-8), p. 1.

7. Latita Kakkar, 'Indian Botanic Garden, Calcutta', *Botanica*, 21 (1971), 95-7.

8. Valentia, *Voyages and Travels in India*, entry for 12 February 1803.

9. *Transactions of the Society for the Encouragement of Arts, Manufactures, and Commerce*, 23 (1805), 407; and 32 (1814), 207.

10. Anon, *The Annual Biography and Obituary of 1816*, pp. 13-14.

11. Archer, *Natural History Drawings in the India Office Library*, p. 23.

12. British Library, OIOC IOR, Board's Collection, F/4/588, contains a number of letters regarding these drawings and their ownership.

13. Grove, *Green Imperialism*, passim, but particularly 399-408; and William Roxburgh, 'Remarks on the landwinds and their causes'.

14. William Roxburgh, 'Remarks on the Land Winds and their Causes'.

15. Bearce, *British Attitudes towards India*, p. 86.

16. Letter from Robert Sealy to G. Pilleri, 17 November 1975, quoted in Pilleri, 'William Roxburgh and Heirich Lebeck', p. 17.

17. Letter from William Roxburgh to Robert Wissett, 24 December 1799, published in *Transactions of the Society for the Encouragement of Arts, Manufactures, and Commerce*, 24 (1806), 143-5.

The following list includes those sources that have been used directly and indirectly in the research. Many have been important in giving background information and a number of others have been used to locate further sources. Roxburgh's own publications have been excluded as they appear in Appendix 1.

1.PRIMARY SOURCES

Works published before 1840, reprints of original works and published manuscript sources.

Manuscript

A full list of all the manuscript sources can be gleaned from the notes. This applies particularly to the sources from the India Office Records and those from the Sutro Library.

Botanical Survey of India, Indian Botanical Garden Calcutta. William Roxburgh drawings.
British Library, Add MSS 28545, 33977, 33978, 33979, 33980, 33981.
British Library, Oriental and Indian Office Collections, India Office Records.
British Library, Oriental and Indian Office Collections. P18, 'A View of the Botanic Garden House and Reach' aquatint by Robert Havell after a watercolour by J. B. Fraser. Plate from *Four Views of Calcutta* (Calcutta, 1824-28).
British Library, Oriental and Indian Office Collections. P1551, 'Madras Landing' aquatint by C. Hunt, May 1856, after a watercolour by J. B. East, of about 1837.
British Museum (Natural History), Botany Library. Buchanan Manuscripts.
British Museum (Natural History), Botany Library. Roxburgh Manuscripts.
Edinburgh Public Library, George IV Bridge, Scottish Room, Parish Registers.
Edinburgh Public Library, George IV Bridge, Scottish Room, IGI Index.
Edinburgh University Library, Special Collections. Register of Matriculations, 1762-1785.
Edinburgh University Library, Special Collections. John Hope, Class Lists.
Linnean Society. *Smith Letters*, volumes 6, and 8.
Mitchell Library, Sydney, New South Wales. Sir Joseph Banks Electronic Archive, 20:83.
National Library of Scotland, MSS Acc 1023, and 7664.
Royal Botanic Garden Edinburgh Library. John Hope's 'Lectures in Botany', 1777-78, 1780 and 1783.
Royal Botanic Garden Edinburgh Library. Roxburgh drawings.
Royal Botanic Garden Edinburgh Library. William Wright's 'Catalogue of the Roxburgh Herbarium' MSS, 1833.
Royal Botanic Gardens Kew Library. Roxburgh drawings, Inwards Book 1805-09, Record Book 1793-1809, and Record Book 1804-26.
Scottish Record Office. Buchanan MSS, GD 161/19/4.
Sutro Library, California. Banks papers.

Printed articles

Anon, *The Lounger*, 1 (1785), 147-51, and 2 (1785), 1-5, 71-80.
Anon, 'Appendix to the History of the Society', *Transactions of the Royal Society of Edinburgh*, 2 (1790), 79.
Anon, 'Appendix to the History of the Society', *Transactions of the Royal Society of Edinburgh*, 3 (1791), 24.
Anon, 'Donations to the Museum of the Linnean Society', *Transactions of the Linnean Society*, 10 (1811), 414.
Anon, *The Edinburgh Advertiser*, 103 (1815), 135.
Anon, *The Edinburgh Evening Courant*, No. 16183 (1815).
Anon, *The Gentleman's Magazine*, 85 (1815), 476.

Anon, 'British Indian Biography, No. 1. William Roxburgh, M.D.', *Asiatic Journal and Monthly Register for British India and its Dependencies*, 1 (1816), 28-30.

Anon, 'Appendix to the History of the Society', *Transactions of the Royal Society of Edinburgh*, 8 (1817).

Asiatick Researches, or, Transactions of the Society, instituted in Bengal, for the inquiring into the History and Antiquities, the Arts, Sciences, and Literature of Asia, 1 (1788).

Asiatick Researches, or, Transactions of the Society, instituted in Bengal, for the inquiring into the History and Antiquities, the Arts, Sciences, and Literature of Asia, 2 (1790).

Asiatick Researches, or, Transactions of the Society, instituted in Bengal, for the inquiring into the History and Antiquities, the Arts, Sciences, and Literature of Asia, 3 (1792).

Asiatick Researches, or, Transactions of the Society, instituted in Bengal, for the inquiring into the History and Antiquities, the Arts, Sciences, and Literature of Asia, 4 (1795).

Asiatick Researches, or, Transactions of the Society, instituted in Bengal, for the inquiring into the History and Antiquities, the Arts, Sciences, and Literature of Asia, 5 (1797).

Asiatick Researches, or, Transactions of the Society, instituted in Bengal, for the inquiring into the History and Antiquities, the Arts, Sciences, and Literature of Asia, 6 (1799).

Asiatick Researches, or, Transactions of the Society, instituted in Bengal, for the inquiring into the History and Antiquities, the Arts, Sciences, and Literature of Asia, 7 (1801).

Asiatick Researches, or, Transactions of the Society, instituted in Bengal, for the inquiring into the History and Antiquities, the Arts, Sciences, and Literature of Asia, 8 (1805).

Asiatick Researches, or, Transactions of the Society, instituted in Bengal, for the inquiring into the History and Antiquities, the Arts, Sciences, and Literature of Asia, 9 (1807).

Asiatick Researches, or, Transactions of the Society, instituted in Bengal, for the inquiring into the History and Antiquities, the Arts, Sciences, and Literature of Asia, 10 (1808).

Asiatick Researches, or, Transactions of the Society, instituted in Bengal, for the inquiring into the History and Antiquities, the Arts, Sciences, and Literature of Asia, 11 (1810).

Asiatick Researches, or, Transactions of the Society, instituted in Bengal, for the inquiring into the History and Antiquities, the Arts, Sciences, and Literature of Asia, 12 (1816).

Bently, John, 'Remarks on the Principal Æras and Dates of the Ancient Hindus', *Asiatick Researches*, 5 (1799), 315-43.

Brown, Robert, 'On the Proteaceae of Jussieu', *Transactions of the Linnean Society*, 10 (1811), 15-226.

Colebrooke, H. T., 'Description of a Species of Ox, named Gayal', *Asiatick Researches*, 8 (1805), 511-24.

Cotton, Joseph, 'Letter to Robert Wissett, et al.', *Transactions of the Society for the Encouragement of Arts, Manufactures, and Commerce*, 33 (1815), 182-95.

Curtis, William, *Botanical Magazine: or, Flower-Garden Displayed* (London, various).

Duncan, Andrew (the Younger), 'On Medical Education', *The Edinburgh Medical and Surgical Journal*, No. 91 (1827), bound in Edinburgh University Library, Special Collections, *Pamphlets*, vol. 69, No. 6, pp. 1-29.

Eyre, J., 'Description of the various Classes of Vessels constructed and employed by the Natives of the Coasts of Coromandel, Malabar, and the Island of Ceylon, for their coasting Navigation', *Journal of the Royal Asiatic Society of Great Britain and Ireland*, 1 (1834), 1-15.

Jones, Sir William, 'Introduction', *Asiatick Researches*, 1 (1788), xiii.

Jones, Sir William, 'On the Gods of Greece, Italy, and India, written in 1784 and since revised', *Asiatick Researches*, 1 (1788), 221-75.

Jones, Sir William, 'The Second Anniversary Discourse', *Asiatick Researches*, 1 (1788), 407-8.

Jones, Sir William, 'The Design of a Treatise on the Plants of India', *Asiatick Researches*, 2 (1790), 345-52.

Jones, Sir William, 'Discourse the Ninth. On the Origin and Families of Nations', *Asiatick Researches*, 3 (1792), 479-92.

Nicholson, William, *A Journal of Natural Philosophy, Chemistry, and the Arts*, various.

Roscoe, William, 'Remarks on Dr. Roxburgh's Description of the Monandrous Plants of India; in a Letter to the President', *Transactions of the Linnean Society*, 11 (1815), 270-82.

Royle, J. Forbes, 'Report on the Progress of the Culture of the China Tea Plant in the Himalayas, from 1935 to 1847', *Journal of the Royal Asiatic Society of Great Britain and Ireland*, 12 (1850), 125-52.

Salisbury, Richard Anthony, 'Species of Erica', *Transactions of the Linnean Society*, 6 (1802), 316-88.

[Taylor, Dr C.], 'William Roxburgh', *Transactions of the Society for the Encouragement of Arts, Manufactures, and Commerce*, 33 (1815), 156-60.

Printed books

Andrews, Henry, *The Botanists Repository Comprising, Colour'd Engravings of New and Rare Plants Only With Botanical Descriptions in Latin and English after the Linnean System*, 10 vols. (London, 1797-1811).

Anon, *The East India Kalendar, or, Asiatic Register ... for the year 1797* (London, 1797).

Anon, *The East India Kalendar, or, Asiatic Register ... for the year 1798* (London, 1798).

Anon, *The East India Kalendar, or, Asiatic Register ... for the year 1799* (London, 1799).

Anon, *The East India Kalendar, or, Asiatic Register ... for the year 1800* (London, 1800).

Anon, *Post Office Directory from Whitsunday 1814 to Whitsunday 1815* (Edinburgh, 1814).

Anon, *The Asiatic Journal and Monthly Register* (London, 1815).

Anon, *Post Office Directory from Whitsunday 1815 to Whitsunday 1816* (Edinburgh, 1815).

Anon, *The Annual Biography and Obituary of 1816* (London, 1816).

Anon, *The Bengal obituary ... being a compilation of monumental inscriptions ... biographical sketches and memoirs* (Calcutta, 1848).

Arnott, Hugo, *The History of Edinburgh* (Edinburgh and London, 1779).

Bauer, Franz, *Delineations of Exotic Plants* (London, 1796).

Beatson, Alexander, *Tracts Relating to the Island of St Helena, written during a residence of five years* (London, 1816).

Berkenhout, John, *Outline of the Natural History of Great Britain and Ireland*, 3 vols. (London, 1769-72).

Blagdon, Francis William, *A Brief History of Ancient and Modern India, From the Earliest Period of Antiquity to the Termination of the late Mahratta War* (London, 1805).

Blume, K. L., *Rumphia, sive commentationes botanicae imprimis de Plantis Indiae orientalis ... quae in Libris ... Roxburghii ... recensentur*, 4 vols. (Leiden, 1835).

Bolts, W., *Considerations on Indian Affairs; Particularly Respecting the Present State of Bengal and its Dependencies* (London, 1772).

Bower, Alexander, *The History of the University of Edinburgh; chiefly compiled from original papers and records, never before published*, 2 vols. (Edinburgh, 1817).

Buchanan, Francis, *A Journey from Madras Through the Countries of Mysore, Canara, and Malabar ...*, 3 vols. (London, 1807).

Burman, Joannes, *Thesaurus Zeylanicus, exhibens plantas in insula Zeylana ... omnia iconibus illustrata ac descriptus cura* (Amsterdam, 1737).

Campbell, Lawrence Dundas, *A Reply to the Strictures of the Edinburgh Review, on the foreign Policy of Marquis of Wellesley's Administration in India; comprising an Examination of the Transactions in the Carnatic* (London, 1807).

Charpentier de Cossigny de Palma, J. F., *Memoir, containing an Abridged Treatise on the Cultivation and Manufacture of Indigo* (Calcutta, 1789).

Curtis, William, *Flora Londinensis* (London, 1775-98).

Dalrymple, Alexander, *Oriental Repertory*, 2 vols. (London, 1793-7).

Dow, Alexander, *The History of Hindoostan*. Volume IV [1772] *From the Death of Akbar, to the Complete Settlement of the Empire under Arungzebe. To which is prefixed. I, A Dissertation on the Origin and Nature of Despotism in Hindoostan. II, An Enquiry into the State of Bengal; with a Plan for Restoring that Kingdom to its former Prosperity and Splendor* (London, 1769-72).

Duncan, Andrew (the Elder), *An Account of the Life, Writings, and Character of the Late Dr. John Hope, Professor of Botany in the University of Edinburgh: Delivered as the Harveian Oration at Edinburgh, For the Year 1788* (Edinburgh, 1789).

Ellis, John, *Directions for Bringing over Seeds and Plants* (London, 1770).

Foote, Samuel, *The Nabob; a comedy in three acts* (London, 1778).

Forbes, James, *Oriental Memoirs: Selected and Abridged from a Series of Familiar Letters Written during Seventeen Years Residence in India: including Observations on parts of Africa and South America, and a Narrative of Occurrences in Four Indian Voyages*, 4 vols. (London, 1813).

Forster, George, *A Journey from Bengal to England through the Northern Part of India, Kashmire, Afghanistan, and Persia, and into Russia, by the Caspian Sea*, 2 vols. (London, 1798).

Graham, Maria (Lady Calcot), *Journal of a Residence in India* (Edinburgh, 1812).

Grant, Robert, *A Sketch of the History of the East India Company, from its First Formation to the Passing of the Regulating Act of 1773: with A Summary View of the Changes which have Taken Place since that Period in the Internal Administration of British India* (London, 1813).

Gupta, P. C. (ed.), *Fort William − India House Correspondence and Other Contemporary Papers Relating Thereto (Public Series), Vol. 3: 1796-1800* (Delhi, 1959).

Hardy, Charles, *A Register of Ships, Employed in the Service of the Honorable The United East India Company, from the Year 1760 to 1810*, revised by H. C. Hardy (London, 1811).

Hickey, William, *Memoirs of William Hickey (1749-1809)*, edited by Alfred Spencer, 4 vols. (London, 1913-25).

Hodges, W., *Travels in India During the Years 1780, 1781, 1782, & 1783* (London, 1793).

Huggins, W., *Sketches in India, Treating on Subjects Connected with the Government; civil and Military Establishments; Characters of the Europeans, and Customs of the Native Inhabitants* (London, 1824).

Hutton, James, *A Theory of the Earth* (Edinburgh, 1785).

Jacquin, Nicolai Josephi, *Fragmenta Botanica, Figuris colorati illustrata, 1800-1809*, 6 vols. (Vienna, 1800-09).

Jacquin, Nicolai Josephi, *Icones Plantarum*, 3 vols. (Vienna, 1786-93).

Jenkinson, Charles, 1st Earl of Liverpool, *A Discourse on the Conduct of the Government of Great Britain in respect to Neutral Nations* (Edinburgh, 1758, new edition 1792, reprinted Edinburgh, 1837).

Kames, Lord, *Sketches of the History of Man*, enlarged edition, 4 vols. (Edinburgh, 1788).

Kay, John, *A Series of Original Portraits and Caricature Etchings ... with Biographical Sketches and Illustrative Anecdotes*, 2 vols. (Edinburgh, 1792, new edition Edinburgh, 1877).

Lambert, Sheila (ed.), *House of Commons Sessional Papers of the Eighteenth Century. George III. Volume 74, 1789-1790* (Wilmington, Delaware, 1975).

Lambert, Sheila (ed.), *House of Commons Sessional Papers of the Eighteenth Century. George III. Volume 84, 1791* (Wilmington, Delaware, 1975).

Lambert, Sheila (ed.), *House of Commons Sessional Papers of the Eighteenth Century. George III. Volume 85, 1792* (Wilmington, Delaware, 1975).

Lambert, Sheila (ed.), *House of Commons Sessional Papers of the Eighteenth Century. George III. Volume 91, 1793-96* (Wilmington, Delaware, 1975).

Lambert, Sheila (ed.), *House of Commons Sessional Papers of the Eighteenth Century. George III. Volume 106, 1796-1797* (Wilmington, Delaware, 1975).

Lambert, Sheila (ed.), *House of Commons Sessional Papers of the Eighteenth Century. George III. Volume 126, 1799-1800* (Wilmington, Delaware, 1975).

Lettsom, J. C., *The Natural History of the Tea-tree* (London, 1772).

Mackenzie, Henry, *The Man of Feeling* (London, 1771, 2nd edition 1773, new edition edited by Brian Vickers, London, 1967).

Maitland, W., *The History of Edinburgh, from its Foundation to the Present Time*, 9 books (Edinburgh, 1753).

Mathison, John and Mason, Alexander Way, *The East India Register and Directory for 1805; ... from Official Returns Received at the East-India House* (London, 1805).

Mathison, John and Mason, Alexander Way, *The East India Register and Directory for 1807; ... from Official Returns Received at the East-India House* (London, 1807).

Mill, James, *The History of British India*, 3 vols. (London, 1813).

Millar, John, *The Origins of the Distinction of Ranks; or, an Inquiry into the Circumstances which give Rise to Influence and Authority, in the Different Members of Society* (London, 1779).

Newte, Thomas, *A Tour in England and Scotland in 1785* (London, 1788).

Newte, Thomas, *Prospects and Observations; on a Tour in England and Scotland: Natural, Œconomical, and Literary* (London, 1791).

Orme, Robert, *A History of the Military Transactions of the British Nation, from the Year MDCCXLV, to which is Prefixed a Dissertation on the Establishment made by the Mahomedan Conquerors in Indostan* (London, 1773-8, 4th edition 3 vols., London, 1803).

Orme, Robert, *Historical Fragments of the Mogul Empires, of the Morattoes, and of the English concerns In Indostan; from the Year M.DC.LIX* (London, 1782, 1805 edition, London).

Piddington, H., *A Tabular View of the Generic Characters in Roxburgh's Flora Indica* (Calcutta, 1836).

Pringle, Sir John, *Six Discourses delivered by Sir John Pringle, Bart. When President of the Royal Society; on Occasion of Six Annual Assignments of Sir George Copley's Medal. To which is prefixed the life of the Author by Andrew Kippis* (London, 1783).

Pulteney, Richard, *Historical and Biographical Sketches of the Progress of Botany in England, from its Origins to the Introduction of the Linnaean System*, 2 vols. (London, 1790).

Robertson, William, *The History of America* (2nd edition, 2 vols., London, 1778).

Robertson, William, *An Historical Disquisition Concerning the Knowledge which the Ancients had of India* (London and Edinburgh, 1791).

Roscoe, William, *Monandrous Plants of the Order Scitamineae, Chiefly Drawn from Living specimens in the Botanic Garden at Liverpool. Arranged according to the System of Linnaeus, with Descriptions and Observations*, 2 vols. bound from 15 fasciculi (Liverpool, 1824-9).

Russell, Patrick, *An Account of Indian Serpents, collected on the Coast of Coromandel; containing Descriptions and Drawings of each species; together with Experiments and Remarks on their several poisons*, 2 vols. (London, 1796-1801).

Scott, Sir Walter, *Guy Mannering; or, The Astrologer* (Edinburgh, 1815, Nelson's New Century Library edition, London, nd).

Scrafton, Luke, Duke of Scrafton, *Reflections of the Government of Indostan, with a Short Sketch of the History of Bengal, from MDCCXXXVIII to MDCCLVI; and an Account of the English Affairs to MDCCLVIII* (London, 1753, reprinted London, 1770).

Scrafton, Luke, Duke of Scrafton, *Observations on Mr. Vansittarts's Narrative* (London, 1767).

Smith, Adam, *An Inquiry into the Nature and Causes of the Wealth of Nations* (London, 1776, 2 vols. Penguin edition, London, 1986).

Smith, Adam, *Lectures on Jurisprudence*, edited by R. L. Mack, D. D. Raphael and P. G. Stein (Oxford, 1978, reprinted Indianapolis, 1982).

Smith, Dr James Edward, *English Botany* (London, 1790-1814).

Spavens, William, *Memoirs of a Seafaring Life: The Narrative of William Spavens*, (London 1796, reprinted London, 2000).

Valentia, Viscount, George Annersley, Earl of Mountnorris, *Voyages and Travels to India, Ceylon, the Red Sea, Abyssinia, and Egypt in the Years 1802, 1803, 1804, 1805, and 1806*, 3 vols. (London, 1802-6).

Verelst, Harry *A View of the Rise, Progress, and Present State of the English Government in Bengal* (London, 1772).

Wight, R., *Contributions to the Botany of India* (London, 1834).

Wight, R., *Icones plantarum Indiae Orientalis*, 6 vols. (London, 1838-53).

Williamson, Peter, *Directory for the City of Edinburgh, Canongate, Leith, and Suburbs* (Edinburgh, 1774), (reprinted Edinburgh, 1889, 1984).

Woodville, William, *Medical Botany, Containing Systematic and General Descriptions, with Plates of all the Medicinal Plants, Indigenous and Exotic, Comprehended in the Catalogues of the Materia Medica, as Published by the Royal Colleges of Physicians of London and Edinburgh*, 3 vols. (London, 1790-3).

2. SECONDARY SOURCES

Articles

Anderson, Thomas, 'Notices of the effects of the cyclone of 5th October 1864 on the Botanic Garden, Calcutta', *Transactions of the Edinburgh Botanical Society*, 8 (1865), 366-75.

Anon, 'The Glenfield Ramblers', *Kilmarnock Standard*, 4 July 1931.

Anon, 'Introduction of Cinchona to India', *Kew Bulletin*, 3 (1931), 113-17.

Anon, 'List of medicinal plants deposited in various Herbaria of the Botanical Survey of India', *Bulletin of the Botanical Society of Bengal*, 2 (1960), 180-273.

Balfour, I. B., 'A Sketch of the professors of Botany in Edinburgh from 1670 until 1887', in F. W. Oliver (ed.), *Makers of British Botany* (Cambridge, 1913), 280-301.

Balfour, I. B., 'Botanical Intelligence', *Transactions of the Edinburgh Botanical Society*, 9 (1862), 162.

[Balfour, I. B.], 'Eighteenth century records of British Plants', *Notes from the Royal Botanic Garden Edinburgh*, 4 (1907), 123-92.

Banerjee, D. N., 'Administration in British Territories in India (1707-1818). Part II. The Madras Presidency down to 1818', in R. C. Majumdar and V. G. Dighe (eds.), *The Maratha Supremacy* (Calcutta, 1977), 614-31.

Basak, R. K., 'The bibliography on the flora and vegetation of Bengal with an introductory note', *Bulletin of the Botanical Society of Bengal*, 15 (1973), 22-38.

Boag Watson, William N., 'Two naval surgeons of the French Wars', *Medical History*, 13 (1969), 213-25.

Boyes, J., 'Sir Robert Sibbald: a neglected scholar', in R. G. W. Anderson and A. D. C. Simpson (eds.), *The Early Years of the Edinburgh Medical School* (Edinburgh, 1976), 19-24.

Britten, James, ' "John" Roxburgh', *Journal of Botany*, 56 (1918), 202-3.

Brown, Stephen W., 'William Smellie and Natural History: Dissent and Dissemination', in Charles W. J. Withers and Paul Wood (eds.), *Science and Medicine in the Scottish Enlightenment* (Edinburgh, 2002), 191-214.

Brown, Stewart J., 'Introduction', in Stewart J. Brown (ed.), *William Robertson*, (Cambridge, 1997), 1-6.

Brown, Stewart J., 'William Robertson (1721-1793) and the Scottish Enlightenment', in Stewart J. Brown (ed.), *William Robertson* (Cambridge, 1997), 7-35.

Bryant, G. J., 'Scots in India in the Eighteenth Century', *The Scottish Historical Review*, 64 (1985), 22-41.

Carnall, Geoffrey, 'Robertson and contemporary images of India', in Stewart J. Brown (ed.), *William Robertson* (Cambridge, 1997), 210-30.

Cater, J. J., 'The Making of Principal Robertson', *Scottish Historical Review*, 49 (1970), 60-84.

Chakraverty, R. K. and Mukhopadhyay, D. P., 'The Great Banyan Tree', *Bulletin of the Botanical Society of Bengal*, 29 (1987), 59-70.

Chatterjee, D., 'Early history of the Royal Botanic Garden, Calcutta', *Nature*, 161 (1948), 362-4.

Christie, J. R. R., 'The origins and development of the Scottish scientific community, 1680-1760', *History of Science*, 12 (1974), 122-41.

Coats, Alice M., 'Henna and Frankincense', *The Gardeners Chronicle*, 160 (16 November, 1966), 14.

Cohen, I. B., 'Stephen Hales', *Scientific American*, 234 (May, 1976), 98-107.

Cohn, Bernard S., 'Structural Change in Indian Rural Society, 1596-1885', in Robert Eric Frykenberg (ed.), *Land Control and Social Structures* (Madison, Wisconsin, 1969), 53-121.

Cowan, J. M., 'The history of the Royal Botanic Garden Edinburgh', *Notes from the Royal Botanic Garden Edinburgh*, 19 (1933), 1-62.

Cowan, J. M., 'The history of the Royal Botanic Garden Edinburgh', *Notes from the Royal Botanic Garden Edinburgh*, 19 (1935), 63-134.

Cowen, D. L., 'The Edinburgh Pharmacopeia', in R. G. W. Anderson and A. D. C. Simpson (eds.), *The Early Years of the Edinburgh Medical School* (Edinburgh, 1976), 24-45.

Crellin, J. K., 'William Cullen: his calibre as a teacher, and an unpublished introduction to his *A Treatise of the Materia Medica*, London, 1773', *Medical History*, 15 (1971), 79-87.

Cunningham, Andrew, 'Medicine to calm the mind: Boerhaave's medical system, and why it was adopted in Edinburgh', in Andrew Cunningham and Roger French (eds.), *The Medical Enlightenment* (Cambridge, 1990), 40-66.

Darwin, Francis, 'A Botanical Physiologist of the Eighteenth Century', *Notes from the Royal Botanic Garden Edinburgh*, 4 (1909), 241-4.

Datta, S. C. and Majumdar, N. C., 'Flora of Calcutta and Vicinity', *Bulletin of the Botanical Society of Bengal*, 20 (1967), 16-120.

Datta, V. N., 'Western Ideas as Reflected in the Official Attitude towards Social and Educational Policy, 1800-1835', in B. Prasad (ed.), *Ideas in History* (London, 1968), 38-53.

Deb, D. B., 'The Indian Botanic Garden', *Bulletin of the Botanical Survey of India*, 19 (1977), 1-4.

Desmond, Ray, 'William Roxburgh's Plants of the Coast of Coromandel 1795-1820', *Hortulus Aliquando*, 2 (1977), 22-41.

Dutta, Navendra Mohan, 'A revision of the Genus *Torenia* Linn. of Eastern India', *Bulletin of the Botanical Society of Bengal*, 19 (1965), 23-7.

Eaves Walton, P. M., 'The early years of the Infirmary', in R. G. W. Anderson and A. D. C. Simpson (eds.), *The Early Years of the Edinburgh Medical School* (Edinburgh, 1976), 71-9.

Embree, Ainslie T., 'Landholding in India and British Institutions', in Robert Eric Frykenberg (ed.), *Land Control and Social Structures* (Madison, Wisconsin, 1969), 33-52.

Emerson, Robert L. and Wood, Paul, 'Science and Enlightenment in Glasgow, 1690-1802', in Charles W. J. Withers and Paul Wood (eds.), *Science and Medicine in the Scottish Enlightenment* (Edinburgh, 2002), 79-142.

Farber, P. L., 'Aspiring naturalists and their Frustrations: the case of William Swainson (1789-1855)', in A. Wheeler and J. H. Price (eds.), *From Linnaeus to Darwin* (London, 1985), 51-9.

Forman, L. L., 'Notes Concerning the Typification of Names of William Roxburgh's Species of Phanerogams', *Kew Bulletin* 52 (1997), 513-34.

Frykenberg, R. F., 'Introduction', in Robert Eric Frykenberg (ed.), *Land Control and Social Structures* (Madison, Wisconsin, 1969), xiii-xxi.

Fynes, Richard, 'Sir William Jones and the Classical Tradition', in Alexander Murray (ed.), *Sir William Jones* (Oxford, 1998), 43-66.

Gage, A. T., 'The Royal Botanic Garden, Calcutta', *Journal of the Royal Horticultural Society*, 51 (1926), 71-81.

Gombrich, Richard, 'Introduction', in Alexander Murray (ed.), *Sir William Jones* (Oxford, 1998), 1-15.

Gosling, W. R. O., 'Leiden and Edinburgh: the seed, the soil and the climate', in R. G. W. Anderson and A. D. C. Simpson (eds.), *The Early Years of the Edinburgh Medical School* (Edinburgh, 1976), 1-18.

Grewal, J. S., 'Ideas. Behind the advocacy of radical social change in India by Utilitarians and the Evangelicals (A Summary)', in B. Prasad (ed.), *Ideas in History* (London, 1968), 54.

Gupta, H. L., 'The Christian Missionaries and their impact on Modern India in the Pre-Mutiny period', in B. Prasad (ed.), *Ideas in History* (London, 1968), 55-65.

Guthrie, Douglas, 'The influence of the Leyden School upon Scottish Medicine', *Medical History*, 3 (1959), 108-22.

Hartley, Stuart, 'Appealing to Nature: Geology in the Field in Late Enlightenment Scotland', in Charles W. J. Withers and Paul Wood (eds.), *Science and Medicine in the Scottish Enlightenment* (Edinburgh, 2002), 280-300.

Hasan, S. Nurul, 'Zamindars under the Mughals', in Robert Eric Frykenberg (ed.), *Land Control and Social Structures* (Madison, Wisconsin, 1969), 17-31.

Hastings, R. B., 'The relationship between the Indian Botanic Garden, Howrah, and the Royal Botanic Garden, Kew, in economic botany', *Bulletin of the Botanical Survey of India*, 28 (1986), 1-12.

Hedge, I. C. and Lamond, J. M., 'Edinburgh's Indian Botanical Connections and Collections', *Bulletin of the Botanical Survey of India*, 29 (1987), 272-85.

Heimann, P. M. and McGuire, J. E., 'Newtonian forces and Lockean powers: concepts of matter in eighteenth century thought', *Historical Studies in the Physical Sciences*, 3 (1971), 233-306.

Holttum, R. E., 'The Historical significance of Botanic Gardens in S. E. Asia', *Taxon*, 19 (1970), 707-14.

Hooker, Sir Joseph, 'A century of Indian orchids, selected from the drawings of plants in the Herbarium of the Royal Botanic Garden, Calcutta', *Annals of the Royal Botanic Garden, Calcutta*, 5 (1895), 1-68 plus 101 Plates.

Howard, Richard A. and Powell, Dulcie A., 'The Indian Botanic Garden, Calcutta and the Gardens of the West Indies', *Bulletin of the Botanical Survey of India*, 7 (1965), 1-7.

Jackson, E., 'From Papyri to Pharmacopeia – the development of standards for crude drugs', in F. N. L. Poynter (ed.), *The Evolution of Pharmacy in Britain* (London, 1965), 151-64.

Jevons, F. R., 'Boerhaave's biochemistry', *Medical History*, 6 (1962), 343-62.

Johri, B. M. and Rau, M. A., 'Plant Sciences in India: Yesterday, Today and Tomorrow', in B. M. Johri (ed.), *Botany in India*, 2 vols. (Lebanon, New Hampshire, 1994), vol. 1, 1-16.

Kakkar, Latita, 'Indian Botanic Garden, Calcutta', *Botanica*, 21 (1971), 95-7.

Karegeannes, Carrie, 'Roxburgh: chronicler of Indian begonias', *The Begonian*, 46 (1979), 261 and 280.

King, Sir George, 'Preface', *Annals of the Royal Botanic Garden Calcutta*, 4 (1893), i-xii.

King, Sir George, 'A brief memoir of William Roxburgh', *Annals of the Royal Botanic Garden Calcutta*, 5 (1895), 1-4.

King, Sir George, 'The early history of Indian botany', *Journal of Botany*, 37 (1899), 454-63.

Lawrence, C. J., 'Early Edinburgh medicine: Theory and Practice', in R. G. W. Anderson and A. D. C. Simpson (eds.), *The Early Years of the Edinburgh Medical School* (Edinburgh, 1976), 81-94.

Mabberley, D. J., 'Francis Hamilton's Commentaries with Particular Reference to Meliaceae' *Taxon*, 26 (1977), 523-40.

Mabberley, D. J., 'William Roxburgh's "Botanical Description of a New Species of *Swietenia* (Mahogany)" and other Overlooked Binomials in 36 Vascular Plant Families', *Taxon*, 31 (1982), 65-73.

Macdonald, Fiona A., 'Reading Cleghorn the Clinician: The Clinical Case Records of Dr Robert Cleghorn, 1785-1818', in Charles W. J. Withers and Paul Wood (eds.), *Science and Medicine in the Scottish Enlightenment* (Edinburgh, 2002), 255-79.

Maheshawari, J. K., 'The food-producing crops in the tropics', *Bulletin of the Botanical Society of Bengal*, 3 (1961), 153-62.

Maheshawari, J. K., 'Taxonomic studies in Indian Guttiferae. II The Genus *Mesua* Linn.', *Bulletin of the Botanical Society of Bengal*, 5 (1963), 335-43.

Majumdar, R. B., 'The Genus *Panicum* Linn. in India', *Bulletin of the Botanical Society of Bengal*, 27 (1973), 39-54.

Matthew, K. M., 'William Roxburgh's Plants of the Coast of Coromandel: an enumeration of species', *Blumea*, 49 (2004), 367-405.

McDougall, Warren, 'Charles Elliot's Medical Publications and the International Book Trade', in Charles W. J. Withers and Paul Wood (eds.), *Science and Medicine in the Scottish Enlightenment* (Edinburgh, 2002), 215-54.

Merrill, Elmer Drew, 'The fugitive place names "Hamoa" and "Romoa" in Roxburgh's *Flora Indica* errors for Honimoa – Sarapua', *Taxon*, 1 (1952), 124-5.

Merrill, Elmer Drew, 'The botany of Cook's voyages', *Chronica Botanica*, 14 (1954), 161-383.

Miller, H. S., 'The herbarium of Aylmer Bourke Lambert, Notes on its acquisition, dispersal, and present whereabouts', *Taxon*, 19 (1970), 489-553.

Morrell, J. B., 'The university of Edinburgh in the late Eighteenth Century: Its Scientific Eminence and Academic Structure', *Isis*, 62 (1971), 158-71.

Morrell, J. B., 'Reflections on the history of Scottish science', *History of Science*, 12 (1974), 81-94.

Morrell, J. B., 'The Edinburgh Town Council and its university, 1717-1766', in R. G. W. Anderson and A. D. C. Simpson (eds.), *The Early Years of the Edinburgh Medical School* (Edinburgh, 1976), 46-65.

Morrison-Low, A. D., 'Feasting my eyes with the view of fine instruments: Scientific Instruments in Enlightenment Scotland, 1680-1820', in Charles W. J. Withers and Paul Wood (eds.), *Science and Medicine in the Scottish Enlightenment* (Edinburgh, 2002), 17-53.

Morton, A. G. and Noble, M., 'Two hundred years of the Botanical Sciences in Scotland: Botany and Mycology', *Proceedings of the Royal Society Edinburgh*, 846 (1983), 65-83.

Morton, C. V., 'William Roxburgh's fern types', *Contributions from the United States National Herbarium*, 38 (1974), 283-396.

Mukherjee, Sunil Kumar, 'Revision of the Genus *Saccharum* Linn.', *Bulletin of the Botanical Society of Bengal*, 8 (1954), 143-9.

Nayar, M. P. and Das, A. R. , 'Glimpse of William Roxburgh through his unpublished letters and his interest in Indian economic botany', *Journal of Economic and Taxonomic Botany*, 5 (1984), 1159-67.

Neale, Walter C., 'Land is to Rule', in Robert Eric Frykenberg (ed.), *Land Control and Social Structures* (Madison, Wisconsin, 1969), 3-15.

Nelson, E. Charles and Rourke, John P., 'James Niven (1776-1827), a Scottish botanical collector at the Cape of Good Hope. His *hortus siccus* at the national Botanic Gardens, Glasnevin, Dublin (DBN), and the Royal Botanic Gardens, Kew (K)', *Kew Bulletin*, 48 (1993), 663-82.

Philips, C. H., 'The East India Company "Interest" and the English Government, 1783-4', *Transactions of the Royal Historical Society*, 20 (1937), 83-101.

Phillipson, Nicholas, 'Commerce and Culture: Edinburgh, Edinburgh University, and the Scottish Enlightenment', in Thomas Bender (ed.), *The University and the City* (New York and Oxford, 1988), 100-16.

Phillipson, Nicholas, 'Providence and Progress: an introduction to the historical thought of William Robertson', in Stewart J. Brown (ed.), *William Robertson* (Cambridge, 1997), 55-73.

Phillipson, Nicholas, 'Politics and politeness in the reigns of Anne and the early Hanoverians', in J. G. A. Pocock (ed.), *The Varieties of British Political Thought* (Cambridge, 1993), 211-45.

Pillerie, G., 'William Roxburgh (1751-1815), Heinrich Julius Lubeck (†1801) and the Discovery of the Ganges Dolphin (*Platanista gangetica* Roxburgh, 1801)', *Investigations on Cetacea*, 9 (1978), 11-21.

Prain, Sir David, ' "John" Roxburgh', *Journal of Botany*, 57 (1919), 28-34.

Price, C. H., 'Medicine and pharmacy at the Cape of Good Hope, 1652-1807', *Medical History*, 6 (1962), 169-76.

R., R. A., 'William Roxburgh', *The Gardeners Chronicle*, 19 (3rd series), (27 January, 1896), 781-2.

Rau, M. A., 'Plant Exploration in India and Floras', in B. M. Johri (ed.), *Botany in India* 2 vols. (Lebanon, New Hampshire, 1994), vol. 1, 17-41.

Raychaudhuri, Tapan, 'Permanent Settlement in Operation: Bakarganj District, East Bengal', in Robert Eric Frykenberg (ed.), *Land Control and Social Structures* (Madison, Wisconsin, 1969), 163-74.

Robinson, C. B., 'Roxburgh's Hortus Bengalensis', *Philippine Journal of Science, C. Botany*, 7 (1912), 411-19.

Samaddar, U. P. and Guha Bakshi, D. N., 'A selective catalogue of the publications on the Indian Botanic Garden', *Bulletin of the Botanical Survey of India*, 29 (1987), 286-91.

Sangwan, Satpal, 'The strength of a scientific culture: Interpreting disorder in colonial science', *Indian Economic and Social History Review*, 34 (1997), 217-50.

Sanjappa, M., Thothathri, K. and Dass, A. R.. 'Roxburgh's Flora Indica drawings at Calcutta', *Bulletin of the Botanical Survey of India*, 33 (1991), 1-232.

Santapau, A., Irani, S. J. and Irani, N. A., 'The Genus *Ceropegia* in Bombay', *Bulletin of the Botanical Society of Bengal*, 12 (1958), 6-17.

Santapu, H., 'The Indian Botanic Garden in the first 175 years', *Bulletin of the Botanical Society of Bengal*, 7 (1965), i-vii.

Santapu, H., 'The story of Tea', *Bulletin of the Botanical Survey of India*, 8 (1966), 103-7.

Sealy, J. Robert, 'The Roxburgh Flora Indica drawings at Kew', *Kew Bulletin*, 2 (1956), 297-348, and 3 (1956), 349-99.

Sealy, J. Robert, 'William Roxburgh's collection of paintings of Indian plants', *Endeavour*, 34 (1975), 84-9.

Sen Gupta, J. C., 'Botanical Survey of India. Its Past, Present and Future', *Bulletin of the Botanical Society of Bengal*, 1 (1959), 9-29.

Sen, J., 'Dr William Roxburgh: The Father of Indian Botany', *Nature*, 207 (1965), 1234-5.

Sen, S. N., 'Marathas and the South Indian States (1772-1799)', in R. C. Majumdar and D. G. Dighe (eds.), *The Maratha Supremacy* (Calcutta, 1977), 422-51.

Shapin, Steven, 'The audience for science in eighteenth century Edinburgh', *History of Science*, 12 (1974), 95-121.

Shapin, Steven, 'Property, patronage, and the politics of science: the founding of the Royal Infirmary of Edinburgh', *British Journal for the History of Science*, 7 (1974), 1-41.

Shapin, Steven and Thackray, A., 'Prosopography as a research tool in history of science: the British scientific community 1700-1800', *History of Science*, 12 (1974), 1-28.

Sinha, N. K., 'Mysore: Haidar Ali and Tipu Sultan', in R. C. Majumdar and V. G. Dighe (eds.), *The Maratha Supremacy* (Calcutta, 1977), 452-71.

Sinha, N. K., 'Progress of the British Power (1785-1798)', in R. C. Majumdar and V. G. Dighe (eds.), *The Maratha Supremacy* (Calcutta, 1977), 472-85.

Sinha, N. K., 'Consolidation of British rule in India (1799-1823)', in R. C. Majumdar and V. G. Dighe (eds.), *The Maratha Supremacy* (Calcutta, 1977), 521-34.

Stewart, I. G. and Ferguson, Joan P. S., 'Rhododendrons, doctors and India, 1780-1860', *Proceedings of the Royal College of Physicians Edinburgh*, 26 (1996), 282-94.

Srinivasan, K. S., 'William Carey: his activities in botanical and horticultural researches in India', *Bulletin of the Botanical Survey of India*, 3 (1961), 1-10.

Steenis-Kruseman, M. J. von, 'Roxburgh, W., *Plants of the Coast of Coromandel*', *Flora Malesiana Bulletin*, 26 (1972), 2018.

Stein, Burton, 'Integration of the Agrarian System of Southern India', in Robert Eric Frykenberg (ed.), *Land Control and Social Structures* (Madison, Wisconsin, 1969), 175-216.

Thomson, Thomas, 'Notes on the herbarium of the Calcutta Botanical Garden, with especial reference to the completion of the Flora Indica', *Journal of Botany and Kew Garden Miscellany* (edited by Sir William Jackson Hooker), 9 (1857), 33-41.

Trautmann, Thomas R., 'The Lives of Sir William Jones', in Alexander Murray (ed.), *Sir William Jones* (Oxford, 1998), 91-122.

Verdoorn, F. (ed.), *Chronica Botanica*, 2 (1936), 206.

Verdoorn, F. (ed.), *Chronica Botanica*, 5 (1939), 116.

Waterston, Charles D., 'Late Enlightenment Science and Generalism: The Case of Sir George Mackenzie of Coul, 1780-1848', in Charles W. J. Withers and Paul Wood (eds.), *Science and Medicine in the Scottish Enlightenment* (Edinburgh, 2002), 301-26.

White, James A., 'Botanical art in the Indian Museum, Calcutta', *Huntia*, 9 (1996), 151-60.

Wight, Robert, 'Illustrations of Indian botany, principally of the southern parts of the Peninsula', *Botanical Miscellany* (edited by William Jackson Hooker), 2 (1831), 90-7.

Withers, Charles W. J., 'Situating Practical Reason: Geography, Geometry and Mapping in the Scottish Enlightenment', in Charles W. J. Withers and Paul Wood (eds.), *Science and Medicine in the Scottish Enlightenment* (Edinburgh, 2002), 54-78.

Withers, Charles W. J. and Wood, Paul, 'Afterword: New Directions?', in Charles W. J. Withers and Paul Wood (eds.), *Science and Medicine in the Scottish Enlightenment* (Edinburgh, 2002), pp. 327-36.

Wood, D., 'Roxburgh's "Plants of the Coast of Coromandel" dates of publication of volume 3', *Notes from the Royal Botanic Garden Edinburgh*, 29 (1969), 211-12.

Woudstra, Jan, 'The use of flowering plants in late seventeenth- and early eighteenth-century interiors', *Garden History*, 28 (2001), 194-208.

Books

Allardyce, Alexander (ed.), *Scotland and Scotsmen in the Eighteenth Century, from the Manuscripts of John Ramsay, Esq. Of Ochtertyre* (Edinburgh, 1888).

Allen, Charles, *The Buddha and the Sahibs. The Men who Discovered India's Lost Religions* (London, 2002).

Anderson, P. J. (ed.), *Marischal College, Officers, Graduates and Alumni* (Aberdeen, 1897).

Anderson, R. G. W. and Simpson, A. D. C. (eds.), *The Early Years of the Edinburgh Medical School* (Edinburgh, 1976).

Anon, *British Museum. General Catalogue of Printed Books*, Photolithographic edition, 263 vols. (London, 1961-6).

Anon, *Catalogue of the Printed Books in the Library of the University of Edinburgh*, 3 vols. (Edinburgh, 1918-25).

Anon, *The National Union Catalog. Pre-1956 Imprints*, 685 vols. (London, 1968-79).

Archer, Mildred, *Natural History Drawings in the India Office Library* (London, 1962).

Armitage, David, *The Ideological Origins of the British Empire* (Cambridge, 2000).

Baker, H. A. and Oliver, E. G. H., *Ericas in Southern Africa* (Cape Town, 1967).

Balfour, J. H., *Class Book of Botany, being an Introduction to the Study of the Vegetable Kingdom* (Edinburgh, 1854).

Ballingall, William (ed.), *Edinburgh Past and Present. Its Associations and Surroundings* (Edinburgh, 1877).

Bayly, C. A., *Indian Society and the Making of the British Empire* (Cambridge, 1988).

Bayly, C. A., *Empire and Information: Intelligence Gathering and Social Communications in India, 1780-1870* (Cambridge, 1996).

Beaglehole, T. H., *Thomas Munro and the Development of the Administration Policy in Madras, 1792-1818. The Origins of the Munro System* (Cambridge, 1996).

Bearce, George D., *British Attitudes towards India, 1784-1858* (London 1961).

Bender, Thomas (ed.), *The University and the City, from Medieval Origins to the Present* (New York and Oxford, 1988).

Beveridge, Henry, *A Comprehensive History of India, Civil, Military, and Social, from the First Landing of the English, to the Suppression of the Sepoy Revolt*, 3 vols. of 9 books (Edinburgh and London, 1858-62).

Biswas, K. (ed.), *The Original Correspondence of Sir Joseph Banks, Relating to the Foundation of the Royal Botanic Garden, Calcutta, and the Summary of the 150th Anniversary Volume of the Royal Botanic Garden, Calcutta* (Calcutta, 1950).

Blunt, Wilfred and Stearn, William T., *The Art of Botanical Illustration*, new edition (London, 1994).

Boase, Frederick, *Modern English Biography Containing Many Thousand Concise Memoirs of Persons who have Died Between the Years 1851-1900 with an Index of the Most Interesting Matter*, vol. VI, supplement to Vol. III, 2nd impression (London, 1965).

Bose, Sugata, *Peasant Labour and Colonial Capital: Rural Bengal Since 1770* (Cambridge, 1993).

Bridson, Gavin D. R., Phillips, Valerie C. and Harvey, Anthony P., *Natural History Manuscript Resources in the British Isles* (London and New York, 1980).

Britten, James and Boulger, George S., *A Biographical Index of Deceased British and Irish Botanists*, 2nd edition revised by A. B. Rendle (London, 1931).

Brown, James, *The Epitaphs and Monumental Inscriptions in Greyfriars Churchyard, Edinburgh* (Edinburgh, 1867).

Brown, Stewart J. (ed.), *William Robertson and the Expansion of Empire* (Cambridge, 1997).

Buchan, James, *Capital of the Mind: How Edinburgh Changed the World* (London, 2003).

Buckland, C. E., *Dictionary of Indian Biography* (London, 1906).

Burke, John and Burke, Sir John B., *A Genealogical and Heraldic History of the Landed Gentry of Great Britain and Ireland*, 3 vols. (London, 1849, 4th edition, 1863).

Burkill, I. H., *Chapters on the History of Botany in India* (Calcutta, 1965).

Campbell-Culver, Maggie, *The Origin of Plants. The People and Plants that have shaped Britain's Garden History since the Year 1000* (London, 2001).

Cannon, Garland (ed.), *The Letters of Sir William Jones*, 2 vols. (Oxford, 1970).

Cannon, Garland, *The Life and Mind of Oriental Jones. Sir William Jones, the Father of Modern Linguistics* (Cambridge, 1990).

Carey, Eustace, *Memoir of William Carey, D.D.* (London, 1836).

Carter, Harold B., *Sir Joseph Banks (1743-1820). A Guide to Biographical and Bibliographical Sources* (London, 1987).

Carter, Harold B., *Sir Joseph Banks, 1743-1820* (London, 1988).

Chambers, R., *A Biographical Dictionary of Eminent Scotsmen*, 2 vols., revised by Thomas Thomson (London, 1875).

Chaudhuri, S. B., *Civil Disturbances during the British Rule in India (1765-1857)* (Calcutta, 1955).

Chitnis, Anand C., *The Scottish Enlightenment. A Social History* (London, 1976).

Coats, Alice M., *The Quest for Plants. A History of the Horticultural Explorers* (London, 1969).

Comrie, John D., *History of Scottish Medicine*, 2nd edition, 2 vols. (London, 1932).

Cotton, Sir Evan, *East Indiamen. The East India Company's Maritime Service*, edited by Sir Charles Fawcett (London, 1949).

Craig, W. S., *History of the Royal College of Physicians of Edinburgh* (London, 1976).

Cranmer-Byng, J. L. (ed.), *An Embassy to China, being the Journal kept by Lord Macartney during his Embassy to the Emperor Chi'en-lung. 1793-94* (London, 1962, reprinted 2004).

Crawford, D. G., *A History of the Indian Medical Service 1600-1914*, 2 vols. (London, Calcutta and Simla, 1914).

Crawford, D. G., *Roll of the Indian Medical Service 1615-1930* (London, Calcutta and Simla, 1930).

Creswell, C. H., *The Royal College of Surgeons of Edinburgh. Historical Notes from 1505 to 1905* (Edinburgh, 1926).

Cronk, Q. C. B., *The Endemic Flora of St Helena* (Oswestry, Shropshire, 2000).

Crowther, J. G., *Scientists of the Industrial Revolution* (London, 1962).

Cunningham, Andrew and French, Roger (eds.), *The Medical Enlightenment of the Eighteenth Century* (Cambridge, 1990).

Dalrymple, William, *White Mughals. Love and Betrayal in Eighteenth-Century India* (London, 2002).

Dalzel, Andrew, *History of the University of Edinburgh from its Foundations*, 2 vols. (Edinburgh, 1862).

Dance, S. Peter, *The Art of Natural History* (London, 1978, reissued London 1989).

Daniels, Stephen, *Humphry Repton. Landscape Gardening and the Geography of Georgian England* (New Haven, Conn., and London, 1999).

Davies, A. M., *Warren Hastings. Maker of British India* (London, 1935).

Davies, C. C., *An Historical Atlas of the Indian Peninsula* (Calcutta, 1965).

Dawson, Warren R., *Catalogue of the Manuscripts in the Library of the Linnean Society of London. Part I. The Smith Papers* (London, 1934).

Dawson, Warren R. (ed.), *The Banks Letters. A Calendar of the Manuscript Correspondence of Sir Joseph Banks Preserved in the British Museum, the British Museum (Natural History) and Other Collections in Great Britain* (London, 1958).

Desmond, Ray, *Dictionary of British and Irish Botanists and Horticulturalists including Plant Collectors and Botanical Artists* (London, 1977).

Desmond, Ray, *The India Office Museum 1801-1879* (London, 1982).

Desmond, Ray, *Wonders of Creation. Natural History Drawings in the British Library* (London, 1986).

Desmond, Ray, *The European Discovery of the Indian Flora* (Oxford, 1992).

Desmond, Ray, *Kew. The History of the Royal Botanic Gardens* (London, 1995).

Desmond, Ray, *Sir Joseph Dalton Hooker. Traveller and Plant Collector* (London, 1999).

Devlin-Thorp, Sheila (ed.), *Scotland's Cultural Heritage. Vol. III, The Royal Society of Edinburgh: 100 Medical Fellows Elected 1783-1844* (Edinburgh, 1982).

Dunthorne, Gordon, *Flower and Fruit Prints of the Eighteenth and early Nineteenth Centuries* (Washington, D.C., 1938, reprinted New York, 1970).

Dyson, K. K., *A Various Universe. A Study of the Journals and Memoirs of the British Men and Women in the Indian Subcontinent, 1756-1856* (Delhi, 1978).

Embree, A. T., *Charles Grant and British Rule in India* (London, 1962).

Fara, Patricia, *Sex, Botany and Empire. The Story of Carl Linnaeus and Joseph Banks* (Cambridge, 2003).

Fry, Howard, *Alexander Dalrymple (1737-1808) and the Expansion of British Trade* (London, 1970).

Frykenberg, Robert Eric (ed.), *Land Control and Social Structure in Indian History* (Madison, Wisconsin, 1969).

Gardner, Brian, *The East India Company* (London, 1971).

Gleig, G. R., *Memoirs of the Life of the Right Hon. Warren Hastings, First Governor-General of Bengal*, 3 vols. (London, 1841).

Gordon, Stewart, *The Marathas 1600-1818* (Cambridge, 1993).

Green, J. R., *A History of Botany in the United Kingdom from the Earliest Times to the End of the Nineteenth Century* (London, 1914).

Grove, R. H., *Green Imperialism. Colonial Expansion, Tropical Island Edens and the Origins of Environmentalism, 1600-1800* (Cambridge, 1995).

Hall, Norman, *Botanists of the Eucalypts* (Melbourne, Australia, 1978).

Harlow, Vincent T., *The Founding of the Second British Empire, 1763-1793. Vol. I Discovery and Revolution. Vol. II New Continents and Changing Values*, 2 vols. (London, 1952 and 1964).

Harris, Stuart, *The Place Names of Edinburgh, Their Origins and History* (Edinburgh, 1996).

Harvey-Gibson, R. J., *Outlines of the History of Botany* (London, 1919).

Heiberg, Knird (ed.), *List of Marriages Registered in the Danish Church Register of Zion Church, Tranquebar, 1767-1845* (Madras, 1935).

Henrey, Blanche, *British Botanical and Horticultural Literature before 1800*, 3 vols. (London, 1975).

Heywood, C. V. (ed.), *Flowering Plants of the World* (Oxford, 1978).

Holmes, Richard, *Sahib, The British Soldier in India* (London, 2005).

Hooker, Sir Joseph, *Flora of British India*, 7 vols. (London, 1872-97).

Howe, W. Stewart, *The Dundee Textiles Industries 1960-1977* (Aberdeen, 1982).

Hudson, D. and Luckhurst, K. W., *The Royal Society of Arts, 1754-1954* (London, 1954).

Hudson, Roger (ed.), *The Raj. An Eye-Witness Account* (London, 1999).

Huxley, Leonard, *Life and Letters of Sir Joseph Dalton Hooker, O.M., G.C.S.I., Based on Materials Collected and Arranged by Lady Hooker*, 2 vols. (London, 1918).

James, Lawrence, *Raj. The Making and Unmaking of British India* (London, 1997).

Johri, B. M. (ed.), *Botany in India. History and Progress*, 2 vols. (Lebanon, New Hampshire, USA, 1994).

Kaye, Sir John William, *Administration of the East India Company: a History of Indian Progress* (London, 1853).

Kincaid, Dennis, *British Social Life in India, 1608-1937* (London, 1938, 2nd edition London, 1973).

Lack, H. Walter, *A Garden for Eternity. The Codex Liechtenstein*, translated by Martin Walters (Berne, Switzerland, 2000).

Lasègne, A., *Musée Botanique de M. Benjamin Delessert* (Paris, 1845).

Lemmon, Kenneth, *The Golden Age of Plant Hunters* (London, 1968).

Love, H. D., *Vestiges of Old Madras 1640-1800. Traced from the East India Company's Records Preserved at Fort St. George and the India Office, and from Other Sources*, 3 vols. (London, 1913).

Lyons, Sir Henry, *The Royal Society, 1660-1940. A History of its Administration under its Charters* (London, 1944).

Lyte, Charles, *The Plant Hunters* (London, 1983).

Mabberley, D. J., *Jupiter Botanicus. Robert Brown of the British Museum* (London, 1985).

Mabey, Richard, *The Flowering of Kew* (London, 1988).

MacMunn, Major G. F., *The Armies of India* (London, 1911).

Majumdar, R. C. and Dighe, V. G. (eds.), *The Maratha Supremacy. Volume VIII of The History and Culture of the Indian People* (Calcutta, 1977).

Marshall, P. J. (ed.), *The British Discovery of Hinduism in the Eighteenth Century* (Cambridge, 1970).

Mason, Philip J., *The Men who Ruled India* (London, 1985).

Mathew, Manjil V., *The History of the Royal Botanic Garden Library, Edinburgh* (Edinburgh, 1987).

McCracken, D. P., *Gardens of the Empire: Botanical Institutions of the Victorian British Empire* (London, 1997).

Metcalf, Thomas R., *Ideologies of the Raj* (Cambridge, 1994).

Miller, P. N., *Defining the Common Good. Empire, Religion and Philosophy in Eighteenth-Century Britain* (Cambridge, 1994).

Morton, A. G., *John Hope. 1725-1786. Scottish Botanist* (Edinburgh, 1986).

Mukherjee, S. N., *Sir William Jones: a Study in Eighteenth Century British Attitudes to India* (London, 1968).

Murray, Alexander (ed.), *Sir William Jones 1746-1794. A Commemoration* (Oxford, 1998).

Musgrave, Toby, Gardner, Chris and Musgrave, Will, *The Plant Hunters. Two Hundred Years of Adventure and Discovery Around the World* (London, 1998).

Noltie, H. J., *Indian Botanical Drawings, 1793-1868, from the Royal Botanic Garden Edinburgh* (Edinburgh, 1999).

O'Brian, Patrick, *Joseph Banks. A Life* (London, 1987).

O'Donoghue, Freeman, *Catalogue of Engraved British Portraits Preserved in the Department of Prints and Drawings of the British Museum*, 6 vols. (London, 1908-25).

Oliver, F. W., *Makers of British Botany. A Collection of Biographies by Living Botanists* (Cambridge, 1913).

Peterkin, A. and Johnston, William, *Commissioned Officers in the Medical Services of the British Army, 1660-1960*, 2 vols. (London, 1968).

Pfeiffer, Louis and Otto, Fr., *Abbildung und Beschreibung Blühender Cacteen*, 2 vols. (Vienna, 1843-50).

Philips, C. H., *The Correspondence of David Scott, Director and Chairman of the East India Company, Relating to India Affairs 1787-1805*, 2 vols. (London, 1951).

Pocock, J. G. A. (ed.), *The Varieties of British Political Thought, 1500-1800* (Cambridge, 1993).

Poggendorff, J. C., *Biographisch-Literaraisches Handwörterbuch zur Geschichteder Exacten Wissenschaften*, 2 vols. (Leipzig, 1863).

Poynter, F. N. L. (ed.), *The Evolution of Pharmacy in Britain* (London, 1965).

Prakash, Om, *European Commercial Enterprise in Pre-Colonial India* (Cambridge, 1998).

Prasad, B. (ed.), *Ideas in History. Proceedings of a Seminar on Ideas Motivating Social and Religious Movements and Political and Economic Policies During the Eighteenth and Nineteenth Centuries in India* (London, 1968).

Pritzel, G. A., *Iconum Botanicarum Index Londinensis*, 6 vols. (Oxford, 1929-31).

Rees, Siân, *The Floating Brothel. The Extraordinary True Story of an Eighteenth Century Ship and Its Cargo of Female Convicts* (London, 2001).

Riddick, John F., *A Guide to Indian Manuscripts. Materials from Europe and North America* (Westport, Conn., and London, 1993).

Rix, Martyn, *The Art of the Botanist* (London, 1981), reissued as *The Art of Botanical Illustration* (London, 1989).

Roselli, John, *Lord William Bentinck. The Making of a Liberal Imperialist, 1774-1839* (London, 1974).

Rourke, J. P., *The Proteas of Southern Africa* (Cape Town, Johannesburg and London, 1980).

Scotland, James, *The History of Scottish Education*, 2 vols. (London, 1969).

Smith, Edward, *The Life of Sir Joseph Banks, President of the Royal Society with some Notices of his Friends and Contemporaries* (London, 1911).

Smith, George, *The Life of William Carey, D.D. Shoemaker and Missionary* (London, 1885).

Smout, T. C., *A History of the Scottish People, 1560-1830* (London, 1969).

Stafleu, Frans A., *Linnaeus and the Linnaeans. The Spreading of Their Ideas in Systematic Botany, 1735-1789* (Utrecht, Netherlands, 1971).

Stafleu, Frans A. and Cowan, Richard S., *Taxonomic Literature. A Selective Guide to Botanical Publications and Collections with Dates, Commentaries and Types*, 2nd edition, 7 vols. (Utrecht, 1976-88).

Stephen, L. and Lee, S. (eds.), *Dictionary of National Biography*, 63 vols. (London, 1885-1900).

Stimson, Dorothy, *Scientists and Amateurs. A History of the Royal Society* (London, 1949).

Stuart, W. J., *The History of the Aesculapian Club* (Edinburgh, 1949).

Taton, René (ed.), *The Beginnings of Modern Science from 1450-1800*, translated by A. J. Pomerans (London, 1964).

Teignmouth, Lord, *Memoirs of the Life and Correspondence of John Lord Teignmouth*, 2 vols. (London, 1843).

Turner, Patrick, *Warren Hastings* (London, 1975).

Turner, William, *The Warrington Academy. With an Introduction by G. A. Carter*, reprinted from *Warrington Monthly Repository*, 8-10 (1813-15), (Warrington, 1957).

Watson, Mark, *Jute and Flax Mills in Dundee* (Tayport, Fife, 1990).

Weld, C. R., *A History of the Royal Society, with Memoirs of the Presidents, Compiled from Authentic Documents*, 2 vols. (London, 1848).

Wheeler, A. and Price, J. H. (eds.), *From Linnaeus to Darwin: Commentaries on the History of Biology and Geology* (London, 1985).

White, Alain C. and Sloane, Boyd L. (eds.), *The Stapelieae*, 2nd edition, 3 vols. (Pasedena, California, 1937).

Whittle, Tyler, *The Plant Hunters* (London, 1970).

Williams, Leighton and Williams, Mornay (ed.), *Serampore Letters Being Unpublished Correspondence of William Carey and Others with John Williams* (New York, 1892).

Wilson, Richard and Mackley, Alan, *Creating Paradise. The Building of the English Country House 1660-1880* (London, 2000).

Wilson, Lt Col. W. J., *History of the Madras Army, compiled by Lieutenant Colonel W. J. Wilson, retired list, Madras Army*, 5 vols. (Madras, 1882-9).

Withers, Charles W. J. and Wood, Paul (eds.), *Science and Medicine in the Scottish Enlightenment* (Edinburgh, 2002).

Wood, P. B., *The Aberdeen Enlightenment: the Arts Curriculum in the Eighteenth Century* (Aberdeen, 1993).

Woodruff, Philip, *The Men Who Ruled India. The Founders* (London, 1953).

Worsdell, W. C. and Hill, Arthur W., *Index Londinensis To Illustrations of Flowering Plants, Ferns and Fern Allies. Supplement for the years 1921-35*, 2 vols. (Oxford, 1941).

Youngson, A. J., *The Making of Classical Edinburgh, 1750-1840* (Edinburgh, 1966).

Unpublished Theses

Anon, 'List of Roxburgh specimens in the von Martius Herbarium, Jardin Botanique National de Belgique', unpublished typescript, Royal Botanic Garden Kew, Library.

Chen, Jeng-Guo, 'James Mill's *History of British India* in its intellectual context', unpublished PhD thesis, University of Edinburgh, 2000.

Cook, Andrew Stanley, 'Alexander Dalrymple (1737-1808), Hydrographer to the East India Company and to the Admiralty, as publisher: A catalogue of books and charts', unpublished PhD thesis, University of St Andrews, 1992.

Leung, Man To, 'Extending Liberalism to Non-European Peoples: A Comparison of John Locke and James Mill', unpublished D.Phil. thesis, University of Oxford, 1998.

Merrill, Elmer Drew, 'A partial list of present locations of authentically named William Roxburgh phanerogams, etc.', unpublished typescript, Royal Botanic Garden Kew, Library.

Robinson, T. F., 'Context for the Study of a Great Botanist, William Roxburgh 1751-1815', unpublished MSc thesis, University of Edinburgh, 2001.

Robinson, T. F., 'William Roxburgh (1751-1815) The Founding Father of Indian Botany', unpublished PhD thesis, University of Edinburgh, 2003.

GLOSSARY

The definitions given below pertain to the time of Roxburgh, and some, particularly units of currency, have changed over the years.

Anna   unit of currency, one sixteenth of a rupee.
Bibi   a European's Indian mistress or a Muslim wife.
Biggah   a unit of area, approximately one sixth of an acre.
Brahmin   member of the first *varma*, traditionally priests and scholars; the highest Hindu caste.
Caste   ascribed and usually hereditary ritual status in the Hindu social hierarchy.
Chunam   lime plaster or stucco that can take a high polish used as a rendering for brick built houses, and is very durable.
Coss   a Mughal unit of distance of about 2 miles.
Cowle   a lease or tenancy agreement.
Dioceious   a plant having male and female flowers borne on separate plants.
Dollar (Spanish)   currency used on some of the East Indian islands, equivalent to 10 rupees.
Factory   a trading port, containing the warehouse and usually a port or harbour.
Ghat   steps beside a river, a crossing point.
Havally lands   a tenancy for which there are no living heirs to take on the hereditary lease, which therefore fell to the rulers to dispose of. In the areas governed by the Company, these areas were taken over by the Company and leased out to their tenants.
Holotype   sole specimen or other element that is either used by or designated by an author as the nomenclatural type of a species when he originally published the description of the taxon.
Homonym   same name given as one that has already been published, which may or may not have been applied to the same species, and which is thus not to be used again.
Isotype   duplicate of a holotype.
Jagary   a strong alcoholic drink.
Lakh   one hundred thousand, often used for 100,000 rupees.
Lectotype   used for describing a specimen or other elements (description, illustration, etc) which has been selected subsequently from the original material on which the taxon was based.
Maund   a unit of weight, equivalent to 80 pounds.
Monoecious   a plant having male and female flowers on the same plant, what Roxburgh often referred to as an 'hermaphrodite' plant.
Munshi   an Indian scholar who was used as a private secretary or tutor by the British when learning native languages, etc., and often a holy man.
Nabob   derived from Hindustani *nawab*, meaning 'deputy', became a term of abuse directed towards the rich and retired old India hands.
Natural system   a method of classification using numerous morphological characters, in contrast to an artificial system, which uses only a few (see Sexual system).
Nopalry   a plantation of *Opuntia*, used for raising the cochineal insect.
Pagoda   the main unit of currency used in the Madras Presidency, equivalent to eight shillings (40p) sterling, and 100 Madras Pagodas (MP) were equivalent to 425 Current Rupees.
Palanquin   an Indian litter, usually carried by four men.
Peon   a guard or native soldier, often wearing a badge and carrying some weapon.
Pundit   a native learned in Sanskrit and Indian law, of the Brahmin caste, often used as a translator by the British.
Quincunx   a mathematical two-dimensional design, where dots (or plants) are arranged as on the five spots on a die.
Rupee   the Rupee used in Bengal was usually referred to as the Sicca or Current Rupee, abbreviated SR or CR, equivalent to two shillings and three pence (slightly over 11p). The Rupee used in Bombay was worth about two shillings and six pence (12½p).
Ryot   a peasant farmer.
Sexual system   an artificial method of classification devised by Linnaeus, whereby plants were grouped by the number of male (stamens) and female (pistils) sexual parts.
Sirdar   Military chief or leader.
Tank   freshwater pond or reservoir.
Type specimen   refers to the original specimen on which a description of a new species is made.
Vissum   a unit of land of nearly two acres, containing 31¼ Countas.
Zemindary   an area of land on control of a Zemindar, who rented the land from a higher holder, and was responsible for the well-being of the people in his area, about the British equivalent of a parish.

Page numbers in *italics* refer to illustrations